AUTOMOTIVE
INTERNETWORKING

Intelligent Transportation Systems

VANET Vehicular Applications and Inter-Networking Technologies
Hannes Hartenstein and **Kenneth P. Laberteaux (Eds.)**

Vehicular Networking: Automotive Applications and Beyond
Marc Emmelmann, Bernd Bochow and **C. Christopher Kellum (Eds.)**

Automotive Internetworking
Timo Kosch, Christoph Schroth, Markus Strassberger and **Marc Bechler**

AUTOMOTIVE INTERNETWORKING

Timo Kosch
BMW Group Research and Technology, Germany

Christoph Schroth
Alumnus of the University of St. Gallen, Switzerland

Markus Strassberger and Marc Bechler
BMW Group Research and Technology, Germany

WILEY

A John Wiley & Sons, Ltd., Publication

Registered Office
John Wiley & Sons Ltd, The Atrium, Southern Gate, Chichester, West Sussex, PO19 8SQ,
United Kingdom

For details of our global editorial offices, for customer services and for information about how to apply for
permission to reuse the copyright material in this book please see our website at www.wiley.com.

Library of Congress Cataloging-in-Publication Data:

Automotive internetworking / Timo Kosch . . . [et al.].
      p. cm.
   Includes bibliographical references and index.
   ISBN 978-0-470-74979-1 (hardback)
   1.   Vehicular ad hoc networks (Computer networks)   I. Kosch, Timo.
   TE228.37.A98   2012
   388.3′12–dc23

                                                      2011042963

A catalogue record for this book is available from the British Library.

Print ISBN: 9780470749791

Set in 10/12pt Times by Aptara Inc., New Delhi, India
Printed and bound in Singapore by Markono Print Media Pte Ltd

Contents

Preface

Tow Stories

Tug-of-war (or towing) games are known in almost every country of the world – in fact, a tug-of-war competition is an ancient team sport, which is still very popular today. In Germany, we have the famous phrase 'den Karren aus dem Dreck ziehen', which literally means to 'tow' something ('a trolley') out of a precarious situation towards a successful destination. The introduction of a new technology into the market can be considered as a *tow story*, too. This is in particular true for cooperative transportation systems based on AutoNets, which may be seen as a tow story of three parts:

1. In the first part of the story, a heavy trolley called 'AutoNet' needs to be defined and constructed.
2. In the main section, this trolley must be towed along a long and challenging path with several hurdles.
3. At the happy end of the tow story, the trolley passes the market introduction successfully.

Interestingly, the early days of cooperative transportation systems already outlined the correlation to tow stories. CHAUFFEUR, one of the first and pathbreaking projects in this field, developed a 'tow bar, which links two trucks electronically'.[1] Whereas the market introduction of a new technology is new to the stakeholders, the rules for a tow story are well understood by the players. Let's see what we can learn from these experiences for our challenging AutoNet tow story:

- Since the AutoNet trolley is pretty heavy, it is not possible for single keyplayers to manage the task. Instead, lots of manpower, expertise and suitable equipment are required to construct the AutoNet trolley and to get it moving. Moreover, a robust tow is essential, which will not break; and all players have to be convinced in the robustness and solidity of the tow.
- Obviously, all players have to push the tow as hard as possible. Otherwise, the movement of the AutoNet trolley will stop immediately. However, physics also teaches us that pure strength is only half the truth: in order to maximise the overall strength, all force vectors have to point (and to tow) in the same direction. Of course, this direction must be defined and agreed among the players in advance.

---

[1] Quoted from http://cordis.europa.eu/telematics/tap_transport/research/projects/chauffeur.html.

- We must not forget that pulling a tow also requires a good portion of skill and interaction rules. Otherwise, it may become inconvenient for players. For example, on sloping roads the players have to cooperate in a fair and foreseeable way to avoid the event that the tow winds around the neck of some players. This is often painful – and may even be dangerous – for the respective players.

Tow stories are also useful for writing a book. It is a hard work to tow a book trolley towards the successful publication of the book. Since the book trolley is also pretty heavy, it is impossible for the authors to tow the trolley towards successful completion without the valuable support and the excellent contribution of many other players. We want to thank all supporters that directly and indirectly helped and supported (and motivated) us during the writing and finalisation process. We are very happy about this great support; their contributions definitely helped to improve the quality of this book, which is one of the most important preconditions for a successful publication. Unfortunately, listing all of the contributors would be difficult: we had a lot of supporters, and we will likely and unintentionally forget some of them. So we prepared another 'towing game': you will find all of our supporters by towing together the letters in the following letter grid. Have fun! :-)

Timo Kosch, Christoph Schroth, Markus Strassberger and Marc Bechler

```
D F G C L I N K O F F R E T T A T S H C M H I S O F R Y G W Q
Q A B G M R E I E S E C A S T R A K A V G I K L M P W E T B S
A R D C T F K E M S L O R E L H C E B M R B Y D C V L T S W S
Z E H I D R F H K L T N N U T T U X G C A W E K D H W B N E A
N O S R E H P C M C A E S C H D C C N H F E L O L V J Z M Y Y
E G I U N Z E S E L E O I H O R S A H E F Y L L E U O H C R S
E I S J I M T O E T C K U N U U C E R O D L I L F V A E Y G E
D E N E C E G K H J S C H R R C H T E K O A T M N M E I V C E
I R N L A N N T O M E O U R - F A M I L I E S E M R I O S C O
O E A E I I A U L S M A H W C C R T M B T R Z O H O E C O X L
P S M K O N L F F U J I S O A Y O G A S Z G N E C I H T E Y M
Y S Y N T R A M J S L H R M R I H M H R E D E R S R L K R B B
D Q E K E S T E I N B E R G S N S C P X R J R H O O L O E M E
F R R P P E U K A F P G B P E O T S C A K E S T N I U R G K D
G L F Y R M S T R A S S B E R G E R S O N A H I E C E S I H R
B L F A I N D M U B R E N M A C L E R S E I B I V N M C E I J
R O O L C H H R O E L L E Q R E L E A S M V E S N A Y H W R U
O K R C F F P E W S A L K I N H Q Y G D B E I S D D R H H T H
I F P R R N S T P I N F I W S E V E M N R Y R R W V F T C T G
Z M D A E L D I E T S M M E R L I N H O A B A R E R E P S E V
G L I B N U F C S C H S R Z V F I H C S M M O E L L I M B U S
R A S S H A R M A M L O H B O F U R E J F N T U H J R S C I A
```

List of Abbreviations

AC	Access Category
ACC	Adaptive Cruise Control
ACID/ACM	Application Class Identifier / Application Context Mark
ACK	Acknowledgement
ADAS	Advanced Driver Assistance Systems
AHP	Analytic Hierarchy Process
AIFS	Arbitration Inter-Frame Space
AKTIV	Adaptive und kooperative Technologien für den intelligenten Verkehr (Adaptive and Cooperative Technologies for the Intelligent Traffic)
ANSim	Ad Hoc Network Simulator
AODV	Ad Hoc On Demand Distance Vector
AP	Access Point
API	Application Programming Interface
ARQ	Automatic Request and Repeat
ASTM	American Society for Testing and Materials
ATCP	Ad hoc Transmission Control Protocol
AUTOSAR	Automotive Open Systems Architecture
BC	Backoff Counter
BLADE	Business Models, Legal Aspects, and Deployment
BMBF	Bundesministerium für Bildung und Forschung (German Federal Ministry of Education and Research)
BMWi	Bundesministerium für Wirtschaft und Technologie (German Federal Ministry of Economics and Technology)
BSA	Basic Set of Applications
BSM	Basic Safety Message
BSS	Basic Service Set
C2C-CC	Car-to-Car Communication Consortium
CA	Certificate Authority
CAL	Communication Adaptation Layer
CALM	Communications Access for Land Mobiles
CAM	Cooperative Awareness Message
CAN	Controller Area Network
CAPEX	Capital Expenditures

CCH	Control Channel
CCMP	Counter Mode with Cipher Block Chaining Message Authentication Code Protocol
CCoA	Co-Located Care-Of Address
CEN	Comité Européen de Normalisation (European Committee for Standardisation)
CEPT	Conference of European Postal & Telecommunications Administrations
CI	Communication Interface
CIDR	Classless Inter-Domain Routing
CIMAE	CI Management Adaptation Entity
CN	Correspondent Node
CoA	Care-of Address
COM2REACT	Cooperative Communication System to Realise Enhanced Safety and Efficiency in European Road Transport
COMCAR	Communication and Mobility by Cellular Advanced Radio
COOPERS	Cooperative Systems for Intelligent Road Safety
COTS	Components Off-The-Shelf
CRL	Certificate Revocation List
CSMA/CA	Carrier Sense Medium Access / Collision Avoidance
CVIS	Cooperative Vehicle-Infrastructure Systems
CW or CWND	Contention Window
DAB	Digital Audio Broadcast
DACL	Discretionary Access Control List
DCF	Distributed Coordination Function
DDT	Distance Defer Transmission
DENM	Decentralised Environment Notification Message
D-FPAV	Distributed Fair Transmit Power Adjustment for Vehicular Networks
DHCP	Dynamic Host Configuration Protocol
DIFS	Distributed Coordination Function Inter-Frame Space
DMB	Digital Multimedia Broadcast
DoS	Denial of Service
DoT	Department of Transportation (USA)
DRiVE	Dynamic Radio for IP Services in Vehicular Environments
DRP	Distributed Revocation Protocol
DSDV	Destination Sequenced Distance Vector
DSR	Dynamic Source Routing
DSRC	Dedicated Short Range Communication
DSSS	Direct Sequence Spread Spectrum
DVB	Digital Video Broadcast
DVDE	Distributed Vehicle Density Estimation
EASIS	Electronic Architecture & System Engineering for Integrated Safety Systems
EC	European Commission
ECC	Electronic Communications Committee
ECC	Elliptic Curve Cryptography
ECDSA	Elliptic Curve Digital Signature Algorithm

ECN	Explicit Congestion Notification
ECU	Electronic Control Unit
EDCA	Enhanced Distributed Channel Access (of IEEE 802.11e)
EDCF	Extended Distributed Coordination Function
EDGE	Enhanced Data Rates for GSM Evolution
EDR	Event Data Recorder
EFCD	Enhanced Floating Car Data
ERM	Electromagnetic compatibility and Radio spectrum Matters
ESA	Enhanced Set of Applications
ESP	Electronic Stability Program
ETC	Electronic Toll Collection
ETSI	European Telecommunications Standards Institute
EU	European Union
FA	Foreign Agent
FACH	Forware Link Access Channel
FCC	Federal Communications Commission (USA)
FCD	Floating Car Data
FCFS	First Come First Served
FEC	Forward Error Correction
FHWA	Federal Highway Administration (USA, DoT)
FIFO	First-In First-Out
FOT	Field Operational Test
FS	Fixed Service
GNSS	Global Navigation Satellite System
GIDAS	German In-Depth Accident Study
GloMoSim	Global Mobile Information Systems Simulation
GLOSA	Green Light Optimal Speed Advisory
GPRS	General Packet Radio Service
GPS	Global Positioning System
GPSR	Greedy Perimeter Stateless Routing
GSM	Global System for Mobile Communications
GST	Global System for Telematics
HA	Home Agent
HMAC	Hashed Message Authentication Code
HMI	Human–Machine Interface
HSCSD	High Speed Circuit Switched Data
HSM	Hardware Security Module
HSPA	High Speed Packet Access (in UMTS networks)
HSDPA	High Speed Downlink Packet Access
HSUPA	High Speed Uplink Packet Access
HTTP	Hypertext Transfer Protocol
HTTPS	Hypertext Transfer Protocol Secure
I-BIA	Intelligent Broadcast with Implicit Acknowledgement
ICMP	Internet Control Message Protocol
ICRW	Intersection Collision Risk Warning
ICT	Information and Communication Technology

ICTSB	Information and Communication Technology Standards Board
ID	Identification
IEC	International Electrotechnical Commission
IEEE	Institute of Electrical and Electronics Engineers
IEEE-SA	IEEE Standards Association
IETF	Internet Engineering Task Force
IFS	Insurance and Financial Services
INVENT	Intelligenter Verkehr und nutzergerechte Technik
	(Intelligent Road Traffic and User-Friendly Technologies)
IP	Internet Protocol
IPR	Intellectual Property Rights
IPSec	Internet Protocol Security
IPv6	Internet Protocol version 6
ISM Band	Industrial, Scientific and Medical Band
ISO	International Organisation for Standardisation
IST	Information Society Technology
ITS	Intelligent Transportation Systems
ITSSG	Intelligent Transport Standards Steering Group
ITU	International Telecommunication Union
IVHW	Inter-Vehicle Hazard Warning
IVI	In-Vehicle Infotainment
LAN	Local Area Network
LCRW	Longitudinal Collision Risk Warning
LDM	Local Dynamic Map
LDW	Local Danger Warning
LIN	Local Interconnect Network
LLC	Logical Link Control
LLCert	Long-Life Certificate
LOS	Line-of-Sight
LTE	Long Term Evolution
M2M	Machine-to-Machine
MAC	Medium Access Control
MAIL	Media Adapted Interface Layer
MAN	Metropolitan Area Network
MBMS	Multimedia Broadcast Multicast Service
MCTP	MOCCA Transport Protocol
MEXT	Mobility Extensions for IPv6
MFR	Most Forward Progress within Radius
MIB	Management Information Base
MN	Mobile Node
MNO	Mobile Network Operator
MNS	Mobile Network Suppliers
MOCCA	Mobile Communication Architecture
MONAMI6	Mobile Nodes and Multiple Interfaces in IPv6
MOST	Media Oriented Systems Transport

MoTiV	Mobilität und Transport im intermodalen Verkehr (Mobility and Transportation in Inter-Modal Road Traffic Scenarios)
NAT	Network Address Translation
NEMO	Network Mobility
NFC	Near-Field Communications
NFP	Nearest with Forward Progress
NHTSA	National Highway Traffic Safety Administration (USA, DoT)
NOW	Network on Wheels
ns2	Network Simulator 2
OBU	On-Board Unit
ODAM	Optimised Dissemination of Alarm Messages
OEM	Original Equipment Manufacturer
OFDM	Orthogonal Frequency Division Multiplexing
OMG	Object Management Group
OPEX	Operational Expenditures
OSEK	Offene Systeme und deren Schnittstellen für die Elektronik im Kraftfahrzeug (Open Systems and the Corresponding Interfaces for Automotive Electronics)
OSGi	Open Services Gateway Initiative
OSI	Open Systems Interconnection
OTCL	Object Tool Command Language
PA	Physical Attack
PAN	Personal Area Network
PATH	Partners for Advanced Traffic Highways
PCF	Point Coordination Function
PDU	Protocol Data Unit
PEP	Performance-Enhancing Proxy
PF	Performance Factor (of IEEE 802.11e)
PKI	Public Key Infrastructure
PND	Personal Navigation Device
PSK	Pre-Shared Key
QoS	Quality of Service
RACH	Random Access Channel
RADIUS	Remote Authentication Dial In User Service
REAR	Reliable and Efficient Alarm Message Routing
RCCRL	Revocation by Compressed Certificate Revocation List
RFC	Request for Comment
RFID	Radio Frequency Identification
RSA	Rivest, Shamir, Adleman
RTO	Retransmission Timeout
RTPD	Revocation of Tamper-Proofed Device
RTS/CTS	Request To Send / Clear to Send
RTT	Round Trip Time
RTTT	Road Transport and Traffic Telematics
SAE	Society of Automotive Engineers
SAP	Service Access Point

SBA	Smart Broadcast Algorithm
SCH	Service Channel
SDK	Software Development Kit
SDO	Standards Development Organisation
SIFS	Short Interframe Space
simTD	Sichere und intelligente Mobilität, Testfeld Deutschland (Safe and Intelligent Mobility, Test Trial Germany)
SLCert	Short-Life Certificates
SOA	Service-Oriented Architecture
SOAP	Simple Object Access Protocol
SPAV	Segment-based Power Adjustment
SPAT	Signal Phase and Timing
SRDoc	System Reference Document
SSL	Secure Socket Layer
ssthresh	Slow Start Threshold
SUMO	Simulation of Urban Mobility
TC	Technical Committee
TC	Traffic Class
TCL	Tool Command Language
TCP	Transmission Control Protocol
TG	Technical Group
TISA	Traveller Information Services Association
TLS	Transport Layer Security
TMC	Traffic Message Channel or Traffic Management Centre
TORA	Temporally Ordered Routing Algorithm
TPC	Transmitter Power Control
TPD	Tamper-Proof Device
TPEG	Transport Protocol Experts Group
TPM	Trusted Platform Module
TRADE	Track Detection
TRRS	Time Reservation-based Relay Node Selecting Algorithm
TTP	Trusted Third Party
UDDI	Universal Description Discovery and Integration
UDP	User Datagram Protocol
UI	User Interface
UMB	Urban Multi-Hop Protocol
UML	Unified Modeling Language
UMTS	Universal Mobile Telecommunications System
UTC	Universal Time Coordinated
V2I	Vehicle to Infrastructure
V2V	Vehicle to Vehicle
VANET	Vehicular Ad-Hoc Network
VDX	Vehicle Distributed Executive
VII	Vehicle Infrastructure Integration
VIN	Vehicle Identificaion Number
VMT	Vehicle Miles Travelled

VPN	Virtual Private Network
VSC	Vehicle Safety Communications
VTS	Vehicle Traffic Simulator
WA	Wireless Attack
WAP	Wireless Application Protocol
WAVE	Wireless Access in Vehicular Environments
WBSS	WAVE Basic Service Set
WG	Working Group
WHO	World Health Organisation
WiMAX	Worldwide Interoperability for Microwave Access
WLAN	Wireless Local Area Network
WME	WAVE Management Entitiy
WPA	Wi-Fi Protected Access
WRAP	Wireless Robust Authenticated Protocol
WS	Wireless Simulator
WSDL	Web Services Description Language
WSM	WAVE Service Management
WSMP	WAVE Short Message Protocol
XFCD	Extended Floating Car Data
XML	Extensible Markup Language
XTR	Efficient and Compact Subgroup Trace Representation

1

Automotive Internetworking: The Evolution Towards Connected and Cooperative Vehicles

Safety on the road is one of the most important aspects of mobility: people want to be highly mobile, but they also have a great interest in travelling safely to their destinations. This is of particular importance since the request for mobility is permanently growing, and thus travelling on the road is getting more and more crowded and dangerous. According to surveys in 2009 of the Federal Statistical Office in Germany (destatis) shown in Figure 1.1, the total mileage of people travelling in Germany continuously increased, and was three times higher in 2008 compared to 1970. At the same time, the number of accidents has also increased by 70% over the past 30 years.

However, Figure 1.1 also shows that the total number of road fatalities – that is severe accidents where death results – continuously decreased in this time interval to about 25% compared to 1970. This is mainly due to the fact that both active and passive safety of the vehicles have been improved and vehicles have become more 'intelligent'. As we will see in the following sections, this intelligence has been mainly driven by the introduction of electronics and software in vehicles, which are basically focused on the 'autonomous perception' of the vehicle itself, that is the vehicle only gets information about its surrounding from its own sensors. In this book, we will address the next logical step where cooperative vehicles exchange information between each other and with their environment. Experts agree that this is one of the key technologies to reduce both fatalities and the number of accidents in everyday traffic.

In the following sections, we will introduce the advances in in-vehicle electronics, and we will explain the idea of connected and cooperative vehicles. We will also define the terminology we use throughout this book, and we will introduce the different stakeholders involved in such a cooperative system. Finally, an outlook about the contents of this book concludes this chapter.

1.1 Evolution of In-Vehicle Electronics

In the 1950s, automobiles were still basically mechanical products. Vehicle development in those days was focused on the mechanical components needed to get vehicles moving quickly

Automotive Internetworking, First Edition. Timo Kosch, Christoph Schroth, Markus Strassberger and Marc Bechler.
© 2012 John Wiley & Sons, Ltd. Published 2012 by John Wiley & Sons, Ltd.

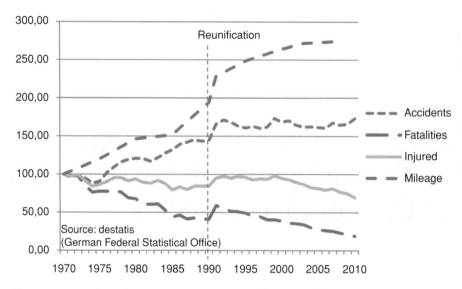

Figure 1.1   Development of accidents in Germany in the context of mileage and road fatalities.

and safely. In the past few decades, electronics has risen as another major component of a vehicle's value, making up around 5% in the early 1970s, growing to 20% in the 1990s and today reaching a mean share of around one third of the value of modern cars. Software has become a major part of this value share, divided approximately into 20% of sensor value, 40% of electronics hardware value and another 40% of software value. This means that the share of software has doubled within the last ten years.

The first generation of vehicle electronics were stand-alone in-vehicle systems that were basically automating or supporting certain driving tasks or comfort features. Prominent examples for systems supporting the driving task are servo (power) steering and anti-lock braking systems. Comfort-oriented systems included the car stereo, electronic windows and automated locking systems. These systems have been implemented with so-called ECUs (electronic control units). The number of ECUs in a typical mid-class vehicle has increased from only a few in the mid-to-late 1990s up to around 50 and more at the end of the first decade of the new millenium. The ECUs control almost every activity in a modern vehicle in order to improve travel comfort and safety and to reduce fuel consumption. Compared to the desktop computer market, the embedded system market has featured a far greater variety of platforms and operating environments. As a move to reduce costs, speed up development and improve quality, French vehicle manufacturers founded VDX (vehicle distributed executive) in 1988, and German vehicle manufacturers and suppliers, together with the University of Karlsruhe, founded OSEK™in 1993[1] in order to establish an industry standard for an open-ended architecture for distributed control units. The two consortia became one joint OSEK/VDX

---

[1] Offene Systeme und deren Schnittstellen für die Elektronik im Kraftfahrzeug (open systems and the corresponding interfaces for automotive electronics). OSEK is a trademark of Continental Automotive GmbH.

group in 1994. With OSEK-OS, a specification for a real-time operating system for ECUs was published. In 2005, it became an international standard as ISO 17356-3.[2]

Besides architecture and system platforms, special communication networks have been developed to interconnect the originally stand-alone systems. For the data exchange of ECUs in vehicles, special vehicle communication systems have been introduced. In 1983, the company Bosch developed the CAN (controller area network) bus which was used commercially in automobiles less than ten years later. Today, it is standardised as ISO 11898. The MOST (media oriented systems transport) [3] bus system defines an optical network and was introduced to satisfy multimedia requirements. Specifications are defined and published by the MOST Cooperation, which was founded in 1998. Around the same time, the LIN consortium (Local Interconnected Network) started as a working group and the first specification of the low-cost solutions-oriented LIN bus was published in 2000. The Flexray consortium was then founded in that same year and has since then defined the FlexRay as a real-time bus system. Many of these activities have been addressed by the AUTOSAR consortium (Automotive Open System Architecture) since 2003. AUTOSAR defines an in-vehicle software architecture with a purpose to support re-use, exchange and integration of software components across platforms and to integrate the different bus systems.

The in-vehicle networks allow the ECUs to exchange data and to realise ever more complex functionality. At the same time, the interface to the driver has become more complex as many of these new functions need to be controlled or interact with the driver. These in-vehicle networks also enable highly sophisticated driver assistance systems that use interconnected sensor technology to constantly monitor the vehicle's environment. Complex sensor-actuator networks have been developed, enabling features such as active cruise control (ACC) , electronic stability programs (ESP) and night vision systems.

While complex in-vehicle networks are state of the art, they are still not or barely connected to systems outside. Hence, a vehicle today is still a relatively autonomous system, because most of the functionality of the vehicle relies on information generated by itself. During the last few years, manufacturers have started to provide connectivity to and from vehicles. BMW's ConnectedDrive technology was the first available system establishing a communication channel between specifically equipped vehicles and a dedicated BMW back-end infrastructure. This connectivity has largely been used for driver-oriented information services like traffic or weather information, emergency call support or mobility assistance through remote vehicle diagnostics.

These features were not originally addressed directly to driving-related tasks; in fact, this type of connectivity was rather viewed as a key technology to introduce new types of applications and services to drivers. Most likely, vehicle connectivity to the outside will be restricted to the non-driving related domain and remain running through the manufacturer's server infrastructure in the near future. Nevertheless, enabling this data exchange between vehicles and dedicated back-end systems can be seen as the origin of a 'connected vehicle'. As new wireless communication standards have emerged, security solutions are on the horizon and hardware prices are decreasing, we presume that the next important step in-vehicle electronics will be the connectivity of vehicles with traffic infrastructure entities and other traffic participants. This connectivity will then be used not only to provide convenience services to the driver,

---

[2] Road vehicles; open interface for embedded automotive applications; Part 3: OSEK/VDX Operating System (OS).

but more importantly to improve their original task: driving. It will not be only the driver information unit, but the vehicle ECUs in general will be networked with other ECUs in other vehicles and with traffic infrastructure controllers.

1.2   Motivation for Connected Vehicles

The main motivation for the application of wireless network technology to road traffic scenarios is to optimise driving with respect to safety and efficiency. Safety has long been one of the main drivers of vehicle communications. At the end of the 20th century, when many of the bigger research activities started in North America as well as in Europe and Japan, the global number of injuries caused by road traffic accidents was close to 40 million people and the number of fatalities was almost 1.2 million people.[3]

Statistics from the United States Department of Transportation's (US DoT) National Highway Traffic Safety Administration (NHTSA)[4] show a total of 37,261 people killed in traffic accidents in 2008 in the United States alone, with about 2.35 million people injured. Estimates of the economic impact (from 2003) state a cost of 230 billion US Dollars related to traffic accidents.

While passive safety systems have been very effective in protecting the passengers, they will typically not help avoid the accident itself. Accident statistics show that the rate of fatalities and injuries over vehicle miles travelled has gone down in many developed countries. In the US, for instance, the injury rate was around 150 per 100 million vehicle miles travelled (VMT) around 1990, decreasing to about 120 at the turn of the millenium and going down further to less than 80 in 2010. Fatalities declined from 1.73 per 100 million VMT in the year 1995 to 1.53 in 2000 and 1.13 in 2009 the US.[5] These numbers show what can be achieved with passive safety, while the total number of accidents has stayed at a high level over the last few years as the following numbers from Germany show: in 2007, 335,845 combined deaths and injuries were caused by 2.33 million accidents while in 2010, 288,297 combined deaths and injuries were caused by 2.41 million accidents.[6] This is why, in addition to the proven effective passive safety systems, research has turned towards active safety systems in order to avoid a crash in the first place. Sensor technology is able to detect objects in front of and around a vehicle. However, current sensors are not able to detect objects around corners or hidden behind other objects. Therefore, data sensed by other traffic participants or roadside infrastructure is very valuable for getting a full picture of the situation a vehicle is facing. How this data can be communicated is an important part of AutoNet technology. The focus of the communication-data based applications have mainly been aimed at driver information rather than vehicle control. While one reason may be that the technological challenges are not as high, there is also justification from accident analysis.

---

[3] In the year 1998 according to the study 'Injury: A leading cause of the global burden of disease', published by the World Health Organisation (WHO) in 2000.

[4] NHTSA homepage: http://www-fars.nhtsa.dot.gov.

[5] Data source: National Highway Traffic Safety Administration (NHTSA) yearly traffic safety report and fatality analysis reporting system at http://www-fars.nhtsa.dot.gov/Main/index.aspx.

[6] Source: http://www.destatis.de.

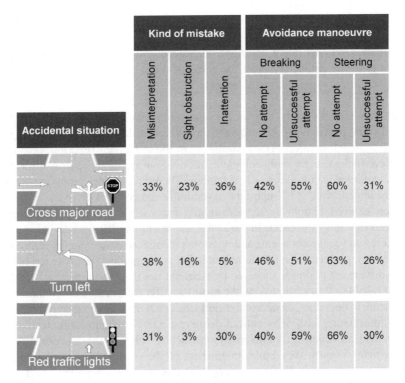

Accidental situation	Kind of mistake			Avoidance manoeuvre			
				Breaking		Steering	
	Misinterpretation	Sight obstruction	Inattention	No attempt	Unsuccessful attempt	No attempt	Unsuccessful attempt
Cross major road	33%	23%	36%	42%	55%	60%	31%
Turn left	38%	16%	5%	46%	51%	63%	26%
Red traffic lights	31%	3%	30%	40%	59%	66%	30%

Figure 1.2   Reasons for crossing traffic accidents in Germany.

Consider three typical accident situations at an intersection as illustrated in Figure 1.2 (for more information on intersection safety, refer to [4]):

1. Turn into (or cross) a major road.
2. Turn left with oncoming traffic.
3. Red traffic lights.

Analysing accidents in each situation, the data shows that there are three main causes: misinterpretation of the situation, sight obstruction and inattention of the driver.[7] Moreover, Figure 1.2 also provides data on the driver's behaviour, showing the rate of drivers reacting not at all or reacting by either braking or steering or both. From these numbers we can draw the following two important conclusions:

1. The three kinds of mistake add up to more than 50% in each scenario; in the first scenario they make a total of 92%. All three mistakes basically occur due to the unavailability of information about the situation. If this information was available for both the vehicle and the driver, these kinds of mistake and accordingly the numbers of accidents might be diminished significantly.

---

[7] Source: Statistisches Bundesamt (destatis) and Database of Accident Research of Dresden and Hannover (GIDAS).

2. On average about 50% of drivers did not react at all to the situation ('no attempt' column). The timely availability of information about the situation and support by vehicle warning and assistance systems might contribute significantly to increasing the share of (proper) reactions to the situation. This way, the accident could be potentially completely avoided or, at least, the consequences made less dramatic.

Gruendl shows in his study that an overwhelming number of accidents are caused by individual drivers' mistakes [2]. According to this reference, the main cause of driving mistakes was that important information on the surrounding traffic and driving situation was not available in time. Nevertheless, even with complete information available, human error cannot be completely excluded in complex traffic situations. Therefore, besides providing the necessary and relevant information, it is important to minimise the complexity of the current driving situation for the driver. Sophisticated foresighted driver assistance and information systems play a crucial role in this context. In particular, the attention of the driver is supposed to be explicitly directed to relevant events and possible potential risks in complex traffic situations.

In this context, powerful wireless communication systems allow the provision of data about a vehicle's driving environment surveyed by other entities that is very precise, timely, of high quality and reliable. They also provide the means to offer information about the vehicles and what they sense to operators, owners and manufacturers. If the vehicles in a traffic environment share their information with other entities using respective communication technologies, we call this system an *AutoNet*. AutoNets systematically enhance the functionality of stand-alone in-vehicle systems, they build an Internet of the in-vehicle networks, and thus we chose the title 'Automotive Internetworking' for this book. Through the ability to receive remote sensing information, AutoNets provide the capability to extend the so-called electronic horizon of vehicles, that is the systems' forecasts about future driving conditions, beyond its own immediate sensing range. The electronic horzion comprises information about the immediate vicinity of a vehicle as well as relevant information at longer distances. In this way, knowing what is lying ahead of them, drivers (and driver assistance functions) gain valuable time to make the right decisions, adapt their controls accordingly and hence reduce driving mistakes. The hardware of wireless communication systems is usually less expensive than sophisticated sensor systems like radar or lidar. Thus, if deployed widely, smaller and less-expensive vehicles (and also motorcycles) which are little equipped with cutting edge sensor technology, would benefit particularly from automotive Internetworking.

As opposed to stand-alone and autonomous systems, the exchange of information between vehicles and their environment enables the deployment of cooperative assistance functions. Vehicles do not act autonomously in such systems; instead, they mutually interact with each other in order to handle a traffic situation in a cooperative way. In such a cooperative system, the quality and effectiveness depends largely on the extent of participation, that is system penetration. The overall quality advances with the increasing penetration rate of communicating entities. In particular, cooperative systems usually require a minimum level of penetration or, to be more precise, a certain minimum number of equipped vehicles travelling in a certain area at the same time. The extent to which this system penetration is sufficient depends on the specific application.

Although traffic safety and efficiency are mostly seen as the key application domains of AutoNets, it is worth mentioning that driving and travelling by car can be made much more pleasant in a connected world too. In addition, vehicle maintenance and management profit from connectivity, and can be seen as commercial enablers to finance the basic infrastructure and penetration before cooperative applications can be offered. Therefore, these applications play an important role with respect to deployment strategies and scenarios.

1.3 Terminology

When talking about communication among vehicles or between vehicles and roadside infrastructure, different communities or institutions from different domains currently use slightly different terminologies and abbreviations, examples are C2C, V2V, Car2X, VSC (vehicle safety communications), DSRC (dedicated short-range communication) or cooperative systems. These terms are often used in a broader sense, even though they clearly refer to different aspects of AutoNets. V2V refers to communication between two vehicles while C2C or Car2Car has a narrower meaning with its reference to cars. Car2X references cars as one endpoint of a communication with an undefined communication partner referred to as 'X'. While these terms tell us nothing about a technology, they are often used to refer to direct vehicle communication, that is with a transceiver in each vehicle and no other communication infrastructure. The term DSRC describes a set of digital communication technologies with a limited communication range. These can be wireless LAN systems or infrared communication systems or others. In the USA, DSRC is often used to refer to a specific communication technology: IEEE 802.11p wireless LAN systems. In Europe, it is sometimes used to refer to toll collection systems based on infrared or other short range communications. VSC is also often used in the USA when talking about vehicle safety systems that use DSRC technology. However, the term itself does not exclude the use of other types of communications. The term 'intelligent transportation system' (ITS) has a far wider meaning and refers, broadly speaking, to all traffic related electronic systems that provide information about traffic, or otherwise influence or control traffic.

With respect to a consistent representation, we use abbreviations with clearly defined semantics. We distinguish communications according to the type of communication partner, as illustrated in Figure 1.3, where the communication partner is a physical rather than a logical entity in order to avoid ambiguity. This also means that we will not use V2I (vehicle-to-infrastructure) in this book even though this is an abbreviation commonly used but with different meanings in different contexts. In the following chapters, we will explain the applications and the necessary technologies for the different communication partners of vehicles. Therefore, we assume the following communication forms in an AutoNet throughout this book (including typical examples to clarify the communication form):

Vehicle-to-vehicle (V2V) refers to communication between any type of vehicle. This is often called C2C or car-to-car in the community, although typically no differentiation is made with respect to different types of vehicles. This way, vehicles may be cars, trucks, motorbikes and so on.

Vehicle-to-mobile-device (V2M) refers to communication between a vehicle and any type of mobile consumer electronics device, such as (cellular) smartphones using respective communication protocols.

Vehicle-to-toadside (V2R) refers to communication between vehicles and roadside traffic infrastructure. Examples include traffic lights or variable message

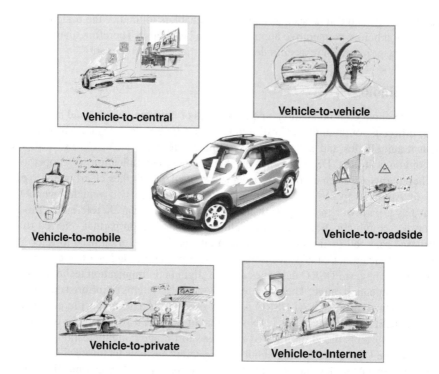

Figure 1.3   Classification of V2X communications. Reproduced by permission of BMW Group.

signs. Often, this type of communication is described by the term vehicle-to-infrastructure (V2I) in literature.

Vehicle-to-central-infrastructure (V2Central)   refers to communication between a vehicle and a traffic management centre. A traffic management centre is typically an institution that controls the traffic flow of a part of the road network. This way, we explicitly distinguish this type of communication form since central infrastructure domains have a significant impact on the traffic flow and thus on traffic efficiency.

Vehicle-to-Internet (V2Internet) refers to any type of communication between a vehicle and an 'accessible' host in the Internet. Hence, V2Internet communication is based on IP communication as used in the Internet. Examples include access to third-party services available on the Internet.

Vehicle-to-private-network (V2Private) refers to communication between vehicles and any type of private network. Here, private networks are networks that limit their access to some vehicles only. A typical example would be communication between a vehicle and a (private) server located inside a home or a company network. Hence, this type of communication also refers to communication between a vehicle and the network of a vehicle vendor, which may provide vendor-specific services to its customers.

For the sake of generality, we use the abbreviation V2X throughout this book in order to refer to vehicle-to-X communication. If we refer to vehicular networks in general, we will either use the term V2X or the term *AutoNet* as an abbreviation for automotive (Inter-) networks. Note there is no explicit vehicle-to-pedestrian communication since this would be covered by the mobile devices pedestrians use. Thus, this communication scenario is covered by V2Mobile. Also, it is important to mention that communication between the different entities is generally bidirectional. For the example of V2R, this communication form comprises communication from a vehicle to a roadside traffic infrastructure unit as well as communication from the roadside traffic infrastructure unit to the vehicle. However, different communication technologies may be used for either direction.

It is also important to mention that V2X communication is not fixed to a particular communication technology or a respective communication protocol. Instead, different wireless communication technologies can be applied to interconnect the entities. They all have their particular properties regarding characteristics like coverage, data rate or latency. Traditionally, broadcast networks have transmitted traffic information to vehicles on radio channels. Analogue broadcast systems have been amended and sometimes replaced by digital broadcast media like DAB (digital audio broadcast), DVB (digital video broadcast), DMB (digital multimedia broadcast) or HDRadio [1]. These systems are unidirectional, usually employ high data rates and cover large areas. Thus, they are useful to distribute information that is of interest to a large number of receivers distributed over a larger area of at least a few dozen square kilometres. In this context, broadcast networks are in particular useful to distribute information of general interest like traffic conditions or weather conditions in the respective area.

To be able to utilise an up-link and also send data out of the vehicles, bidirectional communication systems are required. These enable both the transfer of individual information to single vehicles and the utilising of information that is available in the vehicle for other purposes. Today, typically, cellular systems are used for this kind of communication. Some modern cars feature GSM, GPRS (2.5G), EDGE (2.75G) or UMTS (3G) equipment and offer services like individual location-based information or remote vehicle diagnosis. In addition, many vehicles provide connectivity to mobile phones or other consumer electronic devices via Bluetooth technology, for example. The usage of wireless LAN according to IEEE 802.11 a, b or g within vehicles has also recently been offered by some vehicle manufacturers. In addition to these more general purpose networks, some wireless technologies have been developed specifically for certain usage scenarios in vehicles. Those are infrared technologies and microwave communications for electronic toll collection (ETC) and adapted wireless LAN technologies for vehicle safety and traffic efficiency applications. Wireless LAN technologies specifically adapted for the automotive domain will be called Auto-WLANs in the course of this book. The wireless technologies, their usage for connecting vehicles, their properties in the automotive domain and specific automotive wireless technologies are presented in more detail in Chapter 8.

In this context, we also want to clarify two other terms that we distinguish: *situations* and *events*. Situations describe the environment during a period of time. They usually reflect the existance of specific circumstance. A real-life example would be an area of heavy crosswinds or a construction site along the road. There are also short-lived situations, such as a vehicle engaged in emergency braking. While situations usually describe a certain state of a system, events are well-defined state transitions. In our understanding, events are snapshots of a situation previously unknown to the AutoNet system at a distinct point in time. Thus, the event marks a state transition of the system's knowledge. Vehicles as part of a cooperative traffic situation messaging system, generate events after detecting specific new situations, create messages resembling that information and distribute them. The on-board systems of vehicles receiving these messages calculate their relevance and the confidence they have in the information and then decide on driver notification or autonomous action of the vehicle.

1.4 Stakeholders

Typically, AutoNets comprise a variety of different stakeholders. While it is essential to involve some stakeholders right from the beginning of system development, others may join at a later stage. This greatly depends on the use cases being deployed. The following overview introduces the essential stakeholders of AutoNets together with some examples:

- *Operator organisations* include road operators, network operators and providers and telecommunication providers.
- *Service providers* include public transport providers, emergency service providers, fleet providers, freight service providers, mobility providers (e.g. car sharing), Internet service providers and transportation companies.
- *System users* include end users, public transportation operators, emergency operators, service centre operators, application management centre operators, road network operators, fleet operators and freight operators.
- *Manufacturers and suppliers* include system suppliers, vehicle manufacturers, roadside equipment suppliers and traffic management equipment suppliers.
- *Content providers* include, for example traffic information or navigation map providers.
- *Authorities* include certification authorities, road authorities and other public authorities.
- *Drivers* include private vehicle drivers, freight vehicle drivers, emergency vehicle drivers, hazardous freight vehicle drivers and public transport vehicle drivers.
- *Travellers* include sporadic travellers or frequent travellers.
- *Other stakeholders* include associations of motorists, insurance companies, health service providers, incident managers, town supervisors , traffic managers and many others.

Apparently, the wide variety of stakeholders results in different backgrounds and highly heterogeneous (and often conflicting) aspirations of the system. We will not further discuss the different aspirations and requirements of the stakeholders. In order to realise a respective system architecture, the stakeholders' aspirations have to be mapped onto user requirements. These requirements are the basis to derive the respective functionality of the system architecture in terms of logical and physical subsystems, designs, specifications or implementations.

We will not discuss the stakeholder aspirations in this book. Such issues are important and exhaustive contributions occur in related projects about AutoNets, such as E-FRAME (see Section A.2.9) or COMeSafety (see Section A.2.3).

1.5 Outline of this Book

Authors of books on communication systems always have to decide if they should describe the communication system bottom-up or top-down. Both approaches have their particular strengths (and weaknesses): the bottom-up approach, that is starting at the lowest layer, clearly shows how upper layers deal with deficiencies of lower layers, whereas it is not clear why respective mechanisms are useful for applications or higher layers. Vice versa, top-down approaches – that is starting at the highest communication layer – motivate the requirements for the communication system by examining the uses, while it is not clear for each layer why specific mechanisms are needed. In this book, we decided to follow the top-down approach. This is due to the fact that the design of AutoNets – as well as the development of new technologies for vehicles – is greatly driven by the applications and uses, which are the most important factors for the development of the respective technology. Following this top-down approach, this book basically consists of three main parts:

- *Part I: scenarios and System Architecture.* We outlined in this chapter the development of communicating vehicles, and we introduced the different stakeholders involved in cooperative systems. In order to show the requirements for the system design, Chapter 2 introduces different scenarios and classifies potential AutoNet applications. It also describes the characteristics of the classification scheme, the stakeholders involved and finally the requirements derived for the design of the respective AutoNet communication system. Chapter 3 continues by describing AutoNets from the system architecture's perspective. Starting from a high level domain view, the system's architecture is refined by separating the different domains, their subsystems and their respective mechanisms and relations. This chapter also introduces the *AutoNet Generic Reference Protocol Stack*, which forms the basis for the AutoNet communication system.
- *Part II: AutoNet Communication Layers.* Following the top-down approach, Chapters 4 to 10 describe the different communication layers of the AutoNet Generic Reference Protocol Stack in detail. The application layer in Chapter 4 is most important, because it details exemplary applications of the different classes in our classification scheme, and it derives the requirements necessary for the lower communication layers, especially with respect to the mechanisms required for the different layers. Afterwards, Chapter 5 introduces the application support layer (also known as facility layer), the transport layer and network layer are addressed in Chapters 6 and 7, and the transmission aspects of the physical communication yechnologies are described in Chapter 8. Finally, the two cross-layers for security (Chapter 9) and management (Chapter 10) conclude this part of the book. Throughout this part, we also introduce related work and existing mechanisms and examinations of several aspects of the layers, and we also address some implementation issues for the realisation of some communication layers.
- *Part III: Methodologies and Markets.* In the final part of this book, we provide an overview of the methodologies in Chapter 11, which are used for the development and examination of AutoNets. We also give examples for their realisation and provide some basic results. Additionally, we provide a market survey in Chapter 12, which can be seen as the current state of the art for cooperative systems for vehicles. Finally, Chapter 13 concludes this book and outlines the impact we expect for the market introduction of communicating (and cooperative) vehicles.

In the Appendix, we give a review of past and ongoing activities for AutoNet-based cooperative systems, as well as current standardisation activities. We also provide a summary of already standardised applications, including standardised message formats for the exchange of information among the different entities in an AutoNet.

References

[1] Schiller, J. (2003) *Mobile Communications*, 2nd Edition. Addison Wesley.

[2] Gruendl, M. (2005) Fehler und Fehlverhalten als Ursache von Verkehrsunfällen und Konsequenzen für das Unfallvermeidungspotenzial und die Gestaltung von Fahrerassistenzsystemen [Mistakes and misbehaviour as causes for traffic accidents and consequences for the potential to avoid accidents and design driver assistance systems]. *Dissertation. Regensburg 2005*. (in German).

[3] Grzemba, A. (2011) *MOST. The Automotive Multimedia Network. From MOST25 to MOST150*. Franzis.

[4] Klanner, F., Thoma, S., Winner, H. (2008) *Driver Behaviour Studies and Human-Machine-Interaction Concepts for Intersection Assistance. Proc. of 3rd Conference on Active Safety through Driver Assistance*. Garching / Munich, Germany.

2

Application Classifications and Requirements

Connecting vehicles to different external entities through different network technologies is the enabler of a plethora of novel applications for the vehicular domain, which we call AutoNet applications. These applications can be classified according to different metrics. Very often, they are classified with respect to their intended purpose. Thus, AutoNet applications are typically classified as either:

- traffic safety applications;
- traffic efficiency applications; or
- convenience applications.

While traditionally active safety for accident prevention was the main motivation for developing AutoNets, their potential to improve traffic efficiency has not only added to research activities, but has also spurred the hope for a realistic deployment scenario. Since convenience applications can be offered on an individual basis and do not require standardisation, we have seen the growth of such services in the market over the past few years. Also, the different application types pose different requirements for the wireless communication technologies, as we will see later in Section 2.3. This, however, is an additional reason for convenience applications to appear on the market following mobile service offerings on smartphones. Reflecting this development, it is not uncommon to speak of automotive *apps*. Since latency is more important for active safety applications than throughput, safety systems have not been built on top of cellular technology. Whether this will change with a larger deployment of LTE remains to be seen. Effective communication algorithms for safety applications usually employ data traffic prioritisation schemes to ensure low latency information exchange. We will discuss the potential of cellular technology for safety applications later.

This chapter is partitioned into three sections. In the first, we classify AutoNet applications, describe their basic characteristics and derive implications for information and communication system design. In the second part, we identify key information system properties derived from

Automotive Internetworking, First Edition. Timo Kosch, Christoph Schroth, Markus Strassberger and Marc Bechler.
© 2012 John Wiley & Sons, Ltd. Published 2012 by John Wiley & Sons, Ltd.

the requirements of the application classes described in the first part. Since situation handling is an important feature of AutoNet systems, we describe a methodology for implementing reliable situation estimates as an important prerequisite for information relevance assessment and information filtering in this context. The third part addresses the communication system properties and the communication models relevant for AutoNets. Based on the different characteristics, we show which communication model is relevant for which AutoNet application.

2.1 Classification of Applications and their Implications

You may think of a number of possibilities to classify the plethora of applications that is enabled by connecting the vehicle to different external entities through different network technologies. You've seen a possible classification according to the main purpose of an application above. We understand a classification merely as a means to subdivide the vast field of applications into sections that usually employ certain common characteristics. A classification, however, does not necessarily need to be orthogonal in the sense that there is only one class that a certain application can belong to. Nevertheless, there should be an obvious and unambiguous assignment according to the major characteristics of the application. In the following, we differentiate AutoNet applications along three basic dimensions, as illustrated in Figure 2.1:

> **Driving.** Driving-related applications aim at improving traffic on the road, enhancing both the efficiency and safety of moving vehicles by providing support and assistance to the driver or by improving the cooperation of the traffic participants. These applications may be distributed and include third party services provided by the AutoNet infrastructure domain. Connected navigation or driver warning applications are natural representatives of this class. According to our understanding, new types of traffic light control as infrastructure-oriented applications also belong to this class. As AutoNet applications, they would make use of probe data from vehicles. Since the effect then is clearly on the movement of the vehicles, we classify such applications as driving-related.

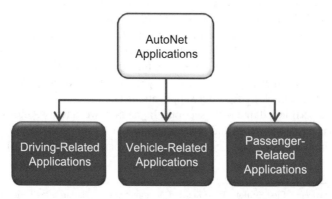

Figure 2.1 AutoNet applications classification.

Passenger. Passenger-related applications are focused on the comfort, convenience and entertainment of the passenger. In our understanding, this includes the driver when taking the role of a passenger. Consider point of interest information (POI) services, for instance. Based on such information, the driver may decide to change their route and thus the information would have an effect on driving. We still classify such POI services as passenger-related because they provide more information than just position for the convenience of the passengers (e.g. pictures, menus of restaurants and the like).

Vehicle. A third class of applications is related to the improved operation, management and simplified configuration of the actual vehicle. Vehicle-related applications are typically implemented for a single or a defined set of vehicles only. They do not have an immediate impact on traffic safety and efficiency. Instead, respective applications have a clear commercial use and business case, and are typically controlled by private service centres. Examples include the connected management of vehicles for fleet operators, or services from automotive manufacturers for their customers like remote software updates. Improved engine and vehicle energy management also belongs to this class if only affecting the operation of the vehicle without influencing the driving behaviour or the movement of the vehicle.

2.1.1   Driving-Related Applications

A large number of the applications that are currently being discussed in the context of intelligent transportation systems are driving-related in the sense described above. These applications usually need to know and have to predict the traffic or driving situation as accurately as possible. Consequently, driving-related AutoNet applications support traffic safety and efficiency by providing more and/or more accurate information about the driving context of the vehicles. Regarding the system requirements that have to be met, two major system aspects deserve a closer look: the impact of the information on the driving task and the dependence on the vehicle context. In the following, we present some more details on these two aspects.

2.1.1.1   The Impact on the Driving Task

Following the three levels of Rasmussen's control model [3], the driving task can be categorised into knowledge-based, rule-based and skill-based levels. Driver assistance systems can be structured according to this logic as follows:

- On a knowledge level, *navigation applications* (e.g. road navigation and traffic information systems) provide knowledge to the driver and facilitate the decision making processes.
- On a rule-based level, *manoeuvering applications* such as active cruise control or lane change assistance help drivers in observing traffic rules.
- On a skill-based level, *stabilising applications* (e.g. anti-lock braking or traction control systems) improve driving safety by controlling the typical actions or driving situations of a vehicle.

Within the driving-related class, a further distinction can be made between driver assistance systems and autonomous vehicle control systems. Autonomous vehicle control is defined as systems that are either making purely autonomous decisions on driving dynamics such as acceleration, braking or steering, or systems that are making autonomous decisions regarding the vehicle operation (e.g. switching gears, engine operation). Usually, both driver assistance and autonomous control systems are targetted towards one or more of the three goals of safety, efficiency and comfort. For example, Active Cruise Control (ACC) is primarily a comfort system but also implicitly improves safety.

Driving-related applications can also be classified along the so-called '4 A axis', emphasising the cognitive interaction with the driver: automation, action, attention and awareness.

Automation refers to autonomous vehicle control systems where direct driver interaction is neither required nor possible. Typically, this refers to driver assistance systems as mentioned before, such as a dynamic stability control. Obviously, such systems require a very high overall reliability. If based on AutoNet communications, such driver assistance systems also require very low communication and processing latency, because of the very limited time frame available for stabilising the vehicle in critical driving situations.

Action refers to situations in which the driver needs to react immediately, for instance in the case of a suddenly hard braking vehicle ahead. Because of the spatial proximity, there is very little time to react. Therefore, the overall AutoNet system again needs to ensure low latency and a high reliability, as well as very intuitive user interaction concepts.

Attention refers to situations that do not require an immediate action from the driver. Instead, the driver should have his or her attention drawn to a specific entity or event, like an approaching emergency vehicle, low grip or an expected spot of black ice on the road segment ahead. Since there is no immediate action required, the latency constraints of the system are usually not as strict as for the Automation and Action class.

Awareness refers to all situations that should not explicitly draw a driver's attention to a specific action, but rather should be subliminally perceived in order to support the driver's cognitive model of the overall traffic condition on the travel route. This can refer, for instance, to an emergency vehicle that is in the proximity, but does not cross one's individual route, or a traffic jam or bad weather conditions on the route ahead at some distance. In such a scenario, obviously there are relaxed latency requirements. Also, the impacts of a false system behaviour are less severe, resulting in soft requirements for system reliability.

The examples mentioned above show, however, that requirements of both the overall in-vehicle system in general and the communication system in particular, do not primarily depend on the specific application. Instead, the requirements are driven to a great extend by the effect on driver interaction and the spatial proximity of the respective event or entity. For example, an application to support the driver in case of a near emergency vehicle can either require a driver's attention, or it can only require his or her awareness. The same is true for all applications that provide information about the route ahead. Depending on the distance to a critical road condition, the required driver interaction is either action (very close ahead), attention (close ahead) or awareness (further away).

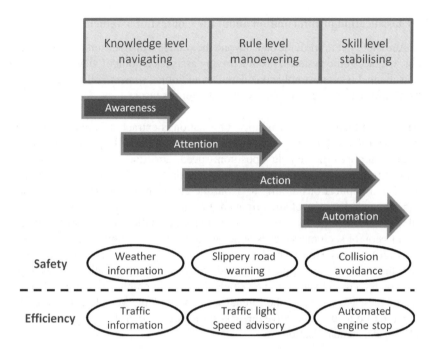

Figure 2.2 AutoNet classification for driving-related applications.

Figure 2.2 illustrates the above mentioned classification of driving-related applications according to the knowledge level, the rule level, and the skill level. In a safety enhancement context, these levels are associated with applications such as weather information (knowledge or awareness, respectively), slippery road warning (rule or attention, respectively), and collision avoidance (skill or action, respectively). In an efficiency improvement context, the levels refer to applications such as traffic information, traffic light speed advisory, and automated engine stop. As an example for application-based classification, consider a parking information system. This would be a driving-related application in the navigation domain with the goal of improving efficiency.

In general, the requirements on AutoNet systems increase from a knowledge-based level to skill-based level, and from an autonomous to an awareness level, respectively. This is in particular true for the overall system latency, availability, fault-tolerance and security.

As we will see in Chapter 7, there are two different approaches to designing respective communication networks. One approach puts each application into a specific application class, depending on the most critical characteristics of the application. For the example of black ice notification, this would result in a classification as an instance of the action class. Correspondingly, data transmission of this application would be handled and prioritised according to the requirements of the application class. The second approach aims at maximising the overall benefit of the available network capacity. Message transmissions in this case are scheduled according to the additional benefit the transmission will actually provide. Following this approach, not all messages of black ice spot notification are treated equally, but rather differently depending on the situation, for example the distance to the reported event.

2.1.1.2   Classification of Information about the Context of a Vehicle

As stated, AutoNet systems provide data beyond the vehicle's own sensors from remote sources to the vehicle's electronic systems. From a driving assistance point of view, an AutoNet communication channel is often simply considered an additional sensor. Following this metaphor, two types of information are provided by this additional sensor:

> **Proximity information.** Proximity information is extracted from a set of basic sensor data that each AutoNet vehicle transmits regularly. In particular, this set of data comprises the vehicle's position, speed, direction and acceleration. The frequent transmission of this kind of information forms the basis for action and automation applications aiming to avoid the collision of two vehicles. Because of the special character of such frequent transmissions, such messages are often called *heartbeat* messages or *here I am* messages. Often it is assumed that the transmission of such messages is done periodically, for example every 500 ms or even every 100 ms. However, like biological heartbeats, there are good reasons to adapt the transmission frequency according to the driving situation, for example based on the current speed of the vehicle. In order to emphasise the fact that this type of message increases the mutual awareness of the vehicles, they are also often referred to as *cooperative awareness messages*, or in short *CAM*, especially in European standardisation.

> **Environmental information.** Environmental information refers to information about the driving environment that may require awareness or attention. In contrast to proximity information, this type does not reflect the state of a (single) vehicle but information about the environment that can be detected by one vehicle or a group of vehicles. A large variety of different context information is covered by this type, ranging from the condition of the road surface, to different weather conditions, black ice or traffic jams. Because of the decentralised character of data acquisition, in Europe, such messages are usually called *decentralised environmental notification* messages, or in short *DENM*. Standardisation activities define common conditions so that receiving vehicles understand the context information. DENMs are usually transmitted whenever a vehicle detects the existence (or absence) of a certain environmental condition.

It is important to note that these types impose very different requirements on the transport and communication layer of AutoNets. As we will see later on, both types are therefore treated very differently by the AutoNet communication system. The reason for this is again mainly correlated to the impact on the driving task and the driver. As menioned, applications that are based on cooperative awareness messages are typically in the domain of automation or action. The obvious reason for this is that the state of a vehicle, that is its speed, position, direction and so on, changes very dynamically. Hence, application of such data is only reliable for a very limited period of time, and thus in turn is only reasonable for vehicles that are in close proximity. Therefore, CAMs are broadcast by each vehicle, but not further distributed throughout a larger area. Because of the comparably high transmission frequency, this also leads to a significant network load. In contrast, applications based on decentralised environmental notification messages mainly focus on early driver attention, and are usually in the domain of awareness or information. In particular, DENMs are not transmitted periodically, but event driven whenever significant situations are detected. In order to ensure an early warning, such information is regularly distributed within a larger area.

Because of these very different requirements on the underlying AutoNet communication, the differentiation of CAMs and DENMs, or the correlated applications respectively, is sometimes also justified by the underlying communication patterns. However, as we have seen, this is only a consequence, not a reason. Consequently, differentiation should be made based on the type of information that is communicated.

2.1.2   Vehicle-Related Applications

With AutoNets, the infrastructure is available to deploy communication-based applications for vehicles and their components, independent of their location and their drivers. Such vehicle-related applications may address a single vehicle or groups of vehicles. Like driving-related applications, vehicle-related applications will typically make use of several communication types in AutoNets, such as V2V and V2R communication, but also V2Internet, V2Private and V2Mobile communication. In contrast to driving-related applications, vehicle-related applications do not have an immediate impact on the traffic situation on the road. As illustrated in Figure 2.3, vehicle-related applications can be differentiated in the following way:

Fleet-based applications. Fleet-based applications are services for a group of vehicles, which form a fleet or which are 'hosted' by one operator.

- *Vehicle business services.* Vehicle fleets are typically integrated into business processes, that is they are an integral part of the business of companies. Especially for fleet operators or logistics companies, fleet management planning

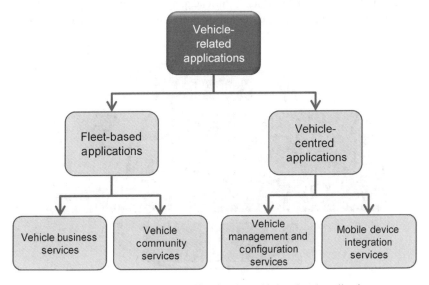

Figure 2.3   Applications classification for vehicle-related applications.

is a mission critical business process for the creation of value. In this context, vehicle business services have to address vehicle planning, vehicle scheduling, goods transportation, and also vehicle maintenance and the re-schedule strategy in case of a vehicle breakdown.

- *Vehicle community services.* Vehicle community applications provide special services for groups of vehicles, for example from one brand or of a certain type. An example for vehicle community services are services provided by vehicle manufacturers for their vehicles, like reminders and automatic appointment arrangements for vehicle inspections based on customer subscriptions.

Vehicle-centred applications. Vehicle-centred applications are services addressing single vehicles, which include the following two types:

- *Vehicle management and configuration services.* Services for vehicle management include remote failure diagnosis, or remote version management and update of non-critical software in the control units of a vehicle. Such services are typically provided by the vehicle manufacturers. BMW TeleServices for instance, as of 2006, enables remote vehicle diagnostics (as illustrated in Figure 2.4). Vehicle management may also comprise V2R AutoNet applications such as a traffic light-controlled dimming of the vehicle head-lights. Here, a traffic light communicates its status to the vehicle. Based on this information, the vehicle decides to dim its lights in order to save energy. Note the similarity to an automated engine start–stop application at a traffic light, triggered by the same information. In contrast to the light dimming, which does not affect the driving, we would classify this application as a driving-related one, because engines may shut down when approaching the traffic light, changing the mode of driving. Vehicle configuration services enable the configuration of vehicles, for example on a personal computer connected to the Internet. This configuration is transmitted to the vehicle, which in turn adapts its configuration parameters accordingly. This enables an easy configuration of the driver's vehicle. Examples

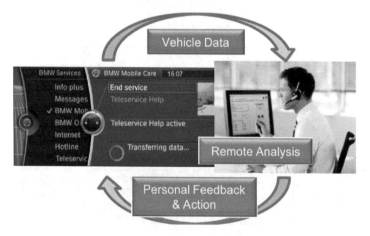

Figure 2.4   TeleServices for remote vehicle diagnostics. Reproduced by permission of BMW Group.

Figure 2.5 Google maps integration in vehicles. Reproduced by permission of BMW Group.

include setting automatic blinking functionality, user preferences of the climate control, or gear shifting characteristics for automatic transmissions.

- *Mobile device integration services.* AutoNets also enable mobile consumer electronics devices to connect to vehicles. Respective services are needed to integrate these devices into the vehicle context. Synchronisation mechanisms are required to combine and to unify mobile device services with the services provided by the vehicles. We therefore classify this integration here as a vehicle-related application. The applications on top of the integration may be classified in either of the three classes. An illustrating application is the integration of smartphones. With MINI Connected,[1] drivers can connect their iPhones to their vehicle. The MINI Connected APP then offers functionality which is driving-related (e.g. the MINIMALISM Analyser which uses vehicle data to provide feedback on an efficient driving style), passenger-related (e.g. news feeds) or vehicle-related by providing status information on the vehicle. Through this channel, MINI also offers a service via this smartphone integration that is also available for other brands in an integrated fashion, called 'Google send to Car'. Here, a driver can choose a location via Google Maps and send this location directly from the Google Maps homepage to their vehicle. The vehicle receives this information either directly via a built-in modem or via a connected cellular phone. The location can then be forwarded to the navigation system, which plans the route to the location (Figure 2.5 illustrates the use of this application).

---

[1] http://www.minispace.com/article/mini-connected/456/.

2.1.3   Passenger-Related Applications

The third application class is focused on the passengers within a vehicle. Such passenger-related applications address individual issues of the driver or passenger, but they neither have any impact on the driving task nor an immediate impact on the traffic situation on the road. As a consequence, such applications have no bearing on intelligent transportation systems, which is an important difference to driving-related applications and vehicle-related applications. However, passenger–related applications are not necessarily limited to local operation within one vehicle only – they may also rely on information from other vehicles or from third party providers. Hence, passenger-related applications may also use AutoNets for their proper operation.

Passenger-related applications have a direct relation with the drivers within a vehicle. Prominent examples of such applications are entertainment or so-called 'infotainment' applications, which make travelling more convenient and enjoyable for the passengers within the vehicle.

Passenger-related applications can be realised by V2V, V2Mobile, V2R, V2Central, V2Internet or V2Private communications. Nevertheless, the most important communication type is V2Internet, since many future entertainment and infotainment applications will be provided in a personalised way via the Internet by their respective content providers.

Besides the traditional entertainment applications available in almost every vehicle (such as CD players or FM radio), there are several more applications that make driving more convenient, pleasurable and comfortable for both the driver and the passengers. We classify passenger-related applications into (ubiquitous) information provisioning, entertainment, communications and comfort. Figure 2.6 illustrates this classification, which will be detailed in the following.

> **Ubiquitous information provisioning.** The main goal of applications in this category is to provide a vehicle's passengers with relevant information, ideally personalised, tailored to the situation and easily accessible. The information itself will be typically offered from third party content providers, either integrated in an overall service offering where the source of the content is not necessarily obvious for the user, or as a clearly labelled third party offering. The information is

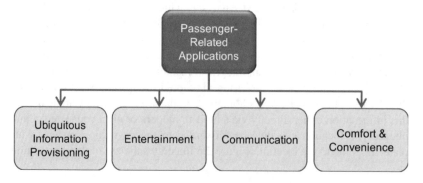

Figure 2.6   Application classification for driver-related applications.

transferred to the vehicle via IP connectivity over an Internet connection, or alternatively, via the portal of the service provider, typically the vehicle manufacturer. Based on the kind of information, we distinguish the following services:

- *General interest information services.* General interest information services provide information which is (supposed to be) relevant or interesting for a large number of recipients. Prominent examples include news-tickers, RSS feeds, or weather or stock market information.
- *Personalised services.* In contrast to general interest, personalised services provide information which is tailored to the specific interest of a driver or passenger. This means that different drivers may get different sets of information depending on their preferences. Examples are Twitter channels or access to personal subscribtions of magazines.
- *Situation-aware services.* Finally, situation-aware information provisioning offers or provides access to information that is pre-filtered, pre-selected or simply pre-organised to serve the information needs in a particular situation. Examples include location-based services or travel directions directly to the next petrol station.

Entertainment. Prominent examples of in-car entertainment are multimedia applications such as audio-based entertainment or (back-seat) video. Sources of entertainment have traditionally been AM/FM radio broadcasts or CDs brought into the car. We have now witnessed the introduction of IP-based entertainment services, enabled by the AutoNet infrastructure. The example of a juke box application within a vehicle demonstrates that modern entertainment applications can of course be realised autonomously within a vehicle without requiring any interaction with other entities outside the vehicle. But obviously the capabilities and performance of such entertainment functions can be substantially improved by AutoNets. Similarly to ubiquitous information provisioning services, entertainment services can be partitioned into the following types:

- *Broadcast services.* Broadcast services include traditional entertainment like radio channels, television, but also digital broadcast applications and IP-based applications like podcasts.
- *Personalised entertainment.* Personalised entertainment items are multimedia applications that are focused on the preferences of the passengers. The most prominent example would be car audio applications, such as players for CD, DVD or MP3, jukebox applications, personalised radio applications or audio books. Another example are so-called music maps [1], which organise either the personal music database or access to a wider database according to a passenger's interest and according to the degree of similarity of the type of music, as illustrated in Figure 2.7.
- *Situation-aware entertainment.* Some modern entertainment features even adapt to the situation at hand. An example are the Dynamic Music and Mission Control features of MINI Connected. Dynamic Music plays specific music themes that adapt to the driving style and thus form a situation-adaptive tune experience. Mission Control features greetings, personal hints or funny remarks virtually

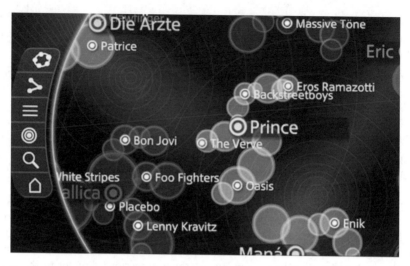

Figure 2.7 Example of personalised entertainment: music maps. Reproduced by permission of BMW Group.

spoken by the car to the driver and passengers, reacting to driving style or environment parameters such as temperature and so on.

- *Communication services.* Probably the best known communication service in vehicles is telephony, supported by built-in microphones, speakers and even voice recognition for hands-free usage. Even deeper integration into the vehicle electronics and infotainment systems enables more value added features for mobile telephony. For example, it is possible to trigger automatic telephone calls to close locations for reservations in restaurants or cinemas, or even to automatically contact a help centre if the vehicle detects a critical emergency situation. Other communication features used by digital natives are also finding their way into the vehicles. These services include instant messaging as well as community platforms.

Comfort and convenience services.   Finally, there are a number of applications improving the comfort and convenience of the passengers while travelling in the vehicle. Example applications are manifold, as illustrated by the following three:

- Pervasive healthcare applications may include monitoring of the passengers' health during travel and possibly allow remote access to these values (from home or for a doctor) – see for example [2].
- Authorised credit cards in combination with respective contracts and applications allow one to pay automatically at petrol stations after filling up the fuel tank ('drive-through refuelling').
- Home services can be used by the passengers in the vehicle to control devices (or to monitor them with cameras) that are at home. This way, the vehicle plays an active role in the so-called 'home environments'.

It is not possible for every application to fit into exactly one application class. For example, consider Facebook access, which is certainly used for communication, but could also be viewed as entertainment, and even considered an information provisioning application (e.g. when receiving messages from celebrities). However, we believe that a differentiation of applications as proposed in this section is helpful since it allows the definition of typical characteristics and requirements for each application class, which is a precondition for the design (and business case calculation) of attractive driver-centred services.

Hundreds of applications have been considered, prototyped and marketed so far for the different application domains. The main focus of this book will be on the driving-related as well as the vehicle-related applications. These applications are those specific to the automotive domain and those arguably posing the biggest challenges to communication technology.

2.2   Requirements and Overall System Properties

An important issue in AutoNets is the distribution and dissemination of context information among the entities participating in the AutoNet. AutoNets may comprise a variety of different communication systems and communication scenarios. Each communication system has its specific characteristics and advantages with respect to the exchange of information in an intelligent transportation system.

Figure 2.8 summarises the impact of the cooperative assistance applications (introduced in Section 2.1.1) on the typical characteristics of a driving task. This figure shows that there is a wide range of different properties in AutoNets, and each of the different properties are more or less important for different applications.

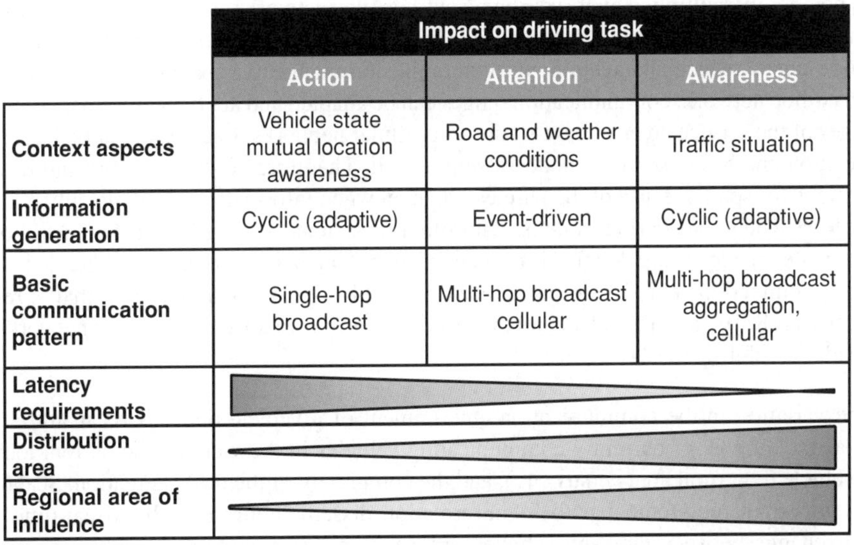

	Impact on driving task		
	Action	**Attention**	**Awareness**
Context aspects	Vehicle state mutual location awareness	Road and weather conditions	Traffic situation
Information generation	Cyclic (adaptive)	Event-driven	Cyclic (adaptive)
Basic communication pattern	Single-hop broadcast	Multi-hop broadcast cellular	Multi-hop broadcast aggregation, cellular
Latency requirements			
Distribution area			
Regional area of influence			

Figure 2.8   Characteristic properties of different driver assistance application classes.

In order to handle information in such heterogeneous AutoNets effectively, the following three requirements have to be met for a successful deployment of applications for intelligent transportation systems:

- An optimal utilisation of the available bandwidth capacity in all driving situations.
- Bandwidth usage according to solidarity and fair rules.
- Fine granular prioritisation of messages.

These requirements are basically focused on ad-hoc networking scenarios in AutoNets, but are also valid for cellular and other types of communication systems. In order to achieve the three requirements, the subsequent properties have to be addressed: scalability, prioritisation, mobility, situation adaptivity and efficient bandwidth utilisation.

Scalability

The density of vehicles on roads is typically subject to high variation, which has an immediate impact on communication, as illustrated by the following two extreme situations:

1. A low density of communicating vehicles means that the possibility of forwarding messages with low latency in an ad-hoc network is limited due to the lower connectivity of the vehicles. In order to maintain regionally important information, vehicles have to store the information physically until they meet another vehicle that incorporates the information (and itself takes care of the dissemination of the information). This principle is called *store and forward* [10, 11]. In cellular networks, this problem is overcome by base stations providing connectivity among all nodes.
2. In scenarios with a high density of vehicles, lots of information will be communicated, which is potentially multiplied by forwarding in ad-hoc type networks. The available bandwidth in this case becomes a limiting bottleneck. The amount of data that has to be handled by each node can be reduced by adapting the communication range to a spatially relevant radius. In an ad-hoc network, communication ranges can be adapted and are typically smaller than the sizes of the service areas of base stations in cellular networks. Each base station has a fixed position and has to serve all nodes within its cell. Therefore, ad-hoc networks allow more options for spatial re-use of the wireless channel when information is only locally relevant, at least when considering systems with omni-directional antenna characteristics. For the interested reader, more detailed information can be found in a number of publications, from early foundations [12, 13] to more recent work [14, 15, 16, 17] and work that is related such as [18, 19, 20]. To understand the mechanisms of the capacity of ad-hoc networks, the reader is referred to [24].

The variation in the communication environment of a vehicle depends on many factors, such as the number of currently communicating vehicles in the vicinity, the current location of a vehicle (e.g. rural area versus city), and the complexity of the traffic situation. Scalability is an important requirement both for information dissemination as well as real-time local data exchange between close-by vehicles. Information dissemination algorithms have to be designed in such a way that the requested reliability of the information is achieved in sparse as well as in highly dense scenarios. Moreover, the heterogeneity of AutoNets has to be utilised appropriately in order to take the best possible advantage of combining the available communication systems in an optimal way. Analyses of the capacity of hybrid cellular and ad-hoc networks provide essential insights to understand this complex topic [21–23].

Prioritisation

A wireless communication channel is usually used by several (and different) applications. AutoNets require that time-critical applications are preferred for channel access. This is necessary in order to reduce their latency as much as possible in order to provide the best conditions for any safety application requiring low latency. However, even time-critical messages require a fine-granular prioritisation strategy since the current criticality of a message depends on both the information the message is carrying and the respective context. This way, a suitable prioritisation strategy has to consider the different application classes (and their requirements) as well as the context of the vehicles. Prioritisation schemes for AutoNets are discussed by [25] or [26].

Mobility

The communication systems as well as the communication protocols have to take into consideration the high mobility of vehicles. For example, the transmission window of two oncoming vehicles on a motorway will be in the order of seconds. In this period of time, it may not be possible to transmit all relevant messages reliably. This is another motivation for the prioritisation introduced in the previous section.

The point in time for both the generation of messages as well as their transmission is of particular importance for AutoNets. Vehicles observing (or predicting) a local danger in their context will warn other vehicles about this event. Since this information will be available for the targeted region for a longer time, the system additionally requires 'revocation messages' in case vehicles cannot verify the local danger (any more). Moreover, it has to be considered that a local danger may be observed by different vehicles in a different way, since the context also differs from vehicle to vehicle. For the example of an aquaplaning local danger warning, the detection of aquaplaning not only depends on the road condition and the amount of water, but also on the speed of a vehicle, type and condition of the tyres, and even the way the vehicle was driven through the aquaplaning area. In the event that a slower vehicle does not detect aquaplaning (due to its better tyres and lower speed profile), it cannot be automatically concluded that the aquaplaning area no longer has any potential danger for other vehicles. More on AutoNet situation handling can be found in [27, 28] and [29].

Situation Adaptivity

As with the mobility of the vehicles, so their situation changes continuously too. This unpredictable behaviour is relevant for safety-relevant weather conditions, as well as for traffic situations or the current condition of the road. The information dissemination strategy has to adapt itself to these changing conditions dynamically. It requires in particular that the distribution area as well as the lifetime of the information is adapted to the current context, while replacing out-dated information with up-to-date information.

Bandwidth Utilisation

In wireless communication networks, the available bandwidth is a rather limited resource and has to be divided among a number of different applications in a fair way. Hence, the available bandwidth should be utilised both to a maximum and efficiently in AutoNets – as well as in other wireless communication scenarios. On the other hand, applications like local danger warning are based on predictions on a partial information base. Their reliability can be improved significantly with the availability of more context information. Potentially available bandwidth can therefore be used for such purposes to a certain degree. In this situation, the communication system always has to ensure that messages with a potentially minor utility for other vehicles do not have a negative effect the transmission and latency reliablilty of more important messages.

2.3   Overview on Suitable Communication Technologies

So far, we have introduced the entities communicating with each other in AutoNets (Section 1.3) as well as a classification scheme for AutoNet applications. We have also shown the cases for different communication systems. We have different types of wireless communication systems, which have their specific characteristics, pros and cons. Hence, they are suitable for some types of AutoNet applications, whereas they cannot be used for other AutoNet applications. In this section, we will map the different communication systems with respect to their suitability for the different AutoNet applications as well as for the different communicating entities participating in the respective AutoNet scenarios. We therefore first introduce the basic communication systems together with their basic characteristics. In the second step, we will show which communication systems can be used for realising the different AutoNet applications (including the communicating entities participating in the respective scenarios). Section 8 will describe the different communication technologies used for different systems.

2.3.1   Communication Technologies

There are different types of wireless communication systems that can be used for information exchange in AutoNets. In general, AutoNets can make use of all kinds of wireless systems: broadcast systems, cellular systems, local wireless systems like WLAN or infrared, and near field communication (NFC) systems.

2.3.1.1   Broadcast Systems

Broadcast communication systems are used to distribute noticeable amounts of information among a group of receivers. Broadcast communication has been used for a long time for information dissemination to vehicles. The analogue FM or AM radio channels have been supplemented and partially replaced by digital radio broadcast services. Broadcast systems are very important for today's vehicular environments since they provide traffic information for vehicles as well as entertainment for the drivers.

Broadcast systems typically allow for unidirection information transfer, that is it is only possible to transmit information from one sender to multiple receivers. This way, it is possible to realise rather high data rates in a rather large cellular size. Depending on the system, one may expect several MB/s in a cellular area of 100 km and above. Broadcast systems also feature a full coverage, that is one may expect to have access to broadcasted information at almost every location in Europe. One also may expect a rather high transmission delay for information sent via broadcast channels. Thus, while broadcast networks work well to widely disseminate information of general interest, shortcomings include the inability to efficiently deliver individual data, potential high latency because messages have to be scheduled in the overall transmission and can only be sent out sequentially, and the lack of an uplink channel.

2.3.1.2 Cellular Systems

Starting with GSM in the early 1990s, UMTS and the upcoming LTE, cellular systems are widely accepted for telephony and data exchange services. Cellular systems are also very important for vehicular environments since they enable vehicles to access personalised information. Today, cellular communication networks support different technologies and are quite ubiquitous. For example, in almost all countries one may expect a full coverage of at least GSM (2G) connectivity, whereas UMTS is nowadays available basically in areas of high population density (like cities) and along populated motorways. With these networks, information can be made available or be distributed easily and quickly in a large geographic area. In contrast to broadcast systems, cellular systems provide bidirectional information exchange, which enables typical communication scenarios relevant for personalised information support. Hence, the available bandwidth within a cell has to be shared among the communicating entities within this cell, also dissemination of information means individual delivery to each recipient and therefore a waste of network resources. Cellular size is far smaller compared to broadcast systems and may be up to 30 km in radius for GSM 2G. Broadcast and multicast services in cellular systems have been standardised but not rolled out. For example, the MBMS (Multimedia Broadcast Multicast Service) in UMTS enables the broadcast of information relevant for all mobile devices within a cell. This feature may become an interesting option in the future. However, today's telecommunication providers do not yet provide it for common use. The operating costs of data exchange using cellular communication systems are still quite noticeable. Also, latency between two vehicles is comparatively high.

2.3.1.3 WLAN Systems

WLAN communication systems based on IEEE 802.11 are the de-facto standard for high speed wireless Internet access in home environments, buildings and public places. For example, almost every notebook computer features a wireless network interface, which enables one to connect the notebook to a base station.

WLAN communication currently plays a minor role in vehicular environments since there is no nation-wide infrastructure of so-called 'hotspots'. Hence, instant access as provided by cellular systems is not available for vehicles. The size of hotspots is rather limited to a reception area of 100 metres and more. However, WLAN systems allow for a high speed information exchange within one cell. Moreover, the delays are rather small, but they may vary significantly depending on the number of communicating systems within one cell. IEEE 802.11 also has an ad-hoc communication mode, where clients can establish links among each other without a central router infrastructure. Such spontaneous networking among vehicles in an ad-hoc fashion avoids the need for costly infrastructure. Therefore, such networks typically have no operational costs. Such systems enable the transmission of information to entities in the direct communication range. This makes them an interesting option for AutoNets since data exchange is always possible between neighbouring vehicles.

The respective handshake protocol of standard WLAN systems is not optimised for low latency, leading to a comparably long handshaking process that is not suitable for environments with frequent changes of participating nodes. The latter, however, is one of the main characteristics of AutoNets. In order to overcome this (and a few more) problems of off-the-shelf WLAN, a dedicated working group of IEEE 802.11 established an AutoNet derivative. We call these particular types Auto-WLANs. Latency for direct communication between neighbouring vehicles is very low in these. However, it takes more time to transfer information over a longer distance due to the short-range nature of the wireless technology used, which requires the forwarding of messages by intermediate nodes. In ad-hoc networks with few vehicles, it may not even be possible to forward messages over longer distances.

Compared to WLAN Systems, Auto-WLAN systems have a slightly higher transmission range, in which the communicating entities may organise themselves in an ad-hoc fashion. Auto-WLAN systems are typically not based on wireless cells, but enable local information exchange only. The data rates are lower than with standard WLANs, trading for higher reliability. Auto-WLAN systems therefore provide a higher communication quality of service, which is necessary for vehicular environments. The transmission delay is very low since no base stations are involved in the communication. In Chapter 8 we will introduce the IEEE 802.11p standard and its specific characteristics in more detail.

Local wireless communication systems using infrared technology have mainly been used and developed for automated road tolling systems [36].

Besides infrared-based communication, radio frequency identidication (RFID) technology has also been used for road tolling applications [37]. The following paragraph covers other uses as well.

2.3.1.4 NFC Systems

Near-field communication (NFC) systems are also widely accepted for short-range communication. Examples include Bluetooth technology or even RFID-based communication, but also the wireless communication between a vehicle and the vehicle keys.[2] Respective systems are an interesting option for the information exchange between mobile (smart) devices and handhelds since they don't consume much energy.

Like Auto-WLAN systems, communication in NFC systems is organised in an ad-hoc fashion. The transmission is short-ranged and typically covers a range from several centimetres up to around 50 metres. Moreover, the data rates of NFC systems are limited and the systems often employ higher delays than specially designed Auto-WLANs. Nevertheless, while Bluetooth is common to connect telephones to vehicles, NFC systems are also under consideration for vehicle safety systems. Vulnerable road users (VRUs) require portable, light, small and extremely energy-efficient devices to participate in AutoNet scenarios. NFC/RFID-based

---

[2] This communication system is based on the frequency bands of 433MHz or 868MHz.

cooperative sensor technology and systems to protect pedestrians and other VRUs have been presented in [30–35].

Both Wireless LANs and NFC systems are sometimes referred to as dedicated short-range communication (DSRC). While this term appears difficult to clearly define, it is often used in conjunction with such systems for AutoNets. An overview of AutoNet relevant DSRC systems is provided in [38].

2.3.2   Suitability for AutoNet Applications

Due to the different strengths and weaknesses of the above mentioned communication systems, they target different AutoNet applications. In this section, we will show which communication system is suitable for which communication requirement of different AutoNet applications. In the figures of this section, the columns reflect the different application types for driving-based, vehicle-based and passenger-based applications, respectively. The rows in the figures represent the different communication types described in Section 1.3, that is vehicle-to-vehicle (V2V), vehicle-to-roadside (V2R), vehicle-to-mobile (V2M), vehicle-to-central-infrastructure (V2Central), vehicle-to-Internet (V2Internet) and vehicle-to-private-network (V2Private). We have clustered them into the four categories, which have similar characteristics with respect to their suitability for the type of communication system:

- *V2V, V2R* will typically both use the same communication technologies – from the communication technology point of view, roadside stations can be considered as 'stationary' vehicles, resulting in the same requirements for the communication technologies.
- *V2M* typically operates either via a local pairing of Bluetooth or NFC, for example, or via an in-vehicle WLAN hotspot. It is worth mentioning that mobile devices comprise common consumer electronics device, but also RFID-based solutions, and will thus also cover the VRU scenarios.
- *V2Central* cannot be clustered with other communication forms.
- *V2Internet, V2Private* are typically realised on IP-based communication protocols, resulting in the same requirements for the communication technologies.

Figure 2.9 illustrates the communication technologies for driving-related communication types. We have already mentioned that there are two major classes of driving-related application, focused on either manoeuvering or operating the vehicle in the current situation (proximity information), or navigating it in a foresighted way (environmental information). As we have seen, these two classes have very different communication requirements with respect to security, latency and geographic coverage. For navigating applications, cellular systems, WLAN and Auto-WLAN systems are suitable for V2V and V2R communication. In general, the distribution of environmental information benefits from the large geographic coverage of cellular systems. In contrast, manoeuvering and safety assistance functions pose very strong performance-related requirements on the AutoNets. Due to the nature of vehicular safety, respective applications require a low-latency, high-availability communication and a high security level. Hence, for manoeuvering and stabilising applications, both cellular and WLAN systems can be used only for applications that do not rely on very low (and robust) latencies, reflected by the braces in Figure 2.9. Nonetheless, as we will see in Chapter 4,

Driving-related applications		
Navigating	**Manoevering**	**Stabilising**
V2V **V2R** Cellular WLAN Auto-WLAN	(Cellular) Auto-WLAN	(Cellular) Auto-WLAN
V2M Cellular WLAN Auto-WLAN NFC	NFC	NFC
V2Central Broadcast cellular	Broadcast cellular	Broadcast cellular
V2Internet **V2Private** Cellular WLAN	Cellular WLAN	Cellular WLAN

Figure 2.9   Technologies for diving-related application communication types.

for some driving-related scenarios a more detailed consideration of the overall system logic
is necessary to deduce the specific requirements and therefore benchmark the most suitable
means of communication.

In this context, V2M is mainly relevant for navigating applications. For manoeuvering and
stabilising applications, so far only the integration of NFC capabilities into mobile devices has
been analysed to offer a potential for the manoeuvering or stabilisiation actions of a vehicle
(in the context of VRUs). V2Central, V2Internet and V2Private are relevant for all three types
of applications since the information collected by the applications may be useful for traffic
management centres, third parties or private service providers. V2Central is typically based
on broadcast services and cellular services, whereas Internet access is typically realised either
via cellular systems or via WLAN systems.

For vehicle-related applications, Figure 2.10 shows the suitable communication technologies
for the respective communication form. V2V and V2R communication is of minor relevance for
fleet-based applications. For vehicle-centred services, they may be based on cellular systems,
WLAN and Auto-WLAN systems. V2M communication is relevant for both types of driver-
related applications: mobile devices may be connected with the vehicle using WLAN systems
or NFC systems in order to interact with the vehicle. In contrast to driving-related applications
as well as passenger-related applications, vehicle-related applications will not be directly
based on V2Central communication. Such information is typically used directly by a fleet
provider, whereas it is of rather limited relevance for vehicle-centred applications according
to our definition above. These providers are typically located in a private network or in the
Internet. Hence, V2Internet and V2Private communication are relevant for both applications,
which will be based either on cellular systems or on WLAN systems.

Finally, Figure 2.11 depicts the relevant communication technologies for passenger-related
applications. For respective applications, almost every communication form may be relevant.

Vehicle-related applications		
	Fleet-based	**Vehicle-centred**
V2V **V2R**		Cellular WLAN Auto-WLAN
V2M	WLAN NFC	WLAN NFC
V2Central		
V2Internet **V2Private**	Cellular WLAN	Cellular WLAN

Figure 2.10   Technologies for vehicle-related application communication types.

Passenger-related applications			
Information	**Enter-tainment**	**Communi-cation**	**Comfort & convenience**
V2V **V2R**		Cellular (WLAN)	Cellular (WLAN)	Cellular (WLAN)
V2M	Cellular WLAN NFC	Cellular WLAN NFC	Cellular WLAN NFC	Cellular WLAN NFC
V2Central	Broadcast cellular	Broadcast cellular		
V2Internet **V2Private**	Cellular WLAN	Cellular WLAN	Cellular WLAN	Cellular WLAN

Figure 2.11   Technologies for passenger-related application communication types.

Readers can easily figure out a scenario for each of the different application types. Since passenger-related applications typically do not have hard real-time requirements, the suitable communication technologies are basically independent of these types. V2V and V2R communication are of minor importance for ubiquitous information support, since this information is likely available on the Internet. For entertainment, communication and comfort services, cellular systems or WLAN systems are typically used. One may even think of highly specialised scenarios, where information sent via Auto-WLAN can be used to realise a respective application. For example, an entertainment application can be customised according to the current traffic situation on the road, which is reflected by the braces in Figure 2.11. V2M communication is typically realised by common communication technologies implemented in mobile devices. Likewise, V2Central communication is typically based on broadcast or cellular systems. It is worth mentioning that V2Central communication is of minor relevance for comfort-based applications. Finally, V2Internet and V2Private communication are important for driver-related applications, typically based either on cellular systems, or on WLAN systems in the case that a vehicle is currently located within a hotspot area.

2.4   Summary

This chapter has focused on AutoNet application scenarios and their requirements. We introduced a classification scheme for AutoNet applications, which basically reflects the different 'stakeholders' involved together with the vehicles. Obviously, driving-related applications are very important for AutoNet since they rely on AutoNet communication as an integral part of their cooperating functionality. Hence, we introduced driving-related applications in more detail by describing their basic functionality and different methodologies for various aspects. We saw that they pose the highest requirements on the respective communication technology being used in AutoNets. In the second part of this chapter, we have focused on situation handling, which is an important component in almost every driving-related AutoNet application. We outlined a feedback principle in order to estimate the current situation in a reliable way with respect to the heterogeneity of information support. Finally, the last part introduced communication systems applicable to AutoNets. We highlighted the typical characteristics of these systems and identified suitable systems for their respective applications and communication forms.

The classification in this section is a precondition for understanding the different design principles in the rest of this book. In this way, we have now laid the necessary foundation to the system architecture of AutoNets, which will be explained in the following chapter.

References

[1] Broy, V., Althoff, F., Klinker, G. (2006) MusicMap – Eine kartographische Metapher für Fahrerinformationssysteme (a cartographic metaphor for driver information systems) *Tagungsband der Fachtagung für Nutzergerechte Gestaltung Technischer Systeme (VDI/VDE USEWARE), atp Automatisierungstechnische Praxis, Heft 10/08.*

[2] Noshadi, H., Giordano, E., Hagopian, H., Pau, G., Gerla, M., Sarrafzadeh, M. (2008) *Remote Medical Monitoring Through Vehicular Ad Hoc Network. Proc. International IEEE Vehicular Technology Conference (VTC 2008-Fall).*

[3] Rasmussen, J. (1983) Skills, rules and knowledge: Signals, signs and symbols, and other distinctions in human performance models. *IEEE Trans. Systems, Man, and Cybernetics* (SMC-13), 257–266.

[4] Dinger, J., Hartenstein, H. (2006) *Defending the Sybil Attack in P2P Networks: Taxonomy, Challenges, and a Proposal for Self-Registration. Proc. International Conference on Availability, Reliability and Security (ARES 2006)*.

[5] Ostermaier, B., Dötzer, F., Strassberger, M. (2007) *Enhancing the Security of Local Danger Warnings in VANETs – A Simulative Analysis of Voting Schemes. Proc. International Conference on Availability, Reliablility and Security (ARES 2007)*.

[6] Weyl, B. (2008) *On Interdomain Security: Trust Establishment in Loosely Coupled Federated Environments*. Verlag Dr. Hut.

[7] Kosch, T. (2004) *Local Danger Warning based on Vehicle Ad-hoc Networks: Prototype and Simulation. Proc. International Workshop on Intelligent Transportation (WIT 2004)*.

[8] Green, P. (1995) *A Driver Interface for a Road Hazard Warning System: Development and Preliminary Evaluation. Proc. World Congress on Intelligent Transportation Systems*, 1795–1800.

[9] Harder, K. A., Bloomfield, J. (2003) *The Effectiveness of Auditory Side and Forward-Collision Warnings in Winter Driving Conditions*. Report from Minnesota Department of Transportation.

[10] Kosch, T. (2004) *Efficient Message Dissemination in Vehicle Ad Hoc Networks. Proc. 11th World Congress on Intelligent Transportation Systems (ITS)*, Nagoya, Japan, October 2004.

[11] Kesting, A., Treiber, M., Helbing, D. (2010) Connectivity statistics of store-and-forward intervehicle communication *IEEE Transactions on Intelligent Transportation Systems*, Volume: 11, Issue: 1, pages 172–181.

[12] Nelson, R., Kleinrock, L. (1984) The spatial capacity of a slotted ALOHA multihop packet radio network with capture *IEEE Transactions on Communications*, Vol. 32 Issue 6.

[13] Kleinrock, L., Silvester, J. (1987) Spatial reuse in multihop packet radio networks. *Proc. of the IEEE*, Volume 75 Issue 1.

[14] Fengji, Y., Su, Y., Sikdar, B. (2003) Improving spatial reuse of IEEE 802.11 based ad hoc networks. *Proc. IEEE Global Telecommunications Conference (GLOBECOM '03)*.

[15] Guo, X., Roy, S., Conner, W. S. (2003) Spatial reuse in wireless ad-hoc networks. *Proc. 58th IEEE Vehicular Technology Conference (VTC 2003-Fall)*.

[16] Santhapuri, N., Nelakuditi, S., Choudhury, R. R. (2008) On spatial reuse and capture in ad hoc networks. *Proc. IEEE Wireless Communications and Networking Conference* (WCNC 2008), Las Vegas, USA.

[17] Toumpis, S., Goldsmith, A. J. (2003) Capacity regions for wireless ad hoc Networks *IEEE Transactions on Wireless Communications* Vol. 2, No. 4.

[18] Scheuermann, B., Lochert, C., Rybicki, J., Mauve, M. (2009) A fundamental scalability criterion for data aggregation in VANETs. *Proc. 15th ACM SIGMOBILE International Conference on Mobile Computing and Networking* (MobiCom 2009), Beijing, China.

[19] Mittag, J., Thomas, F., Haerri, J., Hartenstein, H. (2009) A comparison of single- and multi-hop beaconing in VANETs. *Proceedings of the 6th ACM International Workshop on Vehicular Ad Hoc Networking (VANET)*, pages 69–78, Beijing, China, September 2009.

[20] Schmidt, R. K., Leinmueller, T., Boeddeker, B., Schaefer, G. (2010) Adapting the wireless carrier sensing for VANETs. *Proc. 7th International Workshop on Intelligent Transportation* (WIT 2010), Hamburg, Germany.

[21] Liu, B., Liu, Z., Towsley, D. (2003) On the capacity of hybrid wireless networks. *Proc. 22nd Annual Joint Conference of the IEEE Computer and Communications* (IEEE INFOCOM 2003), San Francisco, CA, USA.

[22] Law, L. K., Krishnamurthy, S. V., Faloutsos, M. (2008) Capacity of hybrid cellular-ad hoc data networks. *Proc. 27th Annual Joint Conference of the IEEE Computer and Communications* (IEEE INFOCOM 2008), Phoenix, AZ, USA.

[23] Wu, H., Qiao, C., De, S., Tonguz, O. (2001) Integrated cellular and ad hoc relaying systems: iCAR *IEEE Journal on Selected Areas in Communications*, Vol. 19, No. 10.

[24] Li, J., Blake, C., De Couto, D. S. J., Lee, H. I., Morris, R. (2001) Capacity of ad hoc wireless networks. *Proc. 7th ACM Annual International Conference on Mobile Computing and Networking* (MobiCom 2001).

[25] Torrent-Moreno, M., Jiang, D., Hartenstein, H. (2004) Broadcast reception rates and effects of priority access in 802.11-based vehicular ad-hoc networks. *Proc. 1st ACM International Workshop on Vehicular Ad Hoc Networks* (VANET'04), New York, NY, USA.

[26] Suthaputchakun, C., Ganz, A. (2007) Priority based inter-vehicle communication in vehicular ad-hoc networks using IEEE 802.11e. *Proc. IEEE 65th Vehicular Technology Conference* (VTC2007-Spring), Dublin, Ireland.

[27] Schoen, T., Sick, B., Strassberger, M. (2007) Hazard situation prediction using spatially and temporally distributed vehicle sensor information. *Proc. IEEE Symposium on Computational Intelligence and Data Mining* (CIDM 2007), Honolulu, HI, USA.

[28] Reiss, M., Sick, B., Strassberger, M. (2006) Collaborative situation-awareness in vehicles by means of spatiotemporal information fusion With probabilistic networks. *Proc. IEEE Mountain Workshop on Adaptive and Learning Systems*, Logan, UT, USA.

[29] Woerndl, W., Eigner, R. (2007) Collaborative, context-aware applications for inter-networked cars. *Proc. 16th IEEE International Workshops on Enabling Technologies: Infrastructure for Collaborative Enterprises* (WETICE 2007), Paris, France.

[30] Lill, D., Schappacher, M., Islam, S., Sikora, A. (2011) Wireless protocol design for a cooperative pedestrian protection system. *Proc. 3rd International Workshop on Communication Technologies for Vehicles* (Nets4Cars 2011).

[31] Lill, D., Gutjahr, A., Schappacher, M., Sikora, A. (2010) Cooperative sensor networks for VRU esafety. *Proc. 5th International Conference on Systems and Networks Communications* (ICSNC 2010), Nizza, France.

[32] Rasshofer, R. H., Sikora, A. (2010) Cooperative sensors using localization and communication for increased VRU safety. *Proc. 4th International Conference on Sensing Technology* (ICST).

[33] Meissner, D., Dietmayer, K. (2010) Simulation and calibration of infrastructure based laser scanner networks at intersections. *Proc. IEEE Intelligent Vehicles Symposium* (IV 2010), San Diego, CA, USA.

[34] Rasshofer, R. H., Schwarz, D., Biebl, E., Morhart, C., Scherf, O., Zecha, S., Gruenert, R., Fruehauf, H. (2007) Pedestrian protection systems using cooperative sensor technology. *Proc. 11th International Forum on Advanced Microsystems for Automotive Applications* (AMAA07), Berlin, Germany.

[35] Andreone, L., Guarise, A., Lilli, F., Gavrila, D., Pieve, M. (2006) Cooperative system for vulnerable road user: the concept of the WATCH-OVER project. *Proc. 13th World Congress on Intelligent Transport Systems and Services*, London, UK.

[36] Shieh, W. Y., Wang, T. H., Chou, Y. H., Huang, C. C. (2008) Design of the radiation pattern of infrared short-range communication systems for electronic-toll-collection applications. *IEEE Transactions on Intelligent Transportation Systems*, Sept. 2008.

[37] Blythe, P. (1999) RFID for road tolling, road-use pricing and vehicle access control. *IEE Colloquium on RFID Technology*, London, UK.

[38] Liu, Y., Dion, F., Biswas, S. (2005) Dedicated short-range wireless communications for intelligent transportation system applications: state of the art *Transportation Research Record: Journal of the Transportation Research Board*, No. 1910, Washington, DC, USA.

3

System Architecture

We have shown that there is research all over the globe on AutoNet communication. However, the system architecture(s) developed for these activities are specifically tailored to the requirements and the goal of the particular project: they address specific aspects of an AutoNet and, thus, their architecture is developed for these aspects. There is still a need for a generic system architecture for AutoNets, which covers the basic mechanisms and principles independently of specific target scenarios.

This chapter introduces AutoNets from the system architecture perspective. We first describe the AutoNet architecture starting from a high-level domain view, followed by a description of the individual domains and their subsystems. The elementary correlations of the different domains and subsystems are essential to gain an understanding of the many varieties AutoNets can take.

Based on our AutoNet generic system architecture, we present standardised architectures that are either as comprehensive as our AutoNet architecture or address certain achitectural parts. We show how these standards map to our generic architecture. To a certain extent, this shows which part of the overall system is addressed by which standard. It also shows how the different architecture standards relate to each other. Therefore, you may interpret the different standards as profiles of our generic AutoNet architecture. We then show how the generic reference architecture is applied to the different AutoNet domains, thus defining respective profiles and going into more detail on the architecture of these domains, that is the vehicle, mobile, roadside infrastructure and central infrastructure domains.

3.1 Domain View of AutoNets

The following description of the AutoNet system architecture is compliant with the IEEE 1471-2000 [3, 4] and ISO/IEC 42010[1] [5] architecture standard guidelines. We assign the entities that cooperate to achieve AutoNet functionalities to three different domains: the AutoNet mobile domain, the AutoNet infrastructure domain and the generic domain (cf. Figure 3.1).

---

[1] ISO/IEC 42010 IEEE Standard 1471-2000 2000 Systems and Software Engineering – Recommended Practice for Architectural Description of Software-Intensive Systems.

Automotive Internetworking, First Edition. Timo Kosch, Christoph Schroth, Markus Strassberger and Marc Bechler.
© 2012 John Wiley & Sons, Ltd. Published 2012 by John Wiley & Sons, Ltd.

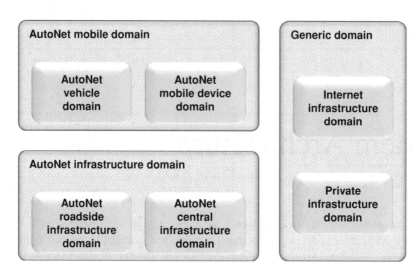

Figure 3.1   AutoNet system domains.

AutoNet-specific components and technologies are required for communication and cooperation in the Autonet mobile and infrastructure domains, whereas services of the generic
domain can be realised without any AutoNet-specific technologies or protocols. However,
AutoNet-specific services may be offered via the generic Internet infrastructure domain as
well as private infrastructure domains.

Within the AutoNet mobile domain, we distinguish the vehicle domain and the mobile
device domain. The vehicle domain comprises all kinds of vehicles such as trucks, buses, cars,
motorcycles and so on. The mobile device domain comprises all kinds of portable devices
like personal navigation devices (PNDs), smartphones and other consumer electronic devices.
Within the AutoNet infrastructure domain, we distinguish the following two subdomains:
the AutoNet roadside infrastructure domain and the AutoNet central infrastructure domain.
The roadside subdomain contains roadside unit entities instantiated, for example, in traffic
lights, variable message signs or road construction sites. The central infrastructure subdomain
contains infrastructure management centres like traffic management centres (TMCs), and
vehicle management centres, such as vehicle service centres or shop floors. The AutoNet central infrastructure entities are thus typically organisational entities, where centrally managed
applications and services are operated.

Figure 3.2 shows the domain entities and their communication relations, that is a link
between two entities means that these two entities may communicate either directly with
each other using one particular wireless or wired communication technology, or indirectly
via one or several communication networks. At this general and abstract level, a link only
means that within an AutoNet system some kind of communication may take place between
any two connected entities. Each link, or interface, is referenced by a simple formalism:
the two domains or entities connected are each denoted by a one or two letter abbreviation
with a hyphen in between, for example V-M denotes the connection between a vehicle and
a mobile device. A link that is connected to a whole domain means that a connection to all
entities within that domain may exist, for example V-I means that a vehicle entity may have

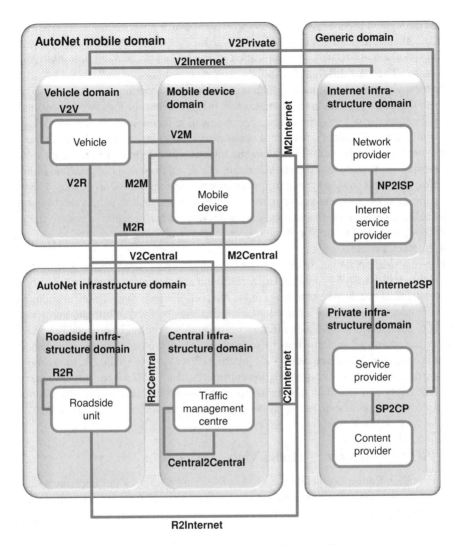

Figure 3.2   AutoNet domain communication architecture.

a connection to either a network provider, an Internet service provider or both.[2] There are also self-referencing links, which mean that, for example, two vehicles may communicate directly with each other. At this stage, we leave it as an exercise for the reader to think about examples for the given relations between the entitites in Figure 3.2. We will reference the

---

[2] The vehicle may only use the services of a network provider to establish a virtual private network connection to a private peer, to the home network or to a company or institution. It may use the services of both a network provider and an Internet service provider to establish a connection to the Internet via a cellular network, for instance. It may only use the services of an Internet service provider if it uses a free access network (this will not be possible if the connections are interpreted on the physical layer, but if they are interpreted on the networking or application layer, for example, the link may denote IP address assignment).

relations throughout the following chapters to explain which relations are addressed by which communication technologies, protocols or standards.

All types of vehicle-related communication introduced in Section 1.3 are reflected in this architecture.[3] There are different viewpoints on system architecture and it is always advisable to define the viewpoint being used, which is a requirement according to ISO/IEC 42010 (see also [6]). The following example illustrates this in more detail. In Figure 3.2 there may be a 'communication party' viewpoint and a 'communication technology' viewpoint, which may differ significantly. Consider a vehicle downloading audio files from the home server of the driver, which is naturally located in the private infrastructure domain. From a communication party viewpoint, the relation is clearly V2Private. However, from a communication technology viewpoint, it depends on the type of communication a vehicle is currently using. The vehicle in the garage can be connected directly to the home network (V2Private), it can be connected via the cellular network of an Internet service provider, which enables access to the private network (V2Internet2Private) or it can be connected via a mobile device to the Internet (V2M2Internet2Private).

While roadside infrastructure entities may feature wireless links to the AutoNet mobile domain, they will usually be connected to infrastructure management centre entities (not shown in Figure 3.2) and may also be connected to the Internet. Hence, a roadside infrastructure entity can communicate with central infrastructure entities, may forward information received from vehicles to central entities and may offer access to AutoNet central infrastructure services. To complete our download scenario, it will also be possible to download the files via a roadside unit into the vehicle, which could be either V2R2Internet2Private or even V2M2R2Internet2Private.

Within the AutoNet-specific domains, the entities contain AutoNet-specific components, and in case the entity consists of an internal network infrastructure, usually a gateway connecting these AutoNet components to legacy systems (vehicle, roadside or central gateway).

3.2   ISO/OSI Reference Model View

From a communication perspective, each AutoNet entity features the implementation of a profile of the generic reference protocol stack depicted in Figure 3.3. This protocol stack follows the ISO/OSI reference model and consists of the following five horizontal layers:

> **Physical communication technologies.** Almost all types of wireless communication technologies can be used for AutoNet applications. We distinguish two types of wireless communication technologies: AutoNet-specific technologies and general technologies. *AutoNet-specific wireless communication technologies* are either created specifically for a certain range of AutoNet applications, or they are derived from a more general wireless communication technology and adapted to the specific requirements of a certain range of AutoNet applications. AutoNet-specific technologies include IEEE 802.11p (derived from IEEE 802.11a) for safety applications, ISO infrared and 5.8GHz DSRC technology for road

---

[3] In fact, there are even more communication types (e.g. NP2ISP, M2M, etc.). They are not introduced in Section 1.3 since they are not focused on vehicles. However, they are also as important for AutoNets as vehicle-centred communication.

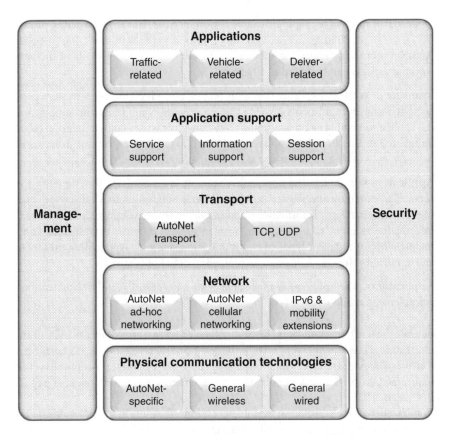

Figure 3.3   AutoNet Generic Reference Protocol Stack.

tolling. *General wireless communication technologies* include broadcast technologies like DAB, DVB, DMB and HDRadio, which are used for traffic information dissemination. Cellular communication technologies like GSM, UMTS, W-CDMA or LTE may be used for a wide range of applications including remote management or Internet access. Finally, local and metropolitan area technologies like several IEEE 802.11 variants or WiMAX are used for local or location-based services or high-speed Internet access, whereas personal wireless technologies like Bluetooth, NFC (near field communication) or RFID are typically used for phone or mobile device connectivity or even for pedestrian detection. The physical communication technologies layer is explained in detail in Chapter 8.

Networking.   Network protocols enable data routing between entities, or forwarding of data to distant entities, that is entities that do not feature a direct communication link. Network protocols are capable of routing data over different types of wireless and wired networks. Again, both general routing protocols as well as AutoNet-specific network protocols are used in AutoNet scenarios. IP needs to be supported for a wide range of applications and also allows the use of higher-layer Internet technologies such as web services. But ad-hoc routing protocols are also required as well as specific optimised variants of them, like specific AutoNet protocols mainly for V2V and V2R communication. Details of the networking aspects can be found in Chapter 7.

Transport. While networking protocols feature an endsystem-to-endsystem connectivity, transport protocols provide true end-to-end communication support for the applications running on a system. Transport layer issues address, among other things, reliability, flow control and congestion avoidance aspects of a network. Both the networking and transport layer are explained in detail in Chapter 6.

Application support.[4] The application support layer contains functionality for service, information and session support. Service support includes discovery and lifecycle mechanisms. Information support comprises messaging capabilities and information storage and aggregation. Session support provides establishment and maintenance of different types of communication sessions. Application protocols and application functionalities are explained in detail in Chapter 4.

Applications. The application functionality of all types of AutoNet applications is implemented on this layer.

In addition, the five communication layers are flanked by two cross-layer planes for management and security. The management layer provides the basic mechanisms to manage a system unit. It also includes cross-layer functionalities. These are important to optimise efficiency, performance and costs of a system unit. The management system component is explained in detail in Chapter 10.

Security is also a crucial issue in AutoNets. Communication in AutoNets deals with the exchange of information relevant for vehicle safety systems' decisions, upon which the safety of human lives depends. The system must therefore take the highest possible economically reasonable security mechanisms to ensure that any kind of tampering with the system or the communication is prevented. To address this, the AutoNet Security system contains the following components: authentication, identity management, privacy, authorisation, key and certificate management, encryption, firewall and intrusion management. The security system is explained in detail in Chapter 9.

3.3   Profiling

So far, we have presented the overall AutoNet system architecture as well as the Generic Reference Protocol Stack. If this Generic Protocol Stack is applied to and instantiated in the different distributed system entities, and if these entities adhere to the specifications set, the entities become able to communicate and even to cooperate. However, different entities have different characteristics, provide or use different services, have different users

---

[4] In the original ISO/OSI reference model, application support consists of the two layers: the presentation layer and session layer. We subsume these two layers into the application support, which indeed covers the aspects of the two layers together with some application-related aspects.

and user interfaces and different communication interfaces. Therefore, the generic architecture needs adaptation to be applied to a specific AutoNet entity. In order to define interoperable configurations for different types of entities and different sets of applications, AutoNets use the concept of *profiles*. Profiles are well-known from existing wireless communication systems like Bluetooth or WiMAX; they are also used for the European ITS Communications Architecture [1] which has been standardised by ETSI as EN 302665 [11].

The profiles concept also allows the handling of the high system complexity with the large number of different entities and the diversity of possible technology components. Profiles typically define a feature or configuration subset of the Generic Reference Protocol Stack. A specific profile incorporates a selection of technology components and, thus, creates a particular variant of the system. A certain profile defines a mandatory set of components (e.g. security components), communication technologies (e.g. wireless technologies), communication protocols (e.g. network protocols) and parameter ranges (e.g. on frequency usage) across the protocol stack. The implementation of a specific profile enables a certain amount of interoperability for the deploying entity. An AutoNet entity may implement one or more profiles.

Profiles are also important for the market introduction of AutoNets. They allow adaptation of the entities to different (regulatory) environments in different markets, and the different feature sets are also useful particularly in the beginning of the deployment phase, since new functionality can be extended easily and interoperably without touching the architecture specification of the overall system. For example, future wireless communication systems can be deployed in an interoperable way for an existing AutoNet by adding a new profile.

3.4 Standardised Architectures

The presented overall AutoNet domain framework architecture, as well as the Generic Reference Protocol Stack, are introduced in this book to present the AutoNet concepts in a generic fashion, abstracting at this level from specific entities, technologies and implementations, but also from specific standards, while aiming to be as comprehensive as possible. They are inspired from the European Intelligent Transport Systems (ITS) Communications Architecture. National and international projects for field operational tests of AutoNets have incorporated this architecture for the development of their system architecture.

Both ETSI and ISO offer standards or draft standards encompassing a comprehensive AutoNet architecture. In the following, we compare our generic AutoNet architecture with these architectures. We also show how they can be mapped. In addition, we outline how other AutoNet-specific architectures addressing certain parts of the system, for example for certain entities, certain system parts or components, protocols or technologies, relate to the presented architecture. We only present AutoNet-specific standards in more detail. This applies for the IEEE 802.11p and IEEE 1609.x family of standards as well as for the draft industry standard published by the Car2Car Communication Consortium (C2C-CC) [2]. Standards from other information and communication technology areas or more general standards for information and communication technology are only referenced but not explained in detail. The reader is kindly requested to consult the existing literature and documentation for these standards (e.g. for TCP/IP including IPv6 and NEMO, GSM, UMTS, DAB, DVB, etc.). AutoNet specific

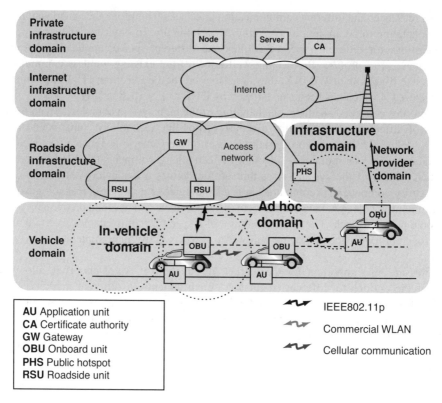

Figure 3.4   Mapping of C2C-CC system view on the AutoNet domain view.

standards that do not address architectural aspects but standardise data formats or protocols are presented in the respective technical chapters throughout the book.

3.4.1   Architecture of the C2C Communication Consortium (C2C-CC)

Figure 3.4 shows how to map the system view of the C2C-CC onto the AutoNet domain view, and Figure 3.5 reflects the mapping of the C2C-CC protocol stack onto the AutoNet Generic Reference Protocol Stack. The AutoNet architecture components are represented as light-grey rectangles with softened edges and are laid over the representation of the C2C-CC system view and protocol stack. For the domain view, the structure of the C2C-CC is quite similar to the AutoNet domain view (aside from a different nomenclature). For the communication reference stack, it is obvious that the C2C-CC stack neither defines a security system in its architecture nor an application support layer (and accordingly is left away in the figure). Also, it only contains wireless communication technologies in the physical communication technologies layer. It can be interpreted as a possible vehicle profile stumb of the AutoNet Generic Reference Protocol Stack. However, without security it would not be complete and therefore needs an additional related security profile. It would also need to be more precise as to which applications out of the mentioned application classes can be implemented on this profile.

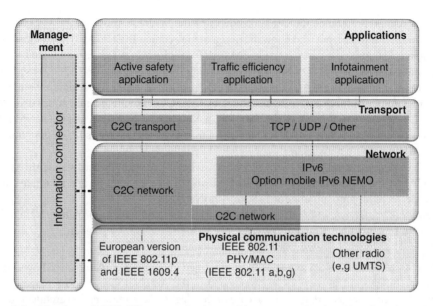

Figure 3.5 Mapping of C2C-CC communication architecture on the AutoNet Generic Reference Protocol Stack.

3.4.2 ISO TC204 CALM Architecture

In ISO, three technical committees are especially relevant for AutoNet systems: TC22 produces standards for road vehicles, TC204 produces standards for intelligent transportation systems and TC211 produces standards for geographic information. Within TC204, Working Group 16 deals with wide-area communications. It has produced a framework and a set of standards for *user transparent continuous mobile packet-switched communication* over a variety of wireless technologies, known as CALM (Communication Architecture for Land Mobile). CALM supports all kinds of ITS applications, ranging from Internet-based services over traffic management services to vehicle safety applications. The standard distinguishes CALM-aware applications and non-CALM-aware applications. The CALM architecture contains four major building blocks:

- The *CALM application block* supports applications that are grouped into the categories safety, non-safety and infotainment. Applications can be either IP-based or non-IP based (the latter relevant especially for safety applications). CALM offers standard Internet interfaces for applications that are not CALM-aware, that is not specifically developed to run on the CALM framework. It provides a special application programming interface to CALM-aware applications with extended functionality.
- The *CALM network block* defines the CALM network layer which is mainly based on IPv6, but also supports dedicated communication protocols, for example to allow for fast safety communications. The IPv6 part was developed in cooperation with IETF.[5] CALM specifically defined the FAST network protocol for vehicle safety communications which

---

[5] IETF: Internet Engineering Task Force; http://www.ietf.org/.

incorporate and reference the IEEE protocols of the WAVE family (wireless access in the vehicle environment), but also allow the usage of the original IEEE WAVE network protocols (compare Sections 3.4.4 and 7.2). CALM FAST additionally offers legacy system support, for example with the CEN DSRC mode, an existing European Norm (EN 13372) used for toll collection systems. The CALM network block also leaves room for vehicle-manufacturer-specific protocols, for example those defined by C2C-CC. For safety applications using IEEE WAVE wireless systems, that is IEEE802.11p technology, a special part of the network layer has been defined that allows a direct binding of the media and bypasses IPv6 communication with safety specific networking.

- The *CALM physical/data link block* encompasses both specific interfaces standardised within the CALM set of standards (CALM-IR, CALM-M5, see Chapter 8) and all other kinds of wireless technologies suitable for ITS, or AutoNet, usage.
- The *CALM management block* is a cross-layer system component, providing system management functionality. It includes a reference to the IEEE 1609.4 standard.

The CALM building blocks can easily be identified in Figure 3.6, which also shows how the CALM architectural framework can be mapped to the AutoNet Generic Reference Protocol Stack. It becomes obvious that, as in the case of the C2C-CC communication architecture, both

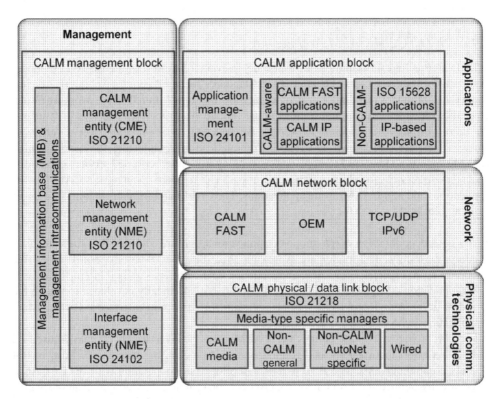

Figure 3.6 Mapping of CALM architectural framework on the AutoNet Generic Reference Protocol Stack.

an explicit application support layer and a security layer are not considered in CALM. Security is handled intrinsically in the other system parts. Within the layers and the management plane, the CALM components can also be mapped to the AutoNet modules. The CALM network management entity maps to the AutoNet network management module and the functionalities of the CALM management entity within the AutoNet reference stack are represented in the respective modules. The AutoNet Generic Reference Protocol Stack does not make an explicit distinction between AutoNet-stack-aware and non-aware applications. The CALM-FAST applications are part of the AutoNet driving-related applications. The other CALM applications are part of either AutoNet passenger-related or vehicle-related applications. CALM FAST is included in the AutoNet ad-hoc networking and the CALM-FAST infrastructure mode is included in the AutoNet cellular networking module. CALM does not provide any particular non-IP networking for cellular networks. IPv6 network mapping is obvious. CALM media and non-CALM AutoNet-specific networks on the physical layer are represented by the AutoNet-specific media within the AutoNet reference stack. Also, both CALM and the AutoNet reference stack support other wireless and wired media technologies. CALM media-type-specific manager functionality is included in the wireless management module in the AutoNet management plane and thus assigned to a different component, being part of the cross-layer optimisation functionality.

An overview of the AutoNet relevant parts of the ISO standardisation organisation and the corresponding standards and work items are provided in Appendix (A). The CALM architecture is standardised as ISO 21217:2010.

3.4.3   ETSI TC ITS Architecture: EN 302 655

As mentioned before, the AutoNet domain and reference stack architectures have largely been inspired by and derived from the European ITS Communications Architecture. ETSI has based the work on its architecture on this harmonised European architecture and published its architecture as a European Norm (EN 302 655). A consolidation of architectures was pushed by the European COMeSafety project (see Section A.2.3) [10], which has convened a task force that has laid the ground for a common understanding of a European framework architecture for AutoNets. The similarity of both the overall European framework architecture (Figure 3.7) with the AutoNet domain architecture and the European ITS station reference architecture with the AutoNet Generic Reference Protocol architecture (Figure 3.8) is evident.

In Figure 3.7 the ITS systems are categorised into four subsystems. In contrast to the AutoNet architecture, there is only one central infrastructure type, the ITS central station. From our perspective, though, it seems important to distinguish between infrastructure management entities and vehicle management entities, which are fulfilling completely different tasks and will therefore feature different system implementations. The figure features examples of possible system implementations of the ITS station reference architecture, that is that the vehicle station consists of a mobile router and a mobile host needs to be understood as a specific implementation of the subsystem. In this case, the mobile router is in charge of the external communication and the applications are deployed on the vehicle host. Also, there is a mobile gateway to the other in-vehicle systems. This is only one deployment alternative. In Section 3.5.1, we provide a more generic vehicle architecture and discuss different possible system implementations. In general, a gateway functionality is part of all subsystems in the

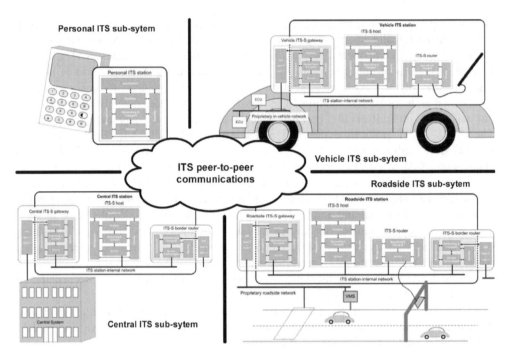

Figure 3.7   European ITS Communications Architecture framework [11]. Reproduced from COMe-Safety project, The project consortium is coordinated by BMW Research and Technology and includes 7 partners: Audi, BMW, Centro Ricerche Fiat, Daimler, ITS Niedersachsen, Renault.

European framework architecture, connecting the ITS system part with the legacy system parts, called vehicle, roadside or central gateway. Also, a host is foreseen for each subsystem. This points to a specific piece of hardware, which is also implicitly stated by using the term 'station' for each subsystem. Rather than the ITS Station Reference Architecture, we use the term *Generic Reference Stack* for our AutoNet architecture and discuss in the following sections how this stack can be implemented and how parts of this stack may be distributed or integrated in legacy systems.

In contrast to the ITS Station Reference Architecture, AutoNet applications are classified differently. Road safety and traffic efficiency map on the driving-related applications, whereas other applications obviously are represented as either passenger-related or vehicle-related classes.

The ITS Station Reference Architecture in Figure 3.8 distinguishes between an ITS network component and a georouting component. Within the AutoNet reference stack, you find the two components AutoNet ad-hoc networking and AutoNet cellular networking instead. This means, that those networking protocols are specifically developed for AutoNets (and presented in detail in Chapter 7). We believe that georouting protocols used for AutoNets should be specifically developed for this purpose. Indeed, a large number of these protocols has been proposed which usually outperform the more generic georouting methods.

Finally, the management and security plane has been structured a bit differently for the AutoNet reference stack. We shifted profile management as a general management functional- ity into the management plane, which usually works together with the authorisation component

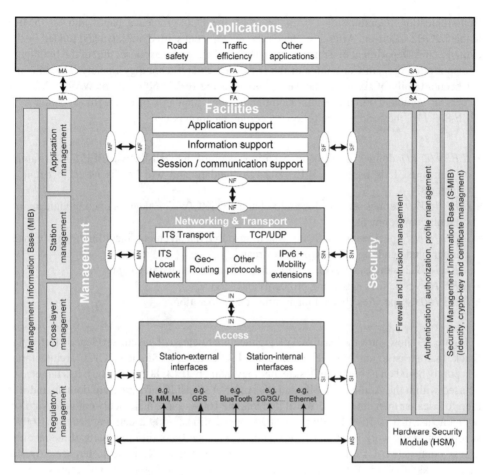

Figure 3.8   European ITS Station Reference Architecture [11]. Reproduced from COMeSafety project, The project consortium is coordinated by BMW Research and Technology and includes 7 partners: Audi, BMW, Centro Ricerche Fiat, Daimler, ITS Niedersachsen, Renault.

of the security system. Instead of a security information base, we move key and certificate management and identity management into single components that offer more than an information base, that is implementations in the infrastructure may deal with distributed certificate grantification and revocation.

3.4.4   IEEE WAVE Architecture Featuring IEEE802.11p and IEEE1609.x Standards

In the United States, work on AutoNet systems has largely had a focus on short range communication at 5.9GHz. This goes back to the development of the initial requirements in the mid 1990s: in 1997, the frequency allocation request was made to the Federal

Communications Commission (FCC). In 1999, the FCC granted a bandwidth of 75MHz in the 5.9GHz spectrum. After that, standards development started, initiated and funded by the United States Department of Transportation (US DoT) under the sponsorship of the Intelligent Transportation System Committee of the IEEE Vehicular Technology Society. The outcome has been a family of six IEEE standards, known as the IEEE 1609 suite of WAVE (Wireless Access in Vehicular Environments) Communications standards (see e.g. [8]. Together with IEEE 802.11p, they address the following communication issues:[6]

- *IEEE 802.11p* defines an AutoNet-specific version of the well-known IEEE 802.11 family of standards. It is based on IEEE 802.11a standard and will be discussed in detail in Chapter 8.
- *IEEE 1609.0* defines services for multi-channel DSRC/WAVE devices to communicate in a vehicular environment.
- *IEEE 1609.1* specifies methods for system resource management, inculding memory management and handling of multiple data streams.
- *IEEE 1609.2* addresses WAVE security and will be presented in Chapter 9.
- *IEEE 1609.3* defines networing protocols and services.
- *IEEE 1609.4* specifies channel management and operation.

Figure 3.9 shows how the IEEE WAVE architecture relates to the AutoNet Generic Reference Stack. In contrast to our AutoNet architecture, ISO CALM, ETSI and C2C-CC architectures, it is purely based on a single wireless technology, namely IEEE 802.11p. Thus, it is targeted to be used within the AutoNet mobile and the AutoNet roadside infrastructure domain and it is therefore leaner than the others. You'll find that the 1609.x parts map nicely onto the AutoNet components, that is 1609.1 defining management entities, 1609.2 defining security, 1609.3 specifying the network layer and 1609.0 and 1609.4 specifying different parts of the medium access control (MAC) layer. Again, we see that the Autonet Generic Reference Protocol Stack works along the ISO/OSI reference model with the management and security system extensions as cross-layer systems. Even though it is based on a single wireless technology, the WAVE architecture supports two kinds of usage with two different network protocols, either UDP/TCP over IPv6 or the so-called WAVE short message protocol (WSMP).

3.5   Subsystem Architectures

In this section, we present the mapping of the AutoNet Generic Reference Stack onto the single AutoNet domains, and what implementation instances of the subsystems can look like. Since the focus of this book is on the vehicles, this will be discussed in more detail than the personal or infrastructure domains.

---

[6] Five of the IEEE 1609.x standards have a direct relation with the system architecture. Besides, IEEE 1609.11 specifies an over-the-air data exchange protocol, which defines services and secure message formats necessary to support secure electronic payments.

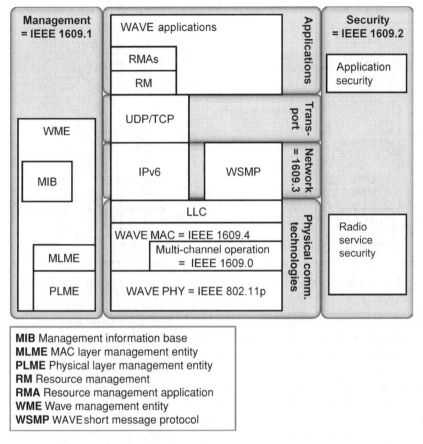

Figure 3.9 Mapping of IEEE WAVE architectural framework on the AutoNet Generic Reference Stack.

3.5.1   Vehicle Architecture

For the in-vehicle architecture definition, we consider first the current situation. The primary motivation for AutoNets is the improvement of vehicle safety and traffic flow, thus from a vehicle-centred perspective the driving-related applications realising information and assistance systems. Current driver assistance systems usually use vehicle integrated sensor and actuator technology and follow the traditional control flow from taking input signals from sensors and the driver, processing these signals and then controlling the system. Typically, this requires coordination of complementing functionality, resulting in either vehicle control via the actuators or driver information via the human–machine interface (HMI) or both. This system behaviour, as outlined in Figure 3.10, needs to go hand in hand with driver behaviour, which follows a similar pattern of perception, information processing with situation detection and assessment, decision making and action. To this control loop, AutoNets basically add two additional logical components for use by the assistance and information systems: an additional input and an additional output component, that is the AutoNet communication capability extends the information base provided by the in-vehicle sensors and driver input.

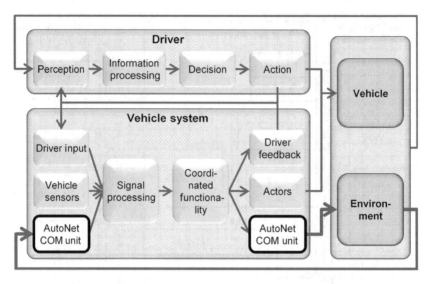

Figure 3.10 AutoNet communications as an addition to vehicle control systems.

This extension potentially reaches far and provides remote input from other sensors which is transmitted to the vehicle by the wireless networks. At the same time, information generated by the in-vehicle systems will be transmitted and provided to other vehicles or infrastructure components, making its own sensor data available to the outside world. Vehicles also offer information that they have deducted from a combined assessment of the vehicle sensor data and data received via the AutoNet communication channels.

If we now take a look at how this control flow is realised in today's vehicles, a very generic architecture looks like the one illustrated in Figure 3.11. Sensors, actuators, driver input and output devices, and processing units called electronic control units (ECUs) are all

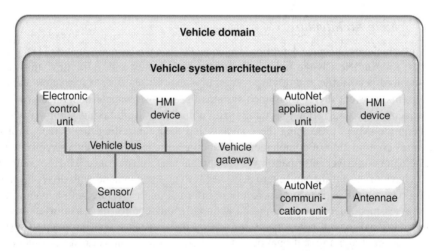

Figure 3.11 In-vehicle generic system architecture.

connected inside the vehicle through – usually wired – in-vehicle communication networks. Typically, these in-vehicle communication networks are different heterogeneous networks depending on their requirements, especially with respect to reliability (for safety systems), latency, real-time capabilities, data rates and cost. In this figure, we just show the very generic architecture, collapsing these heterogeneous networks into one vehicle bus. In reality, networks for multimedia systems like MOST, Flexray, the CAN bus for vehicle control systems like engine management and the LIN bus for simple, low-cost input/output devices like window control systems, and even possibly Ethernet in the future, are all interconnected by one or more gateways. To these in-vehicle networks, AutoNets now open additional channels to a whole world of information providers and consumers outside the vehicle.

To safeguard the in-vehicle safety critical systems from any malfunction or attack through a wireless interface, one may imagine completely decoupling the in-vehicle systems from these wireless networks. In this case, these would be two separate systems rather than a combined system as shown in Figure 3.10. It would mean that only information to the driver could be provided but no input data to the vehicle control systems. In addition, any driver information device can either have a connection to the in-vehicle information providers or the AutoNet applications relying on external information sources.

While this may be a possible architecture for using mobile consumer electronic devices like personal navigation devices (PNDs) inside the vehicle, we will not consider it further but concentrate instead on the integrated system solutions. For these, a gateway needs to connect the in-vehicle networks with the AutoNet wireless networks. We will defer the discussion on protection of the in-vehicle systems from attacks through AutoNets to Chapter 9 and for now assume that an effective protection of the in-vehicle systems is feasible. This gateway may be a special hardware device as illustrated in Figure 3.11, or it may be integrated with either an AutoNet application unit or an AutoNet communication unit. There may be one or more of each of such modules inside the vehicle. Communication is supported by one or more communication modules attached to one or more communication units in charge of any kind of external communication to other vehicles, roadside units or central entities. Applications are supported by one or more AutoNet application units and a number of other dedicated nodes inside the vehicle, namely ECUs. As shown in Figure 3.11, driver input and output devices may be connected to an in-vehicle network or directly to the processing unit running AutoNet applications. An example, as it is implemented today, for the first case is the braking pedal, an example for the second is the navigation information display.

While Figure 3.11 shows the different components for the in-vehicle system from a functional point of view, in a real deployment scenario application units and communication units may, together with the AutoNet gateway, are realised as a single hardware module with connections to the antennae for the wireless systems and driver input/output devices. However, they would usually be implemented as two different devices. This is mainly for cost reasons, because it is sensible to place the communication unit as close as possible to the antennae, since high frequency antenna cable is expensive. Similarly, it is sensible to place the application unit as close as possible to the HMI devices since high quality monitor connections will usually be more costly and also heavier than a network connection between the communication and the application unit. Antennae and HMI devices, however, are physically separated in a vehicle where HMI devices must be within reach of the driver and antennae need to be placed on the outside of a vehicle, today usually on the back part of the vehicle roof or affixed to the vehicle's windows.

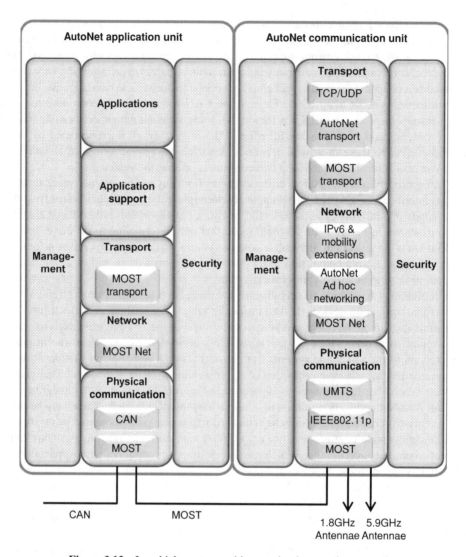

Figure 3.12   In-vehicle system architecture implementation example.

Figure 3.12 shows a specific implementation example of the AutoNet architecture in a vehicle. In this example, the system components are distributed between an AutoNet application unit and an AutoNet communication unit, which are both connected to the vehicle's MOST bus system. The AutoNet application unit in addition is attached to the vehicle's CAN busses. The AutoNet communication unit is equipped with two wireless modules, an UMTS and an IEEE 802.11p module, which both are connected to an antenna cable that ends at the respective antenna placed on the vehicle roof. In this example, all applications and all application support functions only run on the AutoNet application unit.

The AutoNet communication unit implements all ISO/OSI layers up to the transport layer and thus can be regarded as a router. It includes ad-hoc networking capabilities to exchange ad-hoc network packets with other vehicles or roadside units via the 802.11p radio module. It is also capable of IPv6, which is used over the cellular connections established via the UMTS radio module.[7] In the management plane, it implements the respective features necessary to provide the required management functionality (cf. Chapter 10). Vehicle data access is needed, for instance, for position-based routing technologies. Vehicle data access in this case is limited to the vehicle data that can be read from the MOST bus. The cost/benefit module is needed for cost/benefit assessment of channel usage, network beacons or other data packets. These concepts and algorithms will be introduced later. To protect the in-vehicle systems, the communication unit is equipped with firewall functionality. Additionally, it implements privacy mechanisms like pseudonym handling and identity management functionality to hide the vehicle's (and the driver's) identity, and to enable the use of pseudonyms on MAC and network level for ad-hoc communication.

The AutoNet application unit implements the whole stack, but only part of the communication technologies and no wireless network capabilities. Apart from the firewall, which is purely implemented in the communication unit, it is equipped with full security functionality. As an AutoNet-specific network protocol, it implements the MOST network layer and as an AutoNet specific transport protocol, the MOST transport layer. Within the management system, wireless and network management modules are not necessary.

3.5.2   Roadside Architecture

The very basic structure of a roadside architecture is similar to the architecture within the vehicle domain described above. Figure 3.13 shows this general system architecture of a roadside unit. As for vehicles, the functionality of the applications implemented for the roadside unit run on one or more application units, which are called AutoNet roadside application units. Whereas the basic software components of an AutoNet roadside application unit are almost the same when compared to those in AutoNet vehicle application units,[8] the applications will naturally differ significantly. Another almost identical component will be the AutoNet roadside communication unit, which provides the same feature set when compared to an AutoNet vehicle communication unit. Depending on the scenario, the AutoNet roadside communication unit may additionally provide a wired network interface in order to connect the roadside unit to the Internet, for example via a DSL connection (not shown in Figure 3.13.

A roadside unit also may include a gateway, which enables the connection of the roadside unit to a controllable traffic sign, such as a variable message sign or traffic lights. This connection depends on the control unit of the respective traffic sign, and may be realised for example via a simple serial link. The control unit itself finally controls the traffic sign.

In contrast to a vehicle environment, it is worth mentioning that there are no HMI devices in a roadside unit. This is quite obviously because roadside units do not have a direct user interaction.

---

[7] Today, UMTS provides IPv4. However, many operators are planning to upgrade their infrastructure to IPv6. LTE will likely be widely based on IPv6.

[8] To be more precise: in a final deployment of such a system, the basic software components will likely be the same, but they will be profiled according on whether they are working in a vehicle or in a roadside environment.

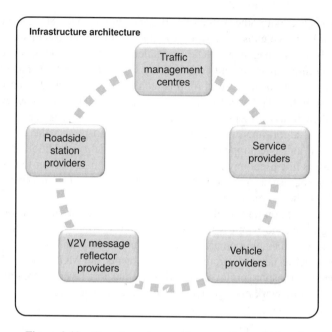

Figure 3.13 Generic system architecture of a roadside unit.

3.5.3 Infrastructure Architecture

The infrastructure-related aspects of AutoNets are somewhat under-represented in current standardisation activities for cooperative intelligent transportation systems. Hence, we will give an overview of the architecture of the infrastructure together with the relevant subsystems, which will have an impact on both the applications and the deployment, without digging too deep into these subsystems. This does not mean that we provide a complete set of relevant subsystems – in fact, there will be additional subsystems in a final deployment scenario. It may be a good exercise for interested readers to figure out which application may require which feature set of which subsystem.

In the following, we will first introduce the relevant subsystems, followed by the components needed for the realisation of a subsystem. Therefore, readers have to keep in mind that – compared to the vehicular environment – many different and heterogeneous stakeholders will be involved in the infrastructure architecture. Several regulatory limitations, heterogeneous processes and even political or tactical decisions and varieties need to be considered in a real-world scenario. We will not address these aspects in the following, but focus on the technical realisation of such a system.

3.5.3.1 Subsystems

In order to substantiate the infrastructure architecture, Figure 3.14 shows the important subsystems for an AutoNet infrastructure domain. From an architecture point of view, the infrastructure for AutoNets are considered as a loosely coupled and interoperable set of distributed subsystems. The different subsystems are able to interact with each other using

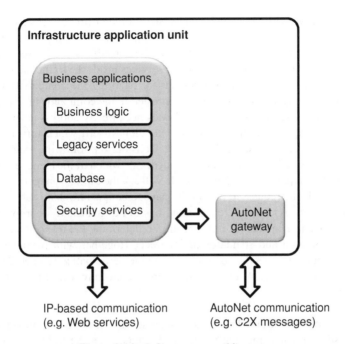

Figure 3.14   Infrastructure architecture.

standardised interfaces. The interfaces have to be designed to provide the exchange of information and control data as well as supporting respective payment and security functionality. Payment support is an important precondition to pave the road for payment services, which itself is an important precondition for sustainable business cases for the operation and deployment of such an infrastructure.

In the final deployment scenario, such an infrastructure architecture will be highly distributed and complex. In order to handle the complexity of such a heterogeneous system, service-oriented architectures (SOA) provide interesting and promising design patterns [9]. Therefore, each subsystem will be considered a service. The information exchange among the subsystems will be based on the Internet Protocol IP.

Although the term suggests that a subsystem corresponds to a single entity, a single operator or stakeholder, neither will be true for a final deployment of the system. Instead (and according the the 'service' definition in SOA), subsystems will form a heterogeneous infrastructure by themselves, consisting of respective services that are more or less coupled with themselves. The following two examples will illustrate this approach:

• Every vehicle manufacturer provides services for its vehicles. Hence, there will be several operators for the vehicle services, which basically will not have any interaction among each other.
• In Europe, there are several (hundreds) of traffic management centres. All of these traffic management centres will form the traffic management centre subsystem, in which the traffic management centres need to cooperate with each other in order to provide a seamless information provisioning and in order to improve the traffic flow on a large scale.

In the following, we will name such a subsystem a *cloud* (of services), which fits better with the purpose of the subsystem. Such a cloud may follow the SOA design pattern too. For example, each traffic management centre may be considered itself a service, which will interact with the other traffic management centre services inside the traffic management centre cloud. Note that this structure is also important for handling the complexity of the overall infrastructure system architecture, which is manifested in the communication patterns among the different clouds. A service within a cloud does not directly address a service in another cloud. Instead, the requesting service contacts the cloud itself, which forwards the request to the service that is responsible for the request. Consider the following example. A vehicle requesting traffic information in its vicinity cannot decide which traffic management centre will be able to handle the request. Instead, the vehicle will ask the traffic management centre cloud, which itself forwards the request to the traffic management centre that is able to provide the respective information. As a result, each cloud has to be 'managed' appropriately in order to provide the respective intelligent functionality to determine the relevant service. This can be achieved by common and standardised interfaces (together with payment and security support), and of course by specific contracts among the participating stakeholders and actors.

Coming back to Figure 3.14, the infrastructure architecture will consist of at least the following clouds:

> **Roadside unit providers.** This cloud comprises all roadside unit providers, that is the institutions that manage the roadside units. Like the traffic management centres cloud, the roadside unit providers cloud may be operated by public or private institutions in a final deployment scenario. For instance, gas stations along motorways as private institutions may think of providing roadside units for their customers and, thus would be included in this cloud. In general, the management of the roadside units themselves does not require any cooperation among the operators. However, the roadside unit providers cloud may offer value added services to the vehicles, which requires a strong cooperation of all operators. For example, one may think of the forwarding of V2V messages by roadside units. A vehicle driving in the south of Europe may send a V2V message to the north of Europe. The message will be forwarded to the next roadside unit, which sends the message via the roadside unit providers cloud to the roadside unit close to the destination. This scenario requires close cooperation of the operators in order to transfer the V2V message to the best possible roadside unit.

> **V2V message reflector providers.** Another interesting service is provided by the V2V message reflector cloud. A V2V message reflector is a service that enables the transmission of V2V messages via cellular networks (as described e.g. in [7]. In an ad-hoc scenario, V2V messages are transmitted directly using respective communication technologies like IEEE 802.11p. However, the vehicle can also send its V2V message to a V2V message reflector, which itself forwards the V2V message to the targetted vehicles by, for example cellular broadcasts (based on the destination position). This is an interesting feature particularly for the following two scenarios: (i) rural areas with a lower penetration of 'AutoNet-enabled' vehicles, and (ii) the beginning of a roll-out phase, in which also only a small percentage of overall vehicles on the road will be equipped with AutoNet

communication technology. A respective solution was developed and used in, for example, the projects CoCar, CoCarX and simTD (see Section A.2). Potential operators of this service could be either public or private institutions, but also (more likely) telecommunication providers, which operate the cellular communication infrastructure. This cloud also requires close cooperation among the operators, since a V2V message sent via the V2V message reflector must be sent to the destination vehicle, independent of the cellular communication technology being used and independent of the operator of the cellular network infrastructure of both vehicles.

Traffic management centres. A traffic management centre is traditionally responsible for providing traffic information, to predict micro- and macro-mobility flows, and to control traffic flow. The latter could be realised either by controlling variable message signs or traffic lights on the road or by sending respective travel direction recommendations to individual vehicles. Moreover, traffic management centres may warn drivers about dangerous situations like objects on the road or motorists driving against the traffic on motorways. For AutoNets, the infrastructure parts of driving-related applications will basically be hosted in this cloud. Traffic management centres are typically operated either by public or by commercial institutions, which have to cooperate very closely in order to generate a maximum of traffic efficiency. Hence, respective information must be exchanged among the different services within this cloud, and the processing and prediction methods need to be harmonised accordingly. Moreover, this cloud has a strong interaction with the other clouds: a traffic management centre may also be able to distribute relevant information for respective vehicles via the roadside unit providers cloud or via the message reflector cloud.

Vehicle providers. The vehicle providers cloud basically manages vehicles and vehicle fleets by providing tailored services for the vehicles. Relevant operators (called vehicle providers in this context) would be vehicle manufacturers or fleet operators. This enables the operators to provide their respective services to their vehicles. For example, a car manufacturer may provide integrated remote diagnostics services for their vehicles. This way, a vehicle will mainly contact its vehicle provider for respective services. A fleet operator, for example, can schedule the respective vehicles to the current delivery situation in order to reduce costs by optimising the travel routes of the vehicles individually. Therefore the different operators within the vehicle providers cloud do not necessarily have to cooperate with each other. However, the different vehicle providers from within this cloud will likely have a strong interaction with the other clouds, for example to collect traffic related (flow) data from traffic management centres. Such data provided to the vehicles is helpful for reducing fuel consumption by adapting the engine or electric powertrain characteristics dynamically to the current (forsighted) traffic situation.

Service providers. Like in the Internet, value added services are provided by third party providers operated by respective service providers. These services are organised in the service providers cloud. Services provided by this cloud may be any kind of business or commercial (or even private) services, which do not necessarily have a relationship to a transportation system. This is an important differentiation to the services provided by the other clouds. But service providers may also provide traffic relevant data directly for respective customers, and some services can be an important input for services in other clouds in order to improve their traffic or vehicle related services. For example, a third party service may provide a detailed weather situation for a large area in real-time. This information can be used either by traffic management centres or by vehicle providers as an important input to derive respective forecasts about the traffic situation in this area or to display the respective weather in the navigation device of the vehicle.

3.5.3.2   Entities of an Infrastructure Subsystem

From an architectural perspective, a service within a cloud runs on one or more application units called *infrastructure application units*. In a typical deployment scenario, such a service can be also considered as a service-oriented architecture, which provides different services to other services in the same cloud or in different clouds, or even directly to vehicles. As illustrated in Figure 3.15, such a service is typically realised by the respective IT business logic, which itself is composed of different internal or external modules. Examples include, among many others, legacy systems still running in the respective business unit, database services, and security services. In SOA-based architectures, such services are typically provided via web services, which themselves are based on common IP protocols. In most cases, nowadays SOA services are provided by SOAP (simple object access protocol) via HTTP over TCP/IP. This way, other services can access the respective service in a common and uniform way. In order to alleviate the cooperation among the services within a cloud, one may think of standardising this interface. Hence, every service participating in a cloud has to implement exactly this standardised interface. This allows an easy introduction of a respective 'service trader', which

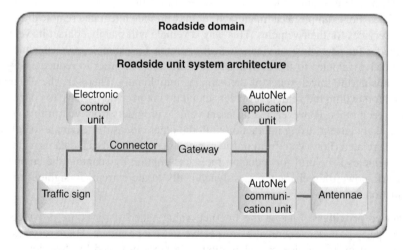

Figure 3.15   Components of an infrastructure subsystem entity.

itself acts as a trading functionality to provide automatically and exactly the service a vehicle currently needs.

In order to integrate a service into the AutoNet infrastructure, an operator has to implement a respective gateway functionality for its service. This gateway functionality enables a service to participate in the AutoNet. This way, the service may get information available in the AutoNet, or it may provide information for the entities participating in the AutoNet. Therefore, the gateway has to be able to generate or process AutoNet messages, which are typically transmitted to vehicles (or even to traffic management centres). In order to link the AutoNet gateway to the business logic, the gateway is either directly connected to the business application, or it can be connected to the business application using IP-based protocols.

3.5.4 Mobile Device Architecture

As we mentioned earlier, mobile devices are also an integral part of an AutoNet. In the context of the AutoNet system architecture, such mobile devices are also called mobile stations. Like for vehicle, roadside and infrastructure systems, the AutoNet-based functionality is implemented on one or more application units, which are called AutoNet mobile device application units in this context. Also, communication with the AutoNet is performed by the communication unit. These components are 'attached' to the mobile device – or, to be more precise, the services provided by a mobile device – via a gateway. Figure 3.16 shows this very basic system architecture of a mobile device.

Obviously the hardware design for the realisation of such a mobile device will look quite different. This is due to the fact that mobile devices have to be produced as cheaply (and smartly) as possible. This way, some of the functional building blocks shown in Figure 3.16 will likely collapse into one single hardware module. For example, the AutoNet mobile device communication unit will be merged into one hardware chip, whereas the AutoNet mobile

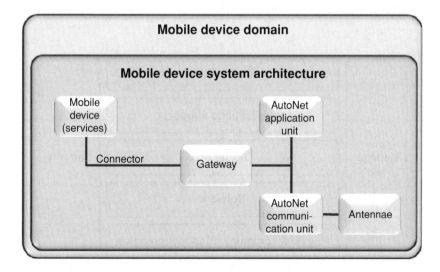

Figure 3.16   Generic system architecture of a mobile device.

device application unit will be integrated into the mobile device's services, for example in a mobile Java environment, whereas the system components required for AutoNets may be integrated into the mobile device's operating system as base functionality.

3.6 Summary

This chapter described the system architecture for AutoNets. Starting with a high-level domain view, we successively refined the different domains into subsystems, and the subsystems into respective components. We paid special attention to the communication reference model, which provides the sound basis for the communication aspects in AutoNets. We provided an overall system architecture for AutoNets covering all aspects that are relevant for both communication and the deployment of such a system. A second focus of this chapter was standardisation. We gave a review of the ongoing standadisation activities in the field of communication architectures in vehicular environments. The results of these activities were mapped onto our AutoNet system architecture. Our AutoNet system architecture consolidates the ongoing standardisation activities in different standardisation boards. We also provided deployment and implementation examples throughout this chapter in order to clarify different mechanisms and principles.

In the following chapters, we will walk through the different layers of the AutoNet generic reference stack shown in Figure 3.17. We will also use this figure in the following chapters to highlight the respective layers and to recall the arrangement of a layer within the reference stack. We will start on top with the applications and run through the stack from top to bottom. Afterwards, we will discuss the cross-layer components security and system management. Finally, we will present methodological issues in research and development and discuss the deployment of AutoNet applications and systems. We will also explain why deployment is difficult for cooperative systems, how markets for cooperative systems could work, which role politics and the public sector may be playing and what impact AutoNet applications for safety and traffic efficiency particularly will have.

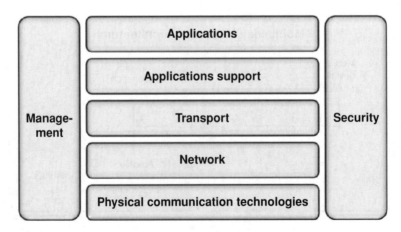

Figure 3.17 Simplified Autonet Generic Reference Protocol Stack used for the following chapters.

References

[1] Kosch, T., Kulp, I., Bechler, M., Strassberger, M., Weyl, B. and Lasowski, R. (2009) Communication architecture for cooperative systems in Europe. *IEEE Communications Magazine*, Vol. 47, No. 5.

[2] Baldessari, R. et al. (2007) *Car2Car Communication Consortium Manifesto*. Available via http://www.car-to-car.org/.

[3] Maier, M. W., Emery, D. and Hilliard, R. (2001) Software architecture: Introducing IEEE standard 1471. *IEEE Computer*, Vol. 34, No. 4.

[4] Maier, M. W., Emery, D. and Hilliard, R. (2004) ANSI/IEEE 1471 and systems engineering. *Systems Engineering*, Vol. 7, No. 3.

[5] Emery, D. and Hilliard, R. (2010) Every architecture description needs a framework: Expressing architecture frameworks using ISO/IEC 42010. *Proc. Joint Working IEEE/IFIP Conference on Software Architecture and European Conference on Software Architecture (WICSA/ECSA)*.

[6] Clements, P., Garlan, D., Little, R., Nord, R. and Stafford, J. (2003) Documenting software architectures: Views and beyond. *Proc. 25th International Conference on Software Engineering (ICSE 2003)*.

[7] Chen, Y., Gehlen, G., Jodlauk, G., Sommer, C. and Goerg, C. (2008) A flexible application layer protocol for automotive communications in cellular networks. *Proc. 15th World Congress on Intelligent Transportation Systems (ITS 2008)*.

[8] Uzcategui, R. and Acosta-Marum, G. (2009) WAVE: A Tutorial. *IEEE Communications Magazine*, Vol. 47, No. 5.

[9] Erl, T. (2005) *Service-Oriented Architecture: Concepts, Technology, and Design*. Prentice Hall.

[10] Kulp, I. et. al. (2010) European ITS Communication Architecture (Version 3.0) *COMeSafety Project, Deliverable D31*, February 2010, available at http://www.comesafety.org.

[11] ETSI EN 302 665 V1.1.1 (2010-09): "Intelligent Transport Systems (ITS); Communications Architecture".

4

Applications: Functionality and Protocols

Communication systems are usually described by different communication layers, which address different aspects, functionalities and protocols for the exchange of data. The topmost layer of all communication systems represents the applications, which use the communication system by exchanging information in order to realise a cooperative functionality. In this chapter, we are focusing on the application layer in AutoNets, which is highlighted in Figure 4.1.

AutoNets are driven by a great variety of applications, in particular applications designed to increase traffic safety and traffic efficiency. In this chapter, we will discuss such applications from different perspectives: with respect to their purpose, foremost their impact on road safety and efficiency, the applications' internal logics and their requirements. At this point we emphasise that at the application layer, the terminology for *application* is used quite heterogeneously. Such terminologies are currently used by different stakeholders in different and sometimes contradictory ways. We feel confident that a well-defined terminology in this specific context neither helps to speed up standardisation, nor does it improve the deployment of AutoNet systems. For the sake of simplicity, we therefore simply stick to the term *application* in this chapter and do not differentiate between terms like applications, functionality, use cases, scenarios or similar expressions. In this sense, an application is simply what the customer of a vehicle would recognise as a feature.

According to the classification scheme presented in Chapter 2, AutoNet-based applications can be clustered into three major categories, which are recalled in Figure 4.2. Based on these categories, we will briefly introduce the respective applications currently examined in related research projects and standardisation activities. Then, we will describe some of the most prominent examples in more detail, namely:

- *(Decentralised) environmental notifications* as an example for a driving-related (foresighted) application, which is primarily based on a shared, common environmental knowledge that is collected by the vehicles.
- *(Intersection) collision avoidance* as another example for a driving-related (critical) safety application, which is based on a mutual cooperative awareness of traffic participants nearby.

Automotive Internetworking, First Edition. Timo Kosch, Christoph Schroth, Markus Strassberger and Marc Bechler.
© 2012 John Wiley & Sons, Ltd. Published 2012 by John Wiley & Sons, Ltd.

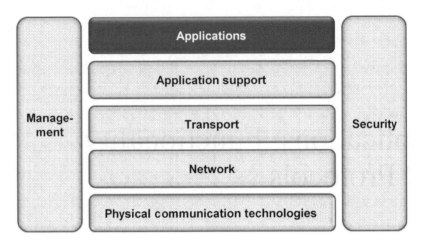

Figure 4.1   Application layer in the AutoNet Generic Reference Protocol Stack.

- *Green light optimal speed advisory* as an example for driving-related applications for roadside-supported traffic efficiency.
- *Insurance and financial appliances* as an example for a typical vehicle-related AutoNet application.

The following sections describe the four application examples. We start with the most prominent but also most challenging class of AutoNet applications, namely driving-related safety applications.

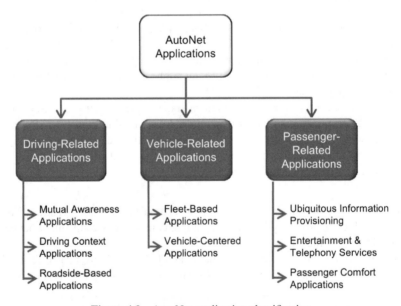

Figure 4.2   AutoNet application classification.

4.1 Foresighted Safety Case Study: Environmental Notifications

Providing drivers with relevant environmental notifications is regarded as one of the most promising AutoNet safety applications. Vehicles share their information about road conditions and dangerous situations. Respective applications are able to warn their drivers about potential upcoming dangers.

Since there is a bit of confusion caused by various terms used for such notifications, it is reasonable to provide a short digression on terminology in the following section. Because of the primary goal to warn the driver about potentially dangerous situations, the example application described in this section is often referred to as *local danger warning*, or *hazardous location warning*. Although this task looks pretty simple and obvious when talking about interconnected vehicles, a closer examination of this application brings up a large number of technological challenges that need to be solved.

The idea of providing up-to-date driving related information in a V2V fashion, that is generated and collected by vehicles rather than by infrastructure sensors, is well known and rather straightforward. For the provision of higher quality traffic information, such *probe vehicle data* has been used since the late 1990s to improve the available traffic information on the infrastructure side and provide the combined overall knowledge about the traffic conditions to feedback to the drivers and in-vehicle systems in order to be able to plan faster routes. Such use of probe information collected by the vehicles is often referred to as *floating car data* (FCD) [20, 21, 22, 23, 24]. In fact, from a technological point of view, this terminology is quite to the point, when used with respect to the generation of probe data by moving vehicles with wireless communication capabilities. However, it is not as accurate anymore and the meaning of this term has sometimes been widened to a synonym for the overall system of collecting and redistributing traffic information via cellular communication, including the refinement of the collected information in a central traffic management centre. Consequently, levers to enrich traffic information with other kinds of environmental information such as local dangers are summarised by the term *extended floating car data* (XFCD), or *enhanced floating car data* (EFCD) [19, 25]. While FCD is usually limited to position and speed data, XFCD encompasses other vehicle sensor data as well like outside temperature or wiper status.

We explicitly emphasise this fact because the driving-related application of environmental notifications described in this section is also based on probe data and, therefore, could be literally referred to as XFCD. However, environmental notifications in our sense are neither limited to probe data and cellular communications only, nor are they dependent on a central traffic management centre. In the absence of a traffic management centre, the application is often also called *decentralised environmental notifications*. The purpose is for driver warnings to improve safety rather than general traffic information.

Having clarified that different stakeholders involved use different terms while having in mind different requirements and expectations, we would also like to mention that environmental notifications cover a great range of environmental parameters with different characteristics, ranging from a blocked road caused by an accident to a forecast of very low friction caused by black ice. As we will see, one immediate consequence is that both the requirements and the resulting reliability of such notifications may vary to a great extent.

According to the current European standardisation activities at ETSI, in the initial deployment phase such environmental notifications should comprise the following so called *basic set of applications*, or in short *BSA* (for further details on standardisation see also [17, 18]), which are the following:

- Stationary vehicle (accident, breakdown) awareness.
- Adverse weather conditions awareness.
- Hazardous location awareness.
- Traffic condition awareness.
- Signal violation awareness.
- Road work awareness.

Along with the further basic applications *emergency vehicle approaching awareness*, *slow vehicle awareness*, *emergency electronic break lights awareness* and *wrong way driving awareness*, these applications are clustered into the so called *cooperative awareness applications (CAA)*.

As stated, environmental notifications are based on an extensive and precise knowledge about the current driving situation of a vehicle. This knowledge comprises the individual driving (current and historical) context, as well as knowledge about objects, hazards, roads and weather conditions being observed by other vehicles or infrastructure entities. Therefore, two of the most critical factors are certainly the acquisition of the respective information and the process of disseminating this data to all entities that could benefit from these information artifacts.

Figure 4.3 illustrates the basic concept of environmental notification. In this example, information about the observation of reduced friction on a slippery road caused by aquaplaning is communicated among the vehicles. In addition, the figure delineates the variety of problem domains inherent to such types of warning applications. In particular, the following challenges need to be addressed:

- Individual observations and data collection.
- Individual situation analysis.
- Cooperative situation analysis.
- Distributed knowledge management.
- Communication and information dissemination.
- Evaluation of individual relevance.
- Prediction of individual occurrence.
- Interaction with the driver.
- Data security and privacy.

In the following, we will address these challenging domains in more detail, pointing out the requirements and expectations according to the classification scheme presented above.

4.1.1 Data Collection and Individual Situation Analysis

The purpose of individual situation analysis is to extract highly precise information from collected data about the current driving context, which is needed to derive valuable information about the current road situation. Sensors in a vehicle provide a large amount of raw data that

Figure 4.3   Environmental notification example: slippery road. Reproduced by permission of BMW Group.

has to be processed. Some of the sensors provide rather direct data about the current context (e.g. a rain sensor, indicating that there is some amount of water, snow or dirt on a small area of the windshield), whereas other sensors can help to derive information about the current context indirectly. For instance, activated fog lights can be used as an indicator for fog, although there is no strong evidence for fog at all. An efficient individual situation analysis can help to reduce the data load to be transmitted for information dissemination, as it is possible to only disseminate significant information within AutoNets. There are two challenges that complicate the individual situation analysis:

- The need for indirect measurement methods.
- The occurrence of imprecise data.

These challenges arise from the fact that environmental data in many situations cannot be directly measured. In those scenarios, it is necessary to determine (or deduce) the situation indirectly and logically from a set of potentially imprecise data.

Modern vehicles are equipped with a variety of different sensors in order to deduce the current driving situation. Examples are lateral and longitudinal accelerometers, or sensors measuring the current wheel speed and yaw rate. While these sensor measurements are typically used to trigger actors to immediately stabilise the vehicle in critical conditions, they can also be used to derive high level information about the potential root cause, such as the existence of water on the road causing aquaplaning. In todays vehicles, such causal data analysis is not done, for a simple reason: since an observing vehicle has already encountered the critical condition, and has hopefully been stabilised, the high-level information is currently not relevant – neither for the vehicle control systems themselves, nor for the driver.

However, with respect to foresighted assistance systems, such information would be highly beneficial for subsequent vehicles. This is in particular true for applications that aim at raising the awareness of the driver to certain events or aspects of the driving situation, such as the just mentioned aquaplaning. Hence, to stabilise a vehicle it is necessary to know the effects of a driving situation. In contrast, to inform a driver adequately in advance about the condition, it is necessary to know the root cause of the event.

Having such a requirement to sense the causal environmental parameter, the typical engineering solution would be to develop a respective sensor system. Unfortunately, in a foresighted system that is based on the sensors of others, it is hard to find a common business case for all involved parties.

Applications that cooperatively detect and forecast critical driving conditions are somewhat altruistic. This means that resources – such as storage capacity, computation capabilities, energy required for transmission, dimensioning of in-vehicle computing platforms or antennae – are consumed without any benefit for one's own vehicle, but with a potential benefit for others. Standardisation of such applications has to ensure cooperative functionality following this altruistic paradigm. However, as we will see in Chapter 12, business considerations may sometimes be in conflict with this paradigm, because obviously there will be a certain trade-off between optimal system performance (functionality) and provided resources (manifested in costs). So far, no mechanisms have been developed to apply market based mechanisms for some kind of pricing for this information. We leave it as an exercise to think of possible ways to market individual vehicle data.

The critical question in this context is: who has to pay for what sensor system? Is it payed by the owner of the vehicle, who in a first step has no benefit from it? Is it commonly paid from all vehicles participating in the network? If you think the latter is fairer: how much organisational overhead is introduced then? And what happens if the sensor is also used for other, individual purposes? These kind of questions are completely new to an automotive industry that has been used to constructing isolated, uninterconnected products. And we therefore feel confident that we are a long way from any commonly accepted solutions.

Anyway, although dedicated sensor systems would ease and/or speed up some technical challenges, we can do a lot with what we already have in almost all modern vehicles. But reliable inferences from such high level information depends on a variety of influencing parameters. For example, it is much more likely that aquaplaning occurs at a certain location if it has been raining a lot at this place in the past hours. In order to take this factor into account, historical knowledge about the amount of rain in a certain area is beneficial too. Another example would be knowledge about specific conditions of the road surface at a certain place. Such knowledge also significantly enhances the reliability of a forecast with respect to aquaplaning.

Another challenge is that the analysis of the situation has to work with imprecise – or even unavailable – information. The collected sensor data may be insufficient and rather limited in terms of accuracy. Existing sensors were typically developed for dedicated purposes, for which their accuracy is sufficient – such as the rain sensor for the wiper interval settings. But with respect to more precise knowledge about the root cause, the derived information will often be imprecise. Inevitably, the data analysis will additionally return probabilities for the existence of certain situations. Hence, further information processing has to deal with both uncertain and imprecise information.

Besides these challenges, time constraints for foresighted systems are not as limited as for scenarios where a vehicle has to be stabilised. Therefore, more complex reasoning algorithms can be applied for foresighted systems. This is especially true for an average consideration. However, it is important to note that the maximum reasonable latency caused by such complex algorithms also depends on the characteristics of a certain situation. The concrete latency constraints primarily arise from the distance to a relevant event and the type of the event: the closer the event, the less time is available to determine the cause, communicate the information and derive appropriate means to interact with the driver. For attention and awareness applications, as classified before, latency constraints are within the magnitude of seconds to minutes, usually providing enough time for complex inference algorithms to be computed. The quality of the available information therefore has significant impact on the required transmission capacity of inter-vehicle communication. Typically, the more information about the driving context can be measured by a vehicle, and the better the quality of this information is, the less communication between the vehicles is required. Following the example of the aquaplaning scenario, it would be sufficient to communicate the readings from a sensor that could measure the water depth beneath the tyres (if such sensors are available – for information on road/tyre friction measurement and especially aquaplaning detection, please refer to [26, 27, 28]). This is because this information is the most important factor to determine whether or not aquaplaning might occur: based on knowledge about the water depth on the road surface, the type and condition of the tyres and the current speed, a reliable and accurate prediction about the risk of aquaplaning at a certain location would be possible. Assuming that such information about the water depth on the road would be continuously measured and communicated by the vehicles passing the dangerous situations on their travel, such reliable prediction of the respective risk potential would be available along the complete route. In particular, it would also be possible even in the event that no vehicle has actually encountered a critical driving condition so far. In other words, it would not be necessary for at least one vehicle to actually encountered aquaplaning, still every vehicle could inform its driver individually about the likelihood of aquaplaning along the route.

In the event that the vehicle does not have such dedicated sensor (as unfortunately is the case for many vehicles today), a likelihood value can be deduced indirectly based on onboard dynamics sensors, indicating the probability of whether or not slippery conditions actually occurred. In this way, besides the fact of slippery conditions at a certain position, the current speed and, in fact, the current vehicle configuration (type and condition of tyres) also have to be shared with the other vehicles.

The aquaplaning example shows clearly that subjectivity and determinism of detection as well as the method of situation analysis significantly influences the amount of data that has to be communicated. In addition, the overall system complexity and reliability also depends on these factors.

4.1.2   Cooperative Situation Analysis

Depending on the current traffic density on the road, a varying number of vehicles may pass the same location within a short period of time. Therefore, the vehicles also encounter the same or

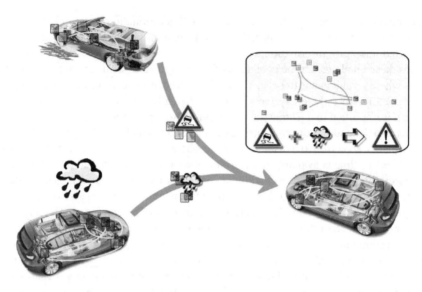

Figure 4.4   Example for cooperative situation analysis. Reproduced by permission of BMW Group.

a very similar driving situation. This fact enables further methods to analyse driving situations through sharing the individual observations, which is called *cooperative situation analysis*. Such cooperative analysis of situations is illustrated in Figure 4.4: The upper vehicle may detect slippery road conditions, whereas the lower vehicle driving at a different (but close) location may detect rainy weather conditions. Both types of information are received by the vehicle on the right-hand side (using AutoNet-based communications), which itself can validate both information items and is able to deduce that the slippery road conditions are caused by aquaplaning. In this way, cooperative analysis can significantly improve both the quality and reliability of the situation evaluation, because the current road condition in many cases cannot be deduced by a single vehicle with sufficient accuracy. Also, cooperative analysis often reduces the overall complexity since the processing needed for the situation analysis can take into consideration the information generated by other vehicles. Another example illustrates this property: let us assume a sufficient density of vehicles participating in a cooperative intelligent AutoNet. Based on position and speed of vehicles driving ahead, or nearby, a vehicle will be able to determine a traffic jam situation with accuracy.

A third example is the cooperative detection of obstacles on the road, based on the analysis of the driving dynamics and trajectories of vehicles. For example, the method presented in [14] makes use of the fact that suddenly appearing obstacles result in characteristic patterns of evasive manoeuvres. Such manoeuvres can be detected by the curve progression of lateral and longitudinal acceleration. Obviously, the occurrence of only one individual evasive manoeuvre is not sufficient evidence for a dangerous obstacle at the respective location. Such a manoeuvre can also be caused by short-lived events that have no enduring risk potential, like for instance the crossing of deer or the temporary inattention or distraction of the driver. Nevertheless, it is a strong evidence that there *might be* a critical situation. Triggered by such manoeuvre, a subsequent cooperative analysis of the driving characteristics and trajectories of the following vehicles provides additional evidence with respect to the existence of an obstacle on the road.

Hence, abrupt braking, steering, or the fact that a certain location on the road is not passed by other vehicles anymore, increases the probability that there may be an obstacle at the road. Vice versa, this also means that vehicles that do not detect unusual driving conditions at the respective location can easily revoke the initial assumption.

Cooperative situation analysis means that additional communication is necessary. Therefore, the required information may be incomplete because of communication constraints. This is particularly true for AutoNets: it cannot be guaranteed that the respective observations of all vehicles are available in time.

In order to compute a highly accurate and reliable prediction of the driving situation, the (cooperative) knowledge must continuously be re-evaluated and adapted in a cooperative way. In contrast to local and autonomous systems, the available data pool may be incomplete. Also, the accuracy of position measurements may not be precise enough to correlate the various observations of different vehicles. Coming back to the example of aquaplaning, the root cause could be, for instance, deep lane grooves. As a result, a few centimetres of variation in the driven route may make the difference between the occurrence of aquaplaning and a normal, uncritical driving condition. Hazards that move or vary in shape and intensity over time (like fog or heavy rain, for instance) impose additional complexity for spatio-temporal reasoning and prediction, leading to more complex inference methodologies that still guarantee reliable results even in the case where a limited amount of information is available.

4.1.3 Distributed Knowledge Management

Cooperative deduction imposes additional requirements on the management of the available knowledge, both measured and inferred either by others or by oneself. Knowledge about the driving situation in this context comprises the following aspects:

- *Domain-specific knowledge* is an important factor, especially with respect to correlations and interactions of different aspects of the driving context. For example, knowledge can be derived from the correlation of rain intensity and wiper settings. Domain knowledge also comprises the spatial impact of areas in the vicinity of a vehicle. In particular, there is a separation of aspects limited to certain road classes, lanes or directions, and of aspects valid in a larger area.
- Knowledge comprises the *ego vehicle's sensor measurements* of several low level aspects of the driving context.
- Individual situation analysis uses *inference of high level aspects of the driving context.*
- In cooperative situations, *sensor measurements of other vehicles* (i.e. several low level aspects about the driving context) are an important factor for knowledge.
- Besides the availability of data or information, *predictions* of both low level and high level aspects of the driving context may be another important source.
- Position (along with the time of the position measurement), speed, direction and acceleration, that is *states of vehicles in the proximity* are important aspects too. As we will see further on in this book, such information is important for routing purposes of data packets in georeferenced AutoNets, as well as for vehicle collision avoidance systems.
- Another important aspect of the knowledge is *an estimation that indicates the importance of the knowledge item for the operation of one's own vehicle.* This value takes into account

the limitations of on-board storage: in case the storage capacity is exhausted, items with the least relevance for the ego vehicle can be removed from the storage.
- Finally, an estimation indicating *the importance* of the knowledge item *for other vehicles* has to be available too. This value takes into consideration that, although not important for the ego vehicle (any more), a piece of information may still be important for other vehicles to predict their driving condition. This is always the case for vehicles travelling close to potentially critical driving situations, but which will not pass this location. This could be true in the case where a vehicle has already passed the location, or the location of the hazard reported previously from another vehicle is not along the route. Such information should still be stored in order to report it to other vehicles that pass the reported location.

A first result of the latter two aspects is the obvious conclusion that not every single piece of information is worth being stored or transmitted. The following example illustrates this. The fact that a vehicle travels along a dry and straight road is not of much interest for driving-related, cooperative AutoNet applications. However, it makes a difference if a vehicle expects wet road conditions because of heavy rain that has been reported previously. This example shows that it depends on the current context (or to be more precise: the particular view of the world based on the current knowledge) of a vehicle rather than on the type of information whether (and how long) or not a certain piece of information needs to be stored by a vehicle. Hence, the utility metric for storing a certain part of the driving context is not only specific to an application, it is also adaptive to the context. Defining such specific metrics for driving-related applications is often a complex task. In short, driving-related and predictive safety applications typically benefit from information about an aspect of the driving context if this aspect may have an impact on the current driving task. The cooperative application also has to consider whether or not the characteristics of this aspect are known in advance, or if the information significantly strengthens, confirms, attenuates or falsifies the belief in certain characteristics.

A second important consideration is the question of whether or not the latter two factors in the enumeration are optional or mandatory for a cooperative safety application. This question is rather of political than of technical nature – it is again a consequence of the altruistic nature of the system.

Low-level sensor information is generated at a high frequency, and is typically neither stored on-board nor arbitrarily transmitted due to the huge amount of data generated. This, in turn, complicates cooperative analysis, because the original measurements may not be available any more, or cannot be shared at all. And it is again a question of cost and fairness for all parties involved, how much (additional) storing capacity a cooperative system should bring along.

A third consideration is the strategy to disseminate the respective information, potentially based on such utility metrics. This aspect will be described in Chapter 7 in more detail.

For the example of sensor measurements, the knowledge items mentioned above are typically correlated with other measured information, describing part of their context:

- Place or position: any measurement in an AutoNet needs to be accompanied by the position of the measurement. In addition, positional information needs to be recorded together with the time of the measurement. This is mainly because current positioning technologies only allow measurements in the magnitude of half a second. For some applications, like a warning about rear end collision accidents, it is crucial to know the exact time of measurement in order to approximate the real position of the respective vehicle when receiving a data packet carrying the information. It should further be noted that plain GPS coordinates are not sufficient for such a purpose. Instead, some applications need additional information such as the road class, information about lane, direction or even street names.

- Time: knowledge must be correlated with the time of the measurement of the respective aspect. Note that this is usually not identical with the time of the position information.
- Accuracy: finally, the accuracy of the measurement must be correlated with the knowledge items.

In the case of inference of high-level aspects of a predicted driving situation, these knowledge items are correlated with the position of the forecast instead of the measurement. In addition, information received from other entities also need to be correlated with a reliability value (or confidence for the correctness of a measurement, respectively), because the different predictions of a cooperative application very often imply uncertainty in the cooperative system.

In contrast to all other knowledge items, domain knowledge is of static nature, that is domain knowledge can be considered as given facts that usually do not change. Changes only appear if the knowledge comprises faults that are corrected, or the domain knowledge is extended based on additional findings. Therefore, apart from some version management, domain knowledge does not require additional meta information.

Finally, the aspects of the stored driving context may vary in size, spatial shape, intensity and location. This determines the optimal strategy to store the different aspects. Thereby, three basic spatial shapes can be distinguished, namely:

- single, isolated points (as may occur in the case of traffic accidents);
- areas with a linear extent along road segments (as may occur in the case of traffic jams); and
- large areas (as may occur in the case of critical weather conditions).

The mutability of a specific aspect of the driving context further determines the optimal storage strategy with respect to their persistence. Positional information of other vehicles in the vicinity is highly volatile compared to georeferenced aspects like road or weather conditions. For this reason, it is sensible to use different storage strategies for the knowledge items described in this section. In particular two methodologies are currently discussed in standardisation: (i) a *neighbourhood table* comprising volatile information of other vehicles, (ii) a *local dynamic map*, comprising georeferenced data. Both methodologies are described in more detail in Chapter 5.

4.1.4   Individual Relevance and Interface to the Driver

Dangerous events in the closer vicinity of a vehicle must be communicated to the driver. However, in each situation it is important to validate whether or not an event is relevant for the driver. This decision is performed by a module we call *relevance checker*, which will be introduced in Chapter 5. The main task of this relevance checker is to decide if one's own vehicle will come across a previously reported event. Therefore, it is not only necessary to know the rough location of such event. It is also important to know how the event affects the traffic situation. An accident in the middle of an intersection, for example, will most likely affect all approaching vehicles, regardless of their direction or lane. Assuming a highway with separated tracks for both driving directions, an accident will – without considering onlookers or rubbernecks – not affect oncoming vehicles. In contrast, upcoming fog will most likely affect both directions of a highway. When designing a reliable foresighted application as described, it would therefore be necessary to know about the road type as well as the characteristics of the cause for critical driving conditions. Being more precise, domain knowledge has to comprise whether the event affects the following four situations:

1. All lanes and driving directions.
2. All, some, or a specific lane of one driving direction, regardless of the road type.
3. All, some, or a specific lane of one driving direction on a road with separated directions.
4. All, some, or a specific lane of one driving direction on a road with non-separated directions.

The relevance checker takes into account all available data in order to compute an optimal warning strategy. For a reliable result, unfortunately, this requires that vehicles are equipped with highly precise digital maps comprising detailed lane information. Additionally, the vehicles must be able to detect their actual driving lane and to locate a detected event with a very high accuracy. From a pragmatic point of view, however, it is still significant progress for traffic safety if the relevance checker is able to decide whether or not the reported event is located on the road ahead, meaning that the remaining situation assessment process is left to the driver – in this case even avoiding the need for a digital map on board all vehicles.

A solution currently being discussed in the respective standardisation bodies was initially introduced in the context of the European research project PReVENT WILLWARN [13]. In this proposal, the idea of *trace chain matching* was born. In short, the key issue is that whenever an environmental notification is created and transmitted, the detecting node additionally transmits a certain number of position markers of its recent trajectory. The linear interpolation of these positions roughly reflects the road geometry. This generated trace chain (often also called path history) is then used by the relevance checker: the higher the overlap of the received trace chain in comparison with the individual trajectory of the receiver, the higher the probability that the reported event is situated along the same road. Because there are some other functionalities that make use of such trace chains, they are considered as part of the application facilities, described in Chapter 5. Although this approach is obviously not as reliable as using digital maps, it is still an effective means to eliminate false alarms. A prominent example of such false alarms are bridges crossing a highway, where it has to be ensured that events that are only relevant for the bridging traffic (e.g. an accident on the bridge) do not result in false alarms for the highway traffic.

In the event that the relevance checker decides that a dangerous event is of relevance for the driver, the respective cooperative application has to ensure that the driver will neither be overstrained nor distracted from the traffic situation. Hence, the user interface (UI) to the driver has to act in a context-adaptive way when informing the driver about the current driving situation. The user interface therefore needs to determine the following two parameters for the representation of the driving situation:

- *When to inform.* The UI has to determine the appropriate moment for informing the driver about the respective event. In the event that there is more than one dangerous event, the UI (or, to be more precise, the relevance checker) has to determine the priority of the warnings. Hence, the UI will display the different warnings in their respective order.

Figure 4.5   Example of a user interface. Reproduced by permission of BMW Group.

- *How to inform.* The UI has to determine in which way a dangerous event will be presented to the driver. The following alternatives exist:
  - *Visual.* Examples include headup display, instrument panel or central information display.
  - *Acoustical.* For example, using the integrated audio system of the vehicle.
  - *Haptic.* Examples include vibrations of the steering wheel or the seat, or even, in urgent cases, automatic straining of the seat belts.

Figure 4.5 shows an example for a potential visual user interface to provide information to the driver of a motorcycle. Here, the biker is informed about fog 750 metres ahead along the route by the display of a respective symbol on the dashboard. The most important factors for effective and specific driver information in this example are: (i) correctness of the existence of the detected situation, (ii) the impact on the driving task and (iii) the consequence of the resulting actions. This example illustrates that the quality of the user interaction greatly depends on the quality of the available context information. Conversely, the ideal information of the driver requires a profound provisioning of relevant context aspects with respect to the actual and future driving environment, condition of the driver and the interpretation of the respective observations. Such issues are discussed extensively in [5, 6].

4.1.5   Data Security and Privacy

An important precondition for user acceptance of cooperative applications is the reliable prediction of the situation. This reliability requires that the available information – provided by other entities in the AutoNets – used for the prediction is reliable, too. Otherwise, the cooperative system can be distracted. Potential attacks range from the dissemination of malicious information, a systematic jamming of the communication channel, or even a manipulation of a vehicle's own sensors or system components in order to generate faulty information. Hence, a cooperative system has to provide protection mechanisms against such malicious systematic attacks, which have to ensure the trustworthiness of the available information or, at least, consider the 'degree of trustworthiness' for the information evaluation.

The detection of such attacks is a serious and challenging task, since it is always hard to detect compromised (own) software or information. A promising approach to address these issues applicable to situational awareness are plausibility checks [2]. Plausibility checks are based on the fact that there are only few attackers in a system, which are facing a large number of genuine participants of the system.[1] Hence, a fault-tolerant system always has to consider different estimations of a situation. Based on these different estimations, the system can distinguish, with the help of additional information, faulty information from a changing situation. However, plausibility checks increase the complexity of both knowledge management (cf. Section 4.1.3) and the deduction mechanisms implemented for the system.

Another crucial acceptance feature is privacy, which has to ensure that vehicles do not disseminate information that can be used to calculate an inference about the customs or habits of the driver. In particular, it must not be possible to deduce motion profiles of a driver, or to monitor speed violations automatically based on the speed readings a vehicle periodically sends to the vehicles in its vicinity. Such private data protection regulations are typically ensured by anonymising strategies using temporary pseudonyms for the vehicles, which are changed periodically and which do not allow a deduction on the vehicle nor the driver [3]. Security-related aspects will be discussed in more detail in Chapter 9.

4.1.6   Reliable Estimation of the Current Driving Condition

The analysis of requirements in the previous sections suggests that the different types of problems and challenges are greatly interwoven with each other. Indeed, this fact needs to be addressed by the overall system design in order to get a proper and reliable estimation of the current situation, resulting in a distributed control loop. Compared to traditional control loops used for the realisation of different functions in a vehicle, this control loop is not limited to individual vehicles. It operates on all entities participating in the AutoNet by taking into account their information and their behaviour. Figure 4.6 illustrates this control loop. It comprises the following processing steps: starting with the interpretation and analysis of the observations, the results will be communicated and propagated if required. Together with a vehicle's own information, this propagated knowledge is analysed and verified in a holistic way. Based on this result, future driving situations are estimated and predicted, which will be confirmed or refuted based on a vehicle's own observations. This result will again be propagated if necessary, which closes the control loop for the reliable situation estimation.

Apparently, the computational complexity needed to evaluate and to verify certain events greatly increases if an event is subjective and affected by non-deterministic driver behaviour. In this case, vehicles may or may not be able to support or to weaken the confidence in a situation that has already been detected and communicated by other vehicles. Putting it differently, the fact that a vehicle could not detect the existence of a certain situation (i.e. support an event

---

[1] As a result, an attacker compromising the majority of participants must be avoided, and the attacker must not be able to impersonate a large number of participants.

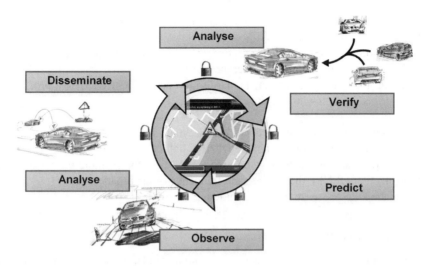

Figure 4.6   Control loop for a reliable situation estimation.

already detected) does not mean that this situation does no longer exist. Consequently, non-deterministic detection impairs the verification of an event, because it reduces the number of vehicles that can actually confirm that event. In addition, more equipped vehicles are necessary to derive situations in a realistic way. In order to calculate the relevance and confidence of a received message properly, subjective events require that the original sender adds information about its configuration. This offers the receiving nodes a more realistic chance to evaluate the message. This additional information has to be evaluated too.

Moreover, the confidence in events that change over time (in a continuous or chaotic fashion) is much more difficult to determine than that in static events, because the potential change or movement has to be taken into account as well. Consequently, whether a situation (which may trigger a specific event) is static, continuous or chaotic has a significant impact on the complexity of calculating the level of confidence in a received event message.

4.1.7   Communication and Information Dissemination

The main purpose of a proper communication protocol is to deliver environmental notifications to all vehicles that are affected by risky environmental situations, which must be identified before that vehicle encounters the situation. Depending on the specific subject of the risk, a respective message is generated if a vehicle detects a specific situation. These messages are called *decentralised environmental notification messages* in the European standardisation community (DENMs).

In order to deliver messages to vehicles that are likely to encounter the same situation, the messages are assigned a target area around the encountered situation. When communicated in an ad-hoc fashion within this area, a DENM is forwarded from one vehicle to another. In order to ensure a reliable delivery in low-density traffic scenarios where vehicles meet each other rarely, DENMs are stored and forwarded as soon as another AutoNet-enabled vehicle comes into communication range. The main challenge for the protocol design therefore is to find a well-balanced trade-off between a fast information dissemination that works for all different traffic densities and a low (ad-hoc) network resource consumption. The protocol has to ensure that more critical messages are propagated faster through the network. Since this is a non-trivial task, we will explain and reference possible approaches in Chapter 7.[2]

According to the current European standardisation activities at ETSI[18], broadcasting a DENM comprises the following architectural domains:

- Detect that the DENM's triggering conditions are met.
- Determine the relevant area in which the DENM shall be disseminated.
- Define the broadcasting conditions and associated rules, for example the latency time to broadcast the first DENM, the number and frequency of the broadcasts.
- Determine the DENM's terminating conditions.

4.1.8 Standardisation Issues

Cooperative applications obviously require common, standardised proceedings in different domains. Cooperation therefore starts with the trivial fact that the application logic spans several vehicles of different vendors. This means that a module from vendor A installed in-vehicle A has to rely on a well known and well defined operation of another module from vendor B deployed in-vehicle B. In order to ensure such common application logic, for each cooperative application there are so-called *application functional requirements*, which are standardised at ETSI too. In order to ensure the correct operation according to these functional requirements, every implementation has to pass through a certification process performed by a trusted certification centre. The in-vehicle system then has to ensure that only certified modules are operated. The application functional requirements also govern the start-up and shut-down behaviour of the application.

In addition to application functional requirements, there are so-called *application operational requirements*. The standardisation of these issues are critical for a proper interaction of all instances. According to ETSI's standardisation activities, the operational requirements comprise the following requirements:

> *Security and Dependability Requirements.* Among others, the current standard in particular enforces that the applications are well protected against abnormal conditions such as the following:

- The absence of formatting DENMs by the facilities layer once started.
- The signalling of a road hazard which does not exist.
- The formatting of false or erroneous DENMs.
- The absence of transmitting the required formatted DENMs.
- The transmitting of DENMs with a frequency that is not in the range of specified DENMs transmission frequencies.
- The transmission of DENMs without being triggered by a specified event.

---

[2] It is worth mentioning that, while DENMs are normally designed for and discussed in the context of ad-hoc networks, the information they carry can also be disseminated by other wireless technologies, cellular networks as well as digital radio broadcasting, for which the TPEG encoding standard offers corresponding methods. The timeliness and granularity of the information will vary though.

- An abnormal silence not informing the driver of a signalled road hazard.
- An abnormal activity informing the driver about a non-existent road hazard.

This protection needs to be done in a *cost effective manner*, indicating that the trade-off of cost versus risk and harm should be thoroughly deliberated, keeping in mind that each system can be broken by just putting enough effort in. In the case of abnormal behaviour being detected, the application is switched off and the driver is informed of the malfunction.

Interoperability requirements. The main purpose of interoperability is that every instance deployed shall be able to transmit, receive, encode and decode DENMs in the specified format, by using the same communication channel, the same medium access protocol, the same networking protocol and the same transport protocol, each as specified in the respective ETSI standardisation documents.

Performance requirements. Performance requirements specify the transmission frequencies of DENMs and the total latency from an end-to-end perspective. Thereby, DENMs that inform of road hazards (awareness and attention) shall be transmitted with a frequency in the range of 1Hz to 10Hz, whereas DENMs with critical content that require a driver alert (action) shall be transmitted with at least 10Hz, while ensuring a first transmission not later than 10 ms after an accident has occurred. The total latency shall be less than 500 ms.

4.2   Active Safety Case Study: Cooperative Collision Avoidance and Intersection Assistance

Avoiding vehicles crashing into one another or into other road users is certainly (along with keeping the car on track) the most important challenge for safe driving and prevention of (fatal) accidents. Not surprisingly, there is widespread belief that AutoNets in future can play a significant role in achieving this goal, assuming at the end of the day that all road users will participate in such cooperative systems.

Reliable collision avoidance systems are very challenging, because they require the cooperation of all nearby vehicles, and every vehicle must have a thorough awareness of the situation. Moreover, such a system must be tolerant to fuzzy or imprecise knowledge: for example, positions of vehicles may have different accuracies, and the predicted behaviour of both the vehicle and the driver may also vary within specific boundaries. Such a system cannot be developed solely based on information generated by a vehicle's autonomous sensors, because individual sensors cannot provide all the relevant information from other vehicles necessary for the correct decisions of the assistance system. In particular, today's sensors all feature line-of-sight characteristics, that is they cannot look around corners or behind objects. AutoNets provide access to such additional information necessary for future collision avoidance systems because they allow the sharing of all information that can be detected by others.

Nonetheless, the vision of using AutoNet communication for collision avoidance introduces very strict requirements for the overall system, in particular with respect to security, latency and reliability.

Vehicles typically crash into one another in either of two ways: laterally, as for example at intersections, or longitudinally, as appears frequently in rear-end or head-on collisions. To introduce some terminology again, the current wording of European standardisation speaks of *intersection collision risk warning (ICRW)* and *longitudinal collision risk warning LCRW*, respectively.

In the introduction to this book, we revealed that intersections are one of the major cause of accidents in everyday traffic situations (see Section 1.2): in 2006, 36 per cent of all accidents with personal damage in Germany occurred at intersections.[3] Further analysis showed that cross-traffic was one of the responsible factors for a crash in 63% of these accidents [9, 11]. This is due to the fact that intersection traffic scenarios are rather complex and often partially unmanageable for drivers, since a lot of information affects them. This is especially true for vehicles that have to wait at the intersection due to crossing traffic. In such situations, the driver of the stopped vehicle has to make their decision within a very short period of time whether or not it is safe to pull into (or cross) the intersection.

Just as we did for environmental notifications, we will describe in the following the most important issues for designing a collision avoidance application with respect to data collection, situation analysis and prediction, knowledge management, communication, security and privacy, as well as driver interaction. Because of their situational complexity and the resulting high number of collisions, we will describe these problem domains in the context of intersection collision avoidance, but also indicating specific characteristics of rear-end or head-on collisions.

4.2.1   Data Collection

The main idea of AutoNet-based collision avoidance is rather simple; at least at first glance. Every traffic participant communicates their own position and velocity to the surrounding vehicles. Having all the information of the surrounding vehicles available, and thus all the relative distances and movement vectors, driver instructions or automatic manoeuvres can avoid critical proximities of the traffic participants.

Both lateral and longitudinal collisions have in common that the point in time where a collision cannot be prevented any more (but can only be mitigated) is surprisingly close to the actual crash. How close mostly depends on one fact: whether or not it is possible in the driving situation to bypass the critical proximity by a steering or evasion manoeuver, or if the braking of one or both vehicles is the only option.

Hence, the pivotal question when designing the application is when to communicate a vehicle's position in order to ensure that the critical proximity can be detected in time. As we have seen, this means that the respective information must be made available in the very small window between the point of no return (when a crash cannot be prevented any more) and the fact of the driving situation becoming critical at all.

The straightforward approach (which has been postulated very often) is to cyclically communicate the vehicle's positions with a rather high frequency. While this leads to a simplified design of the communication protocol, on the other hand it dramatically increases bandwidth consumption, complicates security and privacy measures and increases the computational complexity on the receiver side.

---

[3] This information was generated from data provided by the Federal Statistical Office (http://www.destatis.de.) According to the National Highway Traffic Safety Administration office (NHTSA), the share of intersection accidents in USA is 31%, respectively (http://www.nhtsa.gov).

We would like to mention that we have seen intense discussions about this so called *beaconing* or *heartbeat* strategy in different research works and at different standardisation bodies for several years now. Even with real-world demonstrations from several projects, there is still no real consensus on this issue. We will come back to optimised communication strategies soon in the respective section. However, in order to argue about beaconing strategies, we first want to strengthen the application's point of view and gain a better understanding of the challenges and algorithms to detect such critical proximities.

4.2.2   Situation Analysis and Application Logic

A comprehensive description of a proper cooperative analysis of the driving situation, and in turn a reliable prediction of the collision probability, is beyond the scope of this book. The most detailed application design we know, with a special focus on intersection collisions, is described in [10].

Figure 4.7 depicts an example scenario which we will use in the following for the design of suitable collision avoidance systems with a particular emphasis on intersection assistance. Here, the cube represents the vehicle under consideration for the assistance, which we call *ego vehicle* further on. The balls are the other vehicles that participate in this situation. The arrows mark the direction of the respective vehicles. For intersection assistance, the ego vehicle approaches the crossing and may want to turn left or right, or it may want to cross the intersection. In this situation, the driver of the ego vehicle has to consider the following:

- *Intersection topology and traffic signs.* Traffic signs define the interaction rules among the vehicles. In Figure 4.7, the ego vehicle has to wait at the intersection due to the give way sign. The driver also has to consider the traffic signs that apply to the other vehicles and the topology of the intersection; it must be possible to identify the situation as an 'intersection situation', and on a multi-lane road the driver has to choose the correct lane for a left or right turn.
- *Ego vehicle direction and expected directions of the other vehicles.* Whereas each vehicle may predict its own direction with a high probability, it is almost impossible to determine the turning intention of the other vehicles based solely on autonomous sensors. In the example shown in Figure 4.7, the ego vehicle primarily has to consider the directions from the vehicles of the crossing traffic. If there is enough space to pull into or cross the intersection, the ego

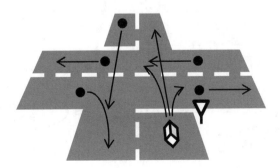

Figure 4.7   Intersection example scenario for intersection assistance.

vehicle additionally has to consider vehicles in front and, depending on its own direction, the oncoming traffic. Furthermore, the performance of the ego vehicle is an important factor: slow or large vehicles (in particular trucks) require more time to pull out into an intersection and therefore have to wait longer for a suitable situation that allows them to pull out safely.

• *Irregularities.* The driver of the ego vehicle has to consider the current 'situation' of the intersection. For example, the driver may have to wait for a long truck that tries to turn into the intersection, there may be road construction at the intersection, or there may be vehicles in front of the driver that may behave irregularly for some unknown reason.

The overall goal is to develop a communication-based assistance system that avoids accidents while crossing or turning into an intersection. In order to avoid conflicts in such situations, the approaching phase of the vehicles must be taken into consideration in order to ensure that the relevant information is available in time for assistance.

From a technological point of view, intersection assistance has to start in the early phase of approaching the intersection by an evaluation of the situation at the intersection. This evaluation is a continuous feedback process as illustrated in Figure 4.8. It basically consists of four steps. First, the system has to predict the driver's behaviour. This behaviour includes whether or not the driver gauged the situation at the intersection correctly. This driver behaviour is used to estimate the direction of the vehicle, in order to predict the trajectory of the vehicle's movement in the second step. Third, the collision probability of the ego vehicle is predicted based on the information received from the vehicles of the crossing traffic and, if necessary, from vehicles of the oncoming traffic. This prediction is based on various information, which is provided by the following sources:

• *Digital map.* A digital map provides information about the topology and geometry of the intersection. This information must be very precise, it has to include the available lanes, stop lines and it may be enriched by information about the driver's route from the navigation system.

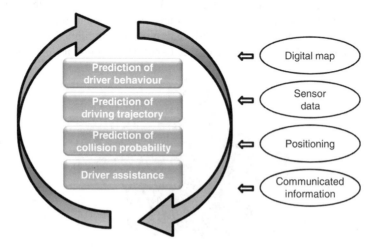

Figure 4.8   Realisation diagram of intersection assistance.

- *Positioning.* The current position of the vehicle must be available, and this position has to be highly accurate. Ideally, the position is matched by the detailed map, including a mapping onto single lanes of the digital map.
- *Sensor information.* Furthermore, the decision has to take into consideration information provided by the vehicle itself. Such information includes the current speed, the status of direction indicators, steering angle, brakes and many other factors.
- *Communicated information from other vehicles.* Information provided by the other vehicles is crucial to determine the collision probability.

Due to the short time window when a warning or automatic manoeuvre makes sense, according to [10] the following parameters have to be considered in the application design.

- Deviation of real and sensed position (translatory error).
- Deviation of real and sensed direction (rotatory error).
- Latency caused by the sensor as well as the communication system.
- In the case of intersection collisions the usual safety gap time before entering the intersection.
- In the case of intersection collisions the fact of whether or not the non-prioritised vehicle has to stop (i.e. there exists a stop sign or a red traffic light).
- In the case of intersection collisions the intentions of all nearby vehicles of where to go (i.e. straight, right or left).

The main principle of calculating the collision probability is to determine the relative distance of the respective vehicles, and the intersection of their extrapolated trajectories, as illustrated in Figure 4.9. Both are distorted by translatory and rotatory errors and the latency caused by the communication system. The latter happens whenever the speed or direction of one of the vehicles changes while the positions are being communicated.

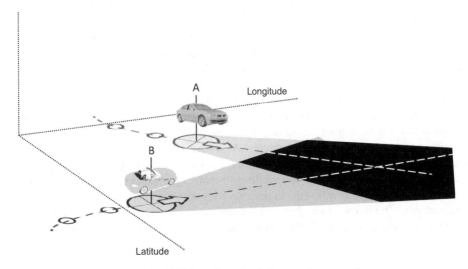

Figure 4.9   Vehicle trajectories in intersection scenarios.

The result of these errors is that it is not possible to compute a precise relative distance or the precise location of the intersection of the trajectories. However, when knowing the current accuracy of the positioning system, starting with the measured position, direction and speed, a fuzzy corridor can be extrapolated for all vehicles. The overlapping area of two of such fuzzy corridors thereby represents a possible collision area. In addition, it also determines the collision probability when set in relation to the remaining relative distance. The impact of the translatory and rotatory errors increases with increasing extrapolation. The obvious result is that, having the same size of overlap of the fuzzy corridors, the collision risk is higher the closer the relative distance of both vehicles is.

Therefore, a meaningful extrapolation can only be computed for a couple of metres, depending on the accuracy of the positioning and direction system. Because of the correlation of speed and distance travelled, a high speed of the vehicle leads to a more fuzzy corridor, assuming a defined look ahead time, as for example 4 seconds. As a consequence, the current speed and acceleration of a vehicle has an influence on the significance of the extrapolation too. From a theoretical point of view, this method of fuzzy extrapolation is usually referred to as *dead reckoning*. As we know from early sea navigation, where dead reckoning was initially used, a new extrapolation is necessary whenever speed or direction changes. And on the other hand it is necessary to take a new measurement from time to time to keep the cumulated errors and thus the fuzzy corridor reasonably small.

When assuming, as we do, that there is a digital map available in the vehicles, the task of a fuzzy extrapolation of the trajectories might be simplified. Since vehicles are supposed to travel on roads, rotatory errors can be neglected. In addition, the fuzzy corridor can be shaped along the road geometry, thus keeping the extrapolated area much smaller in its lateral dimensions. Another important benefit of using digital maps is that the computation of the intersection of the extrapolated trajectories takes into account the road geometry and is not influenced by possible evasion manoeuvers or lane changes. It is important to note that (assuming current location reference technologies) it is a non-trivial task to correlate positions of vehicles that use different digital map data. Another problem when relying on digital maps in this context is the fact that map data can be outdated or faulty, which leads to a wrongly extrapolated corridor and accordingly to a wrong estimation of the collision risk.

Bearing this in mind, an upper bound of the tolerable deviation of real and sensed position can be calculated, assuming that there is no latency caused by the communication system and both affected vehicles know all the parameters. For intersection scenarios, it is necessary to take into account the gap that a non-prioritised vehicle usually leaves for right-of-way vehicles. That is the time the prioritised vehicle needs to pass the intersection before the non-prioritized vehicle can safely enter the intersection. It can be shown by the INVENT project[4] that this safety gap is at least 4 seconds. If this safety gap is undershot, a warning should be generated in the non-prioritised vehicles. Assuming such a safety gap of 4 seconds, the tolerable deviation of real to measured position for a vehicle travelling at 30 metres per second is less than 15 metres. For vehicles travelling 10 metres per second, the tolerable deviation decreases to less than 5 metres. This is because with decreasing speed the distance travelled also decreases, which leads to higher requirements with respect to the positioning accuracy. As one immediate consequence, the position accuracy for scenarios with very low relative

---

[4] INVENT project homepage: http://www.invent-online.de.

speeds, as for example when entering a highway or in very crowded traffic, is significantly higher than the aforementioned 5 metres when approaching an intersection.

The reliability of all these parameters varies in different boundaries, especially with respect to reliability and accuracy of the sensor data, positioning and the information received from other vehicles. For example, different vehicle manufacturers have different representations – and, thus, different resolutions – of sensor data, which must be taken into consideration for the design of a reliable intersection assistance system. A detailed discussion of the reliability aspects of the different parameters can be found in [10].

This leads to the following conclusions relevant for the overall system design.

- The computation of the collision risk is based on extrapolated fuzzy corridors, similar to dead reckoning.
- The size of the fuzzy corridor depends on the accuracy of the position and direction measurements, and the presence or absence of digital maps.
- There is only a small time window, where a warning or autonomous manoeuvre makes sense. The size of this time window depends heavily on whether or not it is possible in the respective situation to perform an evasion manoeuver. In most cases, the traffic characteristics at intersections do not allow for evasion manoeuvers.
- At intersections, the strategy of interaction is also influenced by the so called safety gap, which is typically more than 4 seconds.
- When merging to other lanes, as for example when entering a highway, this safety gap is much smaller. On the other hand, also the relative speed is much lower. As we have seen, one consequence of this is that the accuracy of positioning must be much higher in this scenario.
- The criticality of a relative distance of two traffic participants depends on the relative speed of both vehicles. Therefore, assuming the same relative distance, overlapping lateral trajectories are more critical than from vehicles following one after another (possible rear-end collisions). In contrast, opposing vehicles have the highest relative speed and therefore the highest criticality. It should be noted that such head-on collisions can usually only be prevented by the right (and often rather slight) evasion manoeuvers. As a consequence, the window to react is so small that a driver warning in most cases is not possible. Since autonomous evasion manoeuvres are also not yet reliable, a reasonable warning about a possible head-on collision would rather be restricted to situations when a vehicle can be detected as driving on the wrong lane and (ideally a map-based) correlation with oncoming vehicles.
- The fuzziness of the extrapolated corridor depends on the accuracy of the measurements of speed, direction and acceleration.
- The necessity of communicating position updates depends on the relative speed of the vehicles and the relative direction.
  - Therefore, the accuracy of these measurements has to be communicated, as well.
- Therefore, the fuzzy corridor must only be kept small in the small time window prior to this point of decision.
  - Therefore, there is only the need to recalculate the respective corridors when either there is an update from another vehicle, or one's own position enters an overlapping corridor.
  - Therefore, there is only the need to communicate position updates, if either the vehicle leaves its own fuzzy corridor, or if one's own position enters an overlapping corridor. As a

simplification of the computation of the first issue, a position update might be communicated whenever there is a change of the parameters speed, direction or acceleration within specified boundaries.

– Therefore, besides the current position, it is also necessary to communicate the current speed, direction and acceleration.

Having these issues in mind, we think it is obvious why there have been such intense discussions during the last couple of years about the right strategy to communicate and distribute vehicles' position updates.

4.2.3 Knowledge Management

Having described the major challenges and parameters that have to be taken into account when designing a cooperative collision avoidance application for AutoNet vehicles, we skipped one aspect: the fact that vehicles being within a close mutual proximity need not be travelling on the same crossing or the same road at all. The most prominent and very frequent example are bridges that cross highways. Only taking into account longitude and latitude from an absolute positioning system, as described before, would consequently lead to a high collision probability whenever a vehicle crosses a highway on a bridge. This is the same effect as we have seen in the previous section when talking about environmental notifications. In fact it is worth mentioning here that whenever a vehicle comes very close to a physical object that blocks the road (which has been reported by means of environmental notifications), obviously there is an immanent risk of collision. It therefore has the same impact on the driver as the just described collision avoidance application. Put differently, for such environmental risks there are the same requirements as for a collision avoidance application.

There are three possible ways out of this problem. One option is to use information about the altitude of the vehicles. This would lead to a three-dimensional fuzzy extrapolation of the vehicles' trajectories. The main problem with this is the accuracy of the respective information, which needs to be in a magnitude of a metre.

The most straightforward solution is to make use of digital maps and match the vehicles' positions onto roads. With detailed encoding of street segments and intersections already available, the before described algorithm works fine if it is filtered according to the map information. That is, there is only a potential risk of collisions if either the affected vehicles travel on the same road (oncoming or following), or the risk area of the overlapping extrapolated corridors comprises the location where different roads merge. However, basing the application logic on digital maps requires that all AutoNet equipped vehicles are also equipped with digital maps. Although we strongly believe that in a couple of years this will be true for the great majority of cars anyway, it is a requirement that may increase system costs and therefore complicate market introduction. In addition, map matching algorithms still suffer from low accurate positioning, outdated map data and different data bases, which could occasionally lead to a wrong match of the road. This could again, even if only rarely, lead to wrong and fatal system behaviour.

The third way is again based on a trace chain matching, as described in the previous section. When used in the context of environmental notifications, a match need only be performed between one's own trajectory and the trace chain of the reported events. Since the described events do not move, the latter are static. In contrast, when using trace chain matching in the context of cooperative collision avoidance, a permanent match with all surrounding vehicles has to be performed. This results in a significant computational overhead. In addition, the trace chains of all surrounding vehicles need to be computed and stored locally, assuming that trace chains will not be transmitted with every position update (which is current consensus at all standardisation bodies).

As already mentioned in the previous section, both static georeferenced knowledge as well as dynamic situational knowledge might be stored in a local dynamic map.

Having in mind the algorithmic basis of collaborative collision avoidance, the task of computing the collision risk and the most reasonable time and location of such a potential collision is anything but trivial. Assuming rush hour traffic interchanges with several lanes in all directions, it is pretty obvious that there is a lot work to do for the system. As one consequence, the algorithmic design has to be optimised along with the storage strategy of the nearby vehicles' driving parameters and their predicted travel corridor.

In short, we describe in the following simplification strategies to the above described algorithms that can be used to reduce computational complexity while for most real traffic scenarios still ensuring a sufficient approximation of the collision risk along with the most probable time and location.

Because of the very close proximity, a linear coordinate system rather than a spherical one leads to almost no additional error but significantly reduces computational complexity to compute relative distances. For the same reason, the effects of accelerations and steering angles decreases, which may rationalise to only perform a linear extrapolation and not to take into account accelerations and changes of the steering angles.

Further on, the probability of a crash obviously increases with decreasing relative distances. Having this in mind, calculating the overlapping area of extrapolation driving corridors can be simplified by discretising the extrapolation, as is depicted in Figure 4.10. The discretisation is therefore time-based, meaning that the basis is a look ahead of a specific number of milliseconds, which is transformed in a second step into a location-based look ahead taking

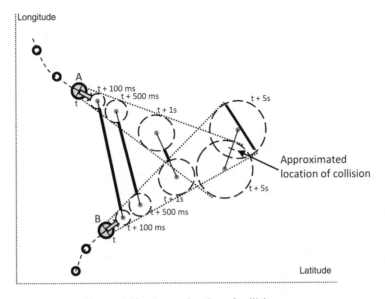

Figure 4.10   Approximation of collision area.

into account heading and speed. Using a time-based look ahead is motivated by the above mentioned safety gap.

Starting with small time interspaces close to the current location, the interspaces can be increased with increasing extrapolation. This simplification is again motivated by the fact that with increasing relative distance the criticality of the situation decreases and therefore the accuracy at this distance can be less. The task of determining the overlap of the corridors can then be interpreted by means of overlaps of discrete circles around the subsequently derived discretised locations, while the circle radius increases along the extrapolation so that it fits right into the boundaries of the virtual corridor. The computation can be further simplified by taking into account the fact that the issue that matters most is the worst case probable relative distance of both vehicles, assuming a dedicated extrapolation time. This is the remaining gap between the virtual corridors at a specific point in time. This gap, however, is again the remaining segment of the straight line of the two respective circle centres, less the sum of both radiuses.

If such gap is negative, then there is an overlap of the extrapolated corridors. In this case, there is a potential risk of collision. If this happens at a very early discretisation step (or putting it differently: close to the current locations), there is an imminent collision probability. The pair with the lowest value of this computation (the highest negative value, to be more explicit) also approximates the most probable collision time and location.

In this way, the problem of computing the collision risk of two vehicles is reduced to a few elementary arithmetic operations. This approach significantly reduces computational complexity of matching the corridors, because only a low number of pairs of discretisation steps need be considered. Therefore, the more discretisation steps used, the more accurate is the warning strategy. In order to make sure that no critical proximity is missed because of computational delays, the sequence of computations should be optimised, for instance by making use of priority queues. Therefore, the computations of the collision probability for all nearby vehicles are ordered according to the above described value, and continuously re-ordered according to each result.

In order to enable such algorithmic optimisations, it is necessary to enhance knowledge management accordingly. This means that it is not sufficient just to store the nearby vehicles' location along with their speed and direction. In addition, the following parameters also have to be stored for each vehicle:

- A time stamp of the latest location measurement, in order to align predicted locations with the size of the look ahead.
- The discretisised future trajectory, which is the computed location of the circle centers and their radiuses.
- The current value of the collision probability.
- The pair of circles with the highest collision probability.
- Optionally a detailed geometry of intersections, in order to enhance the extrapolation of the trajectory for intersection collision avoidance.

4.2.4   Communication

As just seen, the computation of collision avoidance is based on the presented driving dynamic parameters of all nearby vehicles. According to the current work at the standardisation bodies, it is assumed to take into account position, speed, orientation and acceleration, along with a time stamp of the position measurement and its accuracy. In European activities, the respective message that periodically transmits these informations is most often referred to as *cooperative awareness message*, or in short *CAM*, in order to emphasise its main purpose, that is to enable an accurate picture of the nearby traffic participants. While in Europe the terminology emphasises a joint awareness, in the context of US standardisation activities in IEEE and SAE, it is typically referred to as a *basic safety message*, or *BSM* in short.

As mentioned at the beginning, the strategy of frequent position updates is also often called *heartbeat* or *beaconing*, and consequently the respective messages heartbeat or beacon messages.

We think it is worthwhile making a quick side step to clarify the reasons for these different terminologies, because they have an impact on the communication strategy to be standardised. AutoNets, as the name expresses, are driven by two main players: automotive experts and experts for (mobile) computer networks. Position information of vehicles (automotive terminology) or mobile nodes (computer networks terminology) is necessary for both automobile applications and network management. Players from the automobile industry very often focus on the domain of foresighted safety, emphasised by CAM or BSM. Network experts use their common (application agnostic) terminology from mobile networks.

Different terminologies do not at all affect the effectiveness of the system. However, safety applications such as cooperative collision avoidance have different requirements for the position update strategy than networking management has. And more crucially, both issues belong to different layers of the ISO/OSI reference layering of computer networks. In particular, there are differences with respect to message priorisation, accuracy and necessary information artefacts. As one consequence, there is no final consensus within the standardisation bodies whether there should be one (merged) position update strategy, or two individual ones at each respective layer. It is a tricky question because there are pros and cons for each solution. The latter obviously increases the communication overhead, but in contrast simplifies the system layering and therefore the implementation and management of the system. There are dozens of proposals to come to an optimised system performance while fulfilling requirements of both networking and safety applications, ranging from exploiting different channels via adapting the transmission power, to dynamically merging specific information artefacts into one message on demand. It is beyond the scope of this book to discuss the arguments for each proposal in detail. For simplicity reasons, we keep the ISO/OSI reference layering and describe both issues separately throughout this book.

The main challenges for a proper design of the respective position update protocol are to fulfil all position information requirements while

- minimising communication latency;
- ensuring a reliable transmission to all nearby vehicles;
- minimising the communication overhead, that is the consumption of channel capacity and channel idle times;

and ideally to achieve all of this with a simple yet efficient protocol that need not rely on complex information processing or a huge protocol stack.

Focusing on the latter, the straightforward approach (which has been intensively discussed in all related activities) is that every vehicle sends out its CAM or BSM on a dedicated communication channel, with a fixed intermittency and a fixed transmission power, without relying on any implicit or explicit acknowledgment. While this very much meets the goals of simplicity, various improvements have been proposed in order to either improve reliability, latency or the overall channel throughput.

In order to minimise communication latency, messages with critical impact are usually prioritised when accessing the radio channel, that is their waiting times are shorter than those of other messages. For IEEE 802.11 radio systems, the respective channel access mechanisms are described in Chapter 8. Awareness or basic safety messages, respectively, are usually assigned the highest priority because of their important safety impact. However, from an application point of view, there are differences in how critical an awareness message really is. Not surprisingly, this criticality again correlates with the degree of proximity, or to put it differently, whenever there is a high risk of collision caused by a critical proximity, position updates carry a high information entropy and therefore should be transmitted with high priority. Taking this into account, from a theoretical point of view it would be sensible to adapt the priority level that is assigned to each CAM based on the current collision risk of this vehicle. But this would, on the other hand, require a more fine grained medium access strategy that is different from the common approach in IEEE 802.11 MAC, which in turn leads to modifications of the hardware architecture of respective radio modules. From a market perspective the latter is most often considered too expensive and should therefore be avoided. Cellular systems have scarcely been regarded as a means to transmit safety beacons. However, recent analysis provides evidence that they can become an interesting technical alternative or could at least be used as complementary to WLAN or ad-hoc systems. With respect to cost, however, it is unlikely that they will provide the more cost-effective solution for these types of applications.

When using IEEE 802.11 as a communication medium, another important lever to minimise communication based latency is to limit the channel usage. Typically, the overall throughput significantly decreases when too many stations try to access the channel. As a consequence, communication latency increases and transmission reliability decreases.

A simple heuristic to decrease the communication load caused by a fixed periodic transmission of awareness messages is to adapt the intermittency to the vehicles' speed. The obvious idea behind this strategy is that at low speeds only a very small distance is travelled and it takes more time to travel into a critical proximity. This allows us to reduce the position update interval. Assuming several vehicles waiting at a red traffic light illustrates this approach easily. When no one is moving, there is no need to update the positions because everyone is aware of the current positions anyway. This simplified view neglects the fact that other vehicles might approach with high speed, needing to know about the position of the standing vehicles. For the latter reason, current standardisation drafts propose using velocity-based beaconing intervals, ensuring a maximum interval of one second. This approach can be further optimised in two ways:

- A vehicle that detects a possible critical collision path transmits a position update. With this approach, for instance, a standing vehicle might be forced to send an awareness message when another vehicle is approaching, or to be more precise, when there exists a collision risk. In all other cases, a standing vehicle does not need to transmit awareness messages. For urban settings, this would significantly reduce the channel load.
- The update frequency is correlated with the collision risk of a vehicle, complementing the speed-based approach. Doing this in particular increases the reliability, because it advocates the resolution of critical proximities at low or medium speeds also.

4.2.5   Security and Privacy

As for all cooperative applications, ensuring privacy and data security is of crucial importance for both the proper operation and customer acceptance of the overall AutoNetworked application. But when designing a system to avoid collisions, as in this case, handling privacy and security concerns becomes even more challenging, for the following reasons.

- In contrast to a local danger warning scenario, cooperative collision avoidance applications depend on the correctness of the messages of each individual vehicle. Therefore, the application behaviour is designed in a way that assumes this information is trustworthy. In a local danger warning scenario, messages from multiple vehicles can be considered and single attackers are therefore more easily filtered out.
- Most often, location information is still considered information worth being protected. But cooperative collision avoidance as described is explicitly based on publicly sharing such location information in a very precise manner. Therefore, the communication protocol has to include mechanisms to protect peoples' privacy.

As we have seen, cooperative collision awareness requires a very frequent interchange of related messages, leading to a high amount of messages that have to be sent. Assume for instance that a vehicle transmits two position updates per second, in a mutual neighbourhood of 50 vehicles doing the same. In this case, within each second two awareness messages have to be generated by each vehicle, and 98 messages have to be interpreted. From a communication point of view, this seems unchallenging. Neighbourhoods of a few hundred vehicles, however, are also anything but unlikely. Each of them permanently communicating at the same time sending small packets at short intervals does start to pose some challenge on the wireless technology. Neither wireless LANs nor cellular networks have been designed for this type of data traffic.

In addition, securing this amount of messages significantly increases the computational complexity. Securing AutoNets in a cost effective and sustainable way is anything but trivial. Typically, safety applications are based on public messages, that is the message content is not secret and therefore does not have to be protected against unauthorised decoding. But, application design relies on trustworthy messages. Thus, the challenge is to ensure that the message has been generated by a trustworthy network participant, and has not been manipulated during (re-)transmission. There have been dozens of proposals to secure the communication channel in this way. Currently, a system based on a public-key infrastructure is often seen as a suitable solution, despite its rather high communication overhead. Through this, messages are accompanied with a digital certificate that identifies the message originator as an authorised network node and ensures that the message content has not been altered since the origination. The overhead for the message size depends on the length of the cryptographic key and the cryptographic method that is used. As we have seen, cooperative awareness messages only carry some few bytes comprising location, time stamp, speed, direction and acceleration, situationally accompanied by some additional information like turning indicators. In contrast, the size of common certificates is typically in the magnitude of some hundred bytes. This means that the message size is significantly increased, leading to a corresponding increase of the channel consumption. In addition, verifying a high number of certificates requires noticeable computing power or additional dedicated security hardware.

There are two proposals that most likely will become part of the respective standardisation. First, not every awareness message needs to be certified. Assume a stable group of four vehicles for example. Once every vehicle has verified the authorisation of the other three vehicles by checking respective certificates, it is not necessary to transmit further certificates, assuming that is very unlikely that one of the systems is being hacked into just the next second. Awareness messages are not intended to be redistributed or disseminated over a larger area. Thus, once the trustworthiness of a vehicle is verified, all other messages can also be assumed to be trustworthy. In order to reduce the communication overhead, it is only necessary to ensure that, whenever a new vehicle enters a neighbourhood of mutually trusted partners, certificates are shared. While this may look like an easy task, we should have in mind that communication in AutoNets is not uniform or always bilateral, suffering from hidden nodes, message losses and interference.

From an application point of view, the computational overhead can be significantly reduced in a similar way as we have just seen in the context of the computation of the collision risk. It is only necessary to be sure about the trustworthiness of a message whenever there is a (small) risk of collision. In other words, as long as the received messages from another vehicle state that this vehicle is far enough away, everything is fine. Note that this simplification assumes that the only reasonable attack is to pretend to be nearby in order to trigger a false warning for the other driver or a false automated braking manoeuver. But this method is not secure if we assume an attack where a vehicle pretends to be farther away than it actually is, in order to provoke a real accident with the ego vehicle, similar to a suicide attack. Since this simplification is the responsibility of the receiving vehicle, it is up to the manufacturer whether or not the security policy is simplified in that way for the sake of saved resources. In contrast to the certification distribution process, the latter is not necessary to be standardised.

The frequent position updates also lead to a severe threat to drivers' privacy, because peoples' movements can easily be tracked by either the network providers, network participants or an arbitrary attacker. One possible way out of this would be not to use any identifier at any communication layer at all, which would lead to a fully anonymous system. This approach provides the best protection of users' privacy, but is not suitable for several reasons. Even if vehicles send out only some messages within some seconds, the comprised position information or simply the received signal strength make it easy to match those messages to one originator. In order to design a fully anonymous system, this means that there must not be any identifying information in any message. This in turn would mean that it is not possible to deploy bilateral protocols where the originator needs to be known. Further, this also means that messages can neither be encrypted nor certificated as just described, because any signature, encryption or certification requires an identifier of the originator in order to verify the validity of certificate or decrypt the message. The way out is to use pseudonyms for each vehicle. Each vehicle holds an (ideally large) number of identifiers, the so called pseudonyms. From time to time, the pseudonym is changed, to complicate being tracked by others over some time. The actual strategy of when a pseudonym should be changed is tricky, because it has to ensure proper network operation (for instance keep a bilateral communication link in an ad-hoc network) while at the same time minimise the risk of being tracked. Using pseudonyms is therefore a trade-off between privacy protection and network performance. From a networking perspective, a very frequent change of the pseudonym is much like not using any identifier at all, despite the possibility of using security mechanisms to sign or encrypt the messages. On the other hand, this leads to a very high overhead with respect to the management and distribution of identifiers, keys and certificates. In contrast, very infrequent changes raise the risk of being trackable by others. We will describe the approach of pseudonyms, the change strategies and problem domains in more detail in Chapter 9.

Note that the discussed privacy issues are valid for any type of communication system. While largely discussed in the context of ad-hoc networks, the challenges are similar in cellular networks. In addition, each AutoNet participant in a cellular network has to disclose its identity to the network provider.

4.2.6   Driver Interaction

The local proximity in which vehicles can be mutual, while still not encountering any critical situation at all, leads to the second important issue when talking about collision avoidance in vehicles. Such proximity is pretty normal in everyday life traffic and, as long everything goes well and as expected, would be nothing noticeable. Imagine for example two vehicles (A and B) approaching an arbitrary urban intersection at the same time from different directions, where one of the vehicles (A) has to give right of way to the other (B). Assuming a moderate speed due to urban traffic, the brakes are usually activated only a few metres before the stopping line. The fact that vehicle A does not significantly decelerate some 10 metres before the stopping line does not imply that the driver of vehicle A missed the other vehicle's approach. A system that informs or even warns the driver in this situation about a possible collision could quickly become annoying and therefore counterproductive: people would tend to neglect the system, because of too many 'warnings'. This fact is rather tricky, because unfortunately it forces the system to be parameterised for a very late driver interaction (and thus in some circumstances maybe for a too late warning). When having a closer look to this question, in this case we clearly have to distinguish the three usual collision types: head-on, rear-end or lateral collisions:

- The most effective and intuitive warning strategy can be realised with respect to rear-end collisions. Because of the uniform traffic situation and the low relative speeds, it is comparably easy to indicate to the drivers that they are approaching too fast or too close to the vehicle in front. Based on radar sensors, such assistance systems along with automated distance controls are already in the market (usually called active cruise control or dynamic cruise control). Enriching such systems with AutoNet capabilities then allows one to observe vehicles that are not directly in front and therefore help to avoid multiple collisions in a row. However, from the perspective of driver interaction, such extended systems (most often called *extended electronic brake lights* or *forward collision warning*) are very similar.
- Preventing head-on collisions by means of an early driver warning is very complex. Because of the high relative speeds, the driving situation is highly dynamic and the time to react very short. Usually the most suitable manoeuvre is to evade. However, currently, there are no assistance systems in the market that perform an automatic evasion manoeuvre, nor could any driver interaction paradigm so far prove that the driver understands quickly enough to steer right or left in time. Nonetheless we want to emphasise that there is a (prominent) example, where the head-on collision risk can be significantly reduced by means of AutoNet driven driver warnings: against wrong-way drivers on highways.
- As we have seen, AutoNets promise to have great potential with respect to lateral collision avoidance. On the other hand, an effective and intuitive driver interaction strategy is rather

complex. Lateral collisions typically appear at intersections. From the perspective of driver interaction, such networked intersection assistance is typically realised by a three-way assistance system, which assists the driver in three steps: information about the type of traffic control in effect, a warning about a potential collision, and finally active braking.

Information about traffic control. While approaching the intersection, the system should provide the driver at an early stage with information about the intersection, especially with respect to the type of traffic control, in an unobtrusive way. Information on the intersection itself is independent on the current situation, and the driver should be informed about this situation only in case he or she may interpret the situation in the wrong way. With respect to the presentation of this information to the driver, developers must be aware of the fact that this information may be redundant since the driver may already have seized (and interpreted) the situation correctly. This is of particular importance in the early approaching phase, because it is almost impossible for the system to determine whether or not the driver interpreted the situation correctly.

Warning about potential collision. The second step for assistance is to support the driver with a situation-dependent warning: the goal is to warn the driver 'right in time' in order to avoid potential collisions with a vehicle from crossing traffic. The precondition for this warning is that it is possible to mitigate the dangerous intersection situation by an adaptation of the movement of the ego vehicle. An analysis of driver behaviour and corresponding warning strategies can be found in [33].

It should be noted that there are usually two vehicles involved in a possible collision. This means that two drivers may be warned. The warning strategies differ for the two drivers.

Let's first have a look at the situation of the non-right of way vehicle. In this case, an important requirement for the warning is that a warning must be presented to the driver only in the case that the driver misinterpreted the situation, and there is a high probability for collision with a vehicle from the crossing traffic having right of way. Naturally, this warning must be right in time. The term 'right in time' depends on the situation itself: on the one hand, the system should warn the driver as early as possible so that they have enough time to interpret the situation correctly and to react appropriately. On the other hand, the warning has to be presented to the driver rather late in order to avoid a warning being presented to the driver in the case of a conflict-free approach to the intersection. Here we should notice that there exist traffic situations where the usual driver action would appear after the latest possible time to warn. This effect is called the *warning dilemma* [10, 16, 34]. The warning dilemma is delineated in Figure 4.11.

The warning dilemma occurs because, in the case that a warning is triggered, the overall stopping distance is increased by the distance travelled during the human reaction time. The time to react is usually around one second. Thus, the driver has to be alerted one second prior to the latest point in time of the actual braking manoeuvre.

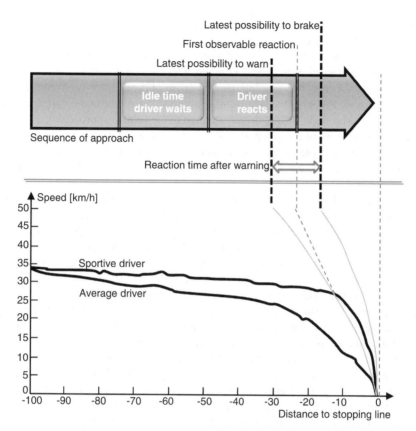

Figure 4.11 The warning dilemma of intersection collision avoidance, based on [10]. Reproduced from a Dissertation by Felix Klanner.

By analysing vehicle dynamics parameters, [16] could show that if we assume fully precise position data, it is possible to make a reliable analysis of whether or not the driver has recognised the actual traffic situation correctly. However, it could also be shown that with imprecise position information – as is the case with current positioning systems – the analysis is not always reliable. As a consequence, the earliest possible time when the driver alert can be triggered has to move closer to the intersection (assuming a behaviour to prevent the system from false warnings in order to prevent customer discontent). The warning dilemma decreases with increasing position accuracy. In turn, this means that the number of required but oppressed warnings also decreases with increasing position accuracy.

The warning strategy for the right-of-way vehicle differs noticeably. The reason is of psychological nature. In the just described scenario, the driver of the non-right-of-way vehicle has their own mental representation of the driving situation. If they have interpreted everything correctly – which is the usual case – they are annoyed by a system insinuating that they did not. To put it differently: the system thinks it is cleverer, but obviously it is not. In contrast, the drivers of the right-of-way vehicle do not know the other driver's mindset. They do not know if and when the other driver slows down and stops. For this reason, a warning can be triggered whenever the non-prioritised vehicle approaches in an unusual way. Even if the vehicle does not cross the stopping line, the driver of the right-of-way vehicle will not be annoyed. Probably the best warning approach would be two-stepped. In the first step there is a light shift of driver attention to the vehicle approaching quickly, so that the driver is aware of a potential risk. If there is a critical proximity and if due to the driving dynamics of the crossing car it is rather likely that it will miss giving way, a warning can be triggered, determining the driver of the right-of-way vehicle to brake.

There is another method to increase the awareness of unattentive drivers or to warn them, that is specially applicable to vulnerable road users on bikes or motorbikes: automatically turning on horns and glinting head-lights in the case that a critical proximity had been detected. Since overlooking (motor)bikes is a major source of fatal accidents, this seems to be a very promising approach to increase traffic safety. Having in mind the discussed warning strategy, providing an automatic external feedback to the driver might also be interesting for vehicle scenarios. This way, the non-right-of-way drivers might be alerted earlier, because in their perception the warning is triggered by the other vehicle (regardless of the technical solution behind it). By doing so, we play a psychological trick, because even if they had read the situation in the correct way, it's the other vehicle's system – or even more likely the other driver – that thinks they have not. Any difference compared to everyday traffic today? None in the perception of the drivers. But still a major step foreward for traffic safety due to the elimination of reaction times. This is a very intuitive way to gain one's attention.

Active braking. The final step of assistance is automated active braking in order to avoid a collision. Active braking should begin *before* the point of no return; until this point in time, a collision can be physically avoided by automated braking, without an additional reaction time. An important precondition for this functionality is that the driver must always have full control over the system. For example, it must always be possible for the driver to overrule the system, for example to accelerate the vehicle although the system would decide for an active braking. This is an important requirement to handle false warnings and for product liability reasons.

4.3   Green Driving Case Study: Traffic Lights Assistance

Besides vehicle-to-vehicle communication, the consideration of communicating roadside infrastructure elements offers interesting opportunities for traffic safety and traffic efficiency. For example, dynamic message signs can communicate the information they currently display immediately to the passing vehicles. Based on this information, drivers can be informed or warned accordingly inside the vehicle. If traffic lights are able to communicate their current status together with their red and green phases, the information is very valuable for both the vehicle and the driver. The following examples demonstrate that this information can be utilised in several ways:

- *Driver support.* A simple case informs the drivers at red traffic lights about the remaining time of the red light. While waiting in such a situation, drivers sometimes bridge the waiting time with other activities, such as checking their emails on their smartphones. Hence, drivers are somewhat distracted from the traffic situation around them. Information about the remaining waiting time – together with a audio-visual feedback – helps to get the driver back into the driving context right in time. Drivers informed about waiting times are typically more relaxed, which also makes the waiting more pleasurable. Moreover, this waiting time can be used to entertain the driver in different ways.
- *Driving assistance.* Information about the behaviour of traffic lights can also be used for driving assistance; especially in the phase while approaching the traffic lights. For example, a 'green light optimal speed advisory' assists drivers in choosing the optimal speed range to pass traffic lights efficiently or to follow a progressive signal system successfully. Such a system contributes to a smoother traffic flow on the road. By following the advice, drivers reduce fuel consumption by avoiding unnecessary acceleration and braking when approaching the traffic lights.
- *Vehicle efficiency.* The vehicle itself can also benefit from this information to further augment the reduction of fuel consumption. This would be made possible by an efficient adaptation of the vehicle's behaviour in this situation. While approaching the traffic lights, there are several options:
  - The power and performance of the engine can be reduced, for example by selectively cutting-off the cylinders of a combustion engine.
  - The gears can be switched optimally to reduce drag torque.
  - The engine can be turned off automatically while the vehicle rolls out.
  - In the case of electric vehicles, the power train can be optimised, for example by increasing recuperation.

  While waiting at the traffic lights, the automatic start-stop system can selectively decide about stopping the engine and knows when to re-start in time. In the case of hybrid vehicles, the power train can be configured accordingly to provide the most suitable – and thus efficient – performance to accelerate the vehicle after the traffic lights have turned green.

In the following, we will emphasise the green light optimal speed advisory as an example to describe the technical prerequisites and challenges for the development of such a system.

4.3.1 Green Light Optimal Speed Advisory

Green light optimal speed advisory (GLOSA) is an assistance system that supports the driver to pass a traffic light in an efficient way. The following scenario shows the basic functionality of such a system: a driver is approaching an intersection with traffic lights. Depending on the status of the traffic lights, the driver has to decide among several alternatives: if the traffic lights show green, the driver may retain or increase his speed to pass the traffic lights before they switch to red, or the driver may decrease his speed since the green phase is not long enough to reach the intersection. Likewise, the driver may needlessly decrease the speed while

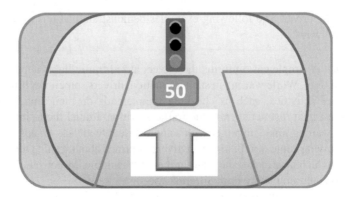

Figure 4.12   Snapshot of a GLOSA HMI in the centre display of an Audi.

approaching traffic lights showing red, although the traffic lights will switch to green within the next second.

The goal of GLOSA is to give the driver advice in such a situation, helping to smooth traffic flow, reduce fuel consuption and – last but not least – inform the driver about the expected situation on the road in the next seconds. Figures 4.12 and 4.13 show two examples of how this information can be presented to the driver. In Figure 4.12 information about the current status of the potential traffic lights, the directions the respective traffic lights regulate, and the required speed to pass the traffic lights is shown in the centre display of an Audi.[5] An alternative representation of this information is depicted in Figure 4.13, which shows a fully programmable prototype dashboard of a BMW. Here, the current phases of the traffic lights are highlighted on the speedometer: when keeping the speed in the green area, the traffic lights will be green before and while passing. Respectively, the driver has to stop at the next traffic lights in the case of the current speed of the vehicle being in the red highlighted area of the speedometer. Additionally, the current phase of the traffic lights is represented in the lower middle of the speedometer in the respective colour, together with a countdown of the remaining time for this phase. This information needs to be updated accordingly, depending on the speed of the vehicle and the status of the traffic lights. As a result, the driver is informed about the expected speed (range) to pass the traffic lights efficiently, and he or she can adapt speed accordingly.

The realisation of such a system requires information about the current status and the expected behaviour of traffic lights. We can think of three possibilities to provide this information to vehicles:

1. Vehicles jointly 'learn' this information autonomously, for example by using a camera system in order to analyse traffic lights that they frequently pass. Assuming the participation of a high number of vehicles passing specific traffic lights, and further assuming that these vehicles are able to share their observations by means of a common data format, a common (distributed) database can be generated and maintained that in many cases will

---

[5] This snapshot was taken from a press release of Autobild (http://www.autobild.de) about the project TRAVOLU-TION. In addition, Audi displays this information in a reduced fashion on the dashboard.

Figure 4.13 Snapshot of a GLOSA HMI in a programmable prototype dashboard of a BMW. Reproduced by permission of BMW Group.

have sufficient accuracy for at least the non-safety critical application domains presented above. Various parameters can be taken into account in order to predict the switching times, for instance waiting times along with the actual distance to the intersection, time of the day or the status of pedestrian signals. This solution is of limited effect, because it can only be used for traffic lights with a deterministic, static control sequence.

2. Information about relevant traffic lights is provided by their respective control or management centres. The vehicles receive this information from respective back-end systems. This approach is very useful for controlled traffic lights with fixed phases. For intelligent traffic lights, which adapt their phases dynamically to the current traffic situation, this solution will only work if the phase information is updated continuously and the round-trip delay is short. Today, traffic lights usually send their information within a couple of seconds of delay to the management centre (if there is one). These do not yet provide interfaces to communicate this information to vehicles.

3. The third solution is to transmit the information by the traffic lights themselves individually via a communication link from each traffic light to the vehicles around it. Avoiding intermediateries and exploiting the local nature of the information, this allows the information to be transferred automatically to all vehicles in the vicinity with small delays.

We will focus on the third solution in the following, because it is the one that has received the most attention in research so far.

4.3.1.1 Realisation of GLOSA

Figure 4.14 depicts the basic scenario for the GLOSA use case, where an ego vehicle approaches the traffic lights. The traffic lights communicate their status periodically to their vicinity via so-called SPAT (signal phase and timing) messages. The message format for SPAT messages is standardised in SAE J2735 [12]. The SPAT message specification is interlaced; several information items in the message are optional and thus need to be customised accordingly. As an example, the message format shown in Figure 4.14 represents the application of SPAT messages used in the German project simTD,[6] which contains a unique intersection identification, the current status of the traffic lights' controller, and a state object for each lane

---

[6] http://www.simtd.de/.

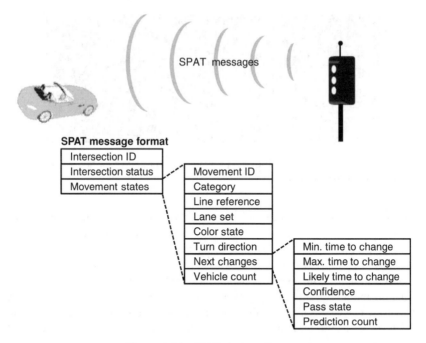

Figure 4.14   SPAT message format.

(also called movement). Of particular importance for GLOSA are the state objects, which provide the following information:

- An optional identification of the movement/lane.
- The traffic category using this lane, which could be individual traffic, public traffic, etc.
- Line reference is a grouping of routes that are generally known to the public by a similar name or number.
- Lane set is a collection of lanes for which the state object is valid.
- Colour state is the current colour displayed by the traffic light for this state object, which could be *green*, *yellow*, *red*, switched off or *flashing*. This information is optional, since a state object may not necessarily be controlled by traffic lights.
- The turn direction represents the direction(s) allowed for the vehicles driving on this lane.
- The 'next changes' information contains the phase information. It comprises the minimum and maximum time-to-change as well as the expected time-to-change (in seconds) to the next colour state. Additionally, there is information about the confidence for the likely time-to-change, the current pass state (i.e. whether or not a vehicle is allowed to pass the traffic lights) and the predicted number of vehicles that may pass the traffic lights.
- Finally, optional information about the current vehicle tailback may be included in the message.

Besides the information provided by SPAT messages, GLOSA also requires sensor data from the ego vehicles. Figure 4.15 summarises potential information necessary for an

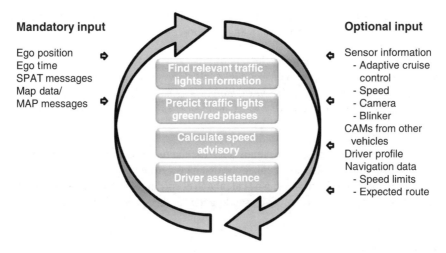

Figure 4.15   GLOSA: functional building blocks for development.

implementation of a robust speed advisory for the driver of the ego vehicle. Mandatory information is the current position of the ego vehicle, the current time and – if available – map data. Depending on the feature set, additional information may be required as shown in Figure 4.15. Based on this input data, the functionality of GLOSA is realised in four steps.

In the first step, relevant SPAT messages for an ego vehicle need to be identified. This is of particular importance since SPAT messages can be transmitted by different (or all) traffic lights at an intersection. Therefore, the relevant SPAT messages need to be found, and the correct information about the relevant lanes needs to be identified. In order to do this, a detailed view of the intersection topology is necessary, comprising in particular the directions of the crossing streets together with information about the respective lanes, that is which lane allows what turnings. In addition, the positioning system has to be accurate enough to distinguish between these lanes.

Second, the phase of the traffic lights with respect to the current position and movement of the vehicle needs to be calculated. In the following, we assume that *MovementState[x]* contains the relevant phase information for the ego vehicle. It is worth mentioning that dynamically controlled traffic lights cannot provide exact information about the time until change of phase. They only provide a likely time-to-change since the phases depend on the current traffic situation in the vicinity. Therefore, the likely time-to-change of a phase at time t needs to be corrected by a function $f(t, a, b, \ldots)$. Potential input factors for f could be the confidence of the phase, or the maximum/minimum time-to-change. The corrective function f also enables the customisation of GLOSA to the different requirements of the vehicle manufacturers.

In the following, we assume for the sake of clarity that the traffic lights switch between green and red only, without any yellow phases. We also assume that the current status of the traffic lights is 'green'.[7] Therefore, the respective values for the end of the current green phase

---

[7] Ambitious readers may formulate the algorithm for the red traffic lights as an exercise.

(end_{green}), the end of the next red phase $end_{nextRed}$ and the end of the following green phase $end_{nextGreen}$ can be calculated by the following pseudo-code:

if (MovementStates[x].ColorState == 'green')
    end_{green} ← *MovementStates[x].NextChanges[0].LikelyTTC* $* f(t, a, b, \ldots)$
    $end_{nextRed}$ ← *MovementStates[x].NextChanges[1].LikelyTTC* $* f(t, a, b, \ldots)$
    $end_{nextGreen}$ ← *MovementStates[x].NextChanges[2].LikelyTTC* $* f(t, a, b, \ldots)$
fi

Based on these calculations, a speed advisory for the driver can be computed in the third step. To do so, GLOSA first has to calculate the distance d from the vehicle to the traffic lights. Then, the speed interval boundaries can be calculated accordingly based on the simple physical formula $speed = \frac{distance}{time}$. The respective pseudo-code for the speed advisory calculation is the following:

$d = |pos_{ego} - pos_{trafficLights}|$
$speed_{minPass}$ ← d/end_{green}
$speed_{maxWait}$ ← $d/end_{nextRed}$
$speed_{minWait}$ ← $d/end_{nextGreen}$
// determine speed corridors
if ($speed_{minPass} > speed_{allowed}$)
        $pass_{currentGreen}$ ← *null #Impossible to pass the current green phase*
else
        $pass_{currentGreen}$ ← $[speed_{minPass}; speed_{allowed}]$
fi
$wait_{nextRed}$ ← $[speed_{minWait}; speed_{maxWait}]$
$pass_{nextGreen}$ ← $[0; speed_{minWait}]$

The results of these calculations are so-called speed corridors, in which the driver may pass (or has to wait at) the traffic lights. If possible, $pass_{currentGreen}$ defines the minimum and maximum speed needed to pass the current green phase, $wait_{nextRed}$ is the speed interval in which the vehicle has to stop at the traffic lights. If the vehicle moves in the speed corridor of $pass_{nextGreen}$, it may pass the traffic lights at the next green phase.

In the final step, the result needs to be presented to the driver. While approaching the traffic lights, the speed corridors can be presented to the driver as illustrated in Figures 4.12 and 4.13. If the vehicle has to wait in front of a red traffic light, GLOSA may also display the estimated remaining waiting time, which can be generated easily from the SPAT messages. The optimal speed advisory for the driver must be updated periodically since several input parameters may change while approaching the traffic lights. For example, the traffic lights may switch to red earlier due to the current traffic situation and the ego vehicle's speed will likely vary since vehicles in front may break or accelerate. This way, the four steps described in this section need to be repeated until the ego vehicle passes the traffic lights.

4.3.1.2 Challenges

Although the physical fundamentals are quite simple and straightforward, there are several challenges that need to be solved for the development of a green light optimal speed advisory system ready for real-world usage. Such aspects of system design are important for the differentiation of the functionality in different vehicle brands. We will only emphasise three important challenges without providing solutions for them, because such solutions greatly depend on the requirements of the individual manufacturers.

As already touched on in the previous section, GLOSA has to handle the inaccuracy of several input factors in order to predict a suitable speed advisory. Examples include the inaccuracy of the ego vehicle's position, variations in speed while approaching the traffic lights, the inaccuracy of available map information, or variations in the time-to-change values of the traffic lights. The latter is of particular importance for demand-actuated traffic lights, which react, for example, to pedestrians crossing the street. Moreover, the system has to cope with potentially inconsistent information. The ego vehicle may not receive all messages due to transmission errors, or the vehicle may receive information from other vehicles describing a different context. Such inaccurate – and sometimes incomplete – information needs to be evaluated and estimated by the system to support the driver with robust and reliable information. Therefore, the system may also consider additional sensor information from the vehicle, and – if available – from communicating vehicles in the vicinity.

A second challenge is the identification of information about relevant traffic lights. Therefore, two calculations and predictions are necessary:

- The route of the ego vehicle needs to be predicted. This can be realised by using route information provided by the navigation unit, or it could be a learning-based system predicting the expected route of the ego vehicle (some methods can be found in [29, 30, 31, 32]. This is necessary to identify the traffic lights relevant for the movement of the ego vehicle.
- Based on the intersection ID in the SPAT message, the position of the different traffic lights need to be mapped either onto the digital map, or onto MAP message[8] information describing the intersection. This way, GLOSA is able to access the relevant SPAT messages, which describe the state of the relevant traffic lights for the ego vehicle.

Third, a major concern is the presentation of the speed advisory to the driver. This human–machine interface (HMI) is the most important factor for how the driver experiences GLOSA. The HMI therefore has to consider the different varieties and look-and-feel strategies of the vehicle brands, but also the cooperation of GLOSA with other AutoNet applications. In the beginning of this section, we showed two possible HMIs. The AUDI solution shown in Figure 4.12 displays the minimum (or maximum) speed to pass the green traffic lights, which is enhanced by a 'speed carpet' shown in the dashboard. In contrast, the BMW solution depicted in Figure 4.13 maps the speed advisory onto the speedometer by highlighting the different speed corridors in their respective colours. This way, the driver can immediately see if the current speed is sufficient to pass the traffic lights without stopping, or if he or she has to increase or decrease speed in order to pass the traffic lights in the current or next green phase.

The green light optimal speed advisory described so far is focused on one vehicle only. However, at most intersections most of the day there is more than just one vehicle approaching at a time. Imagine an intersection with dedicated traffic lights (and short lanes) for different

---

[8] MAP messages (sometimes also referred to as TOPO message) have been defined to provide an accurate layout of a signalised intersection to AutoNet vehicles by the intersection itself or by a corresponding service provider.

directions, for instance one lane to go straight and right, and a dedicated lane to turn left. Imagine a further two vehicles approaching. The preceding vehicle wants to go straight. The respective traffic light is red, and the calculated speed advisory is 30 km/h so that the traffic lights have turned to green just before the vehicle reaches the intersection. However, there is a second vehicle following the first one. This vehicle would like to turn left. The respective traffic light is currently green, and the calculated speed advisory is at least 45 km/h in order to reach the intersection in time. Unfortunately, it is not possible to continue at 45 km/h because the leading vehicle decelerates to 30 km/h as it has been advised to.

There are three possible reactions of the driver of the following vehicle with respect to the missed green phase to turn left. First, they know that the vehicle in front is equipped with a GLOSA system. In this case, they might complain about the stupidity of the system. Second, they are not aware of the GLOSA system in the vehicle in front. In this case, they might be angry with the driver in front. Without any speed advise, typically drivers do not decrease their speeds too much when approaching a red light. This is likely to change when a GLOSA system is available. With an increasing penetration rate of such a system, such scenarios will appear more often. Finally, the driver knows that the vehicle in front is equipped with a GLOSA system, but understands both the unfairness (not to say the simplicity) of the system and that the driver in front does not care. The first two reactions should be avoided for the sake of customer acceptance of the system. Thus, one has to think of technical solutions to alleviate the problem.

We see that one of the greatest challenges of a green light optimal speed advisory is to find ways to resolve possible conflicts. As a way out, a system following an overall optimisation strategy may calculate a joint optimal speed advisory in a way that means that either the overall waiting time or the overall fuel consumption of all vehicles that approach an intersection at one time is minimised. By doing so, the HMI concept is more complicated too, because ideally it should also present the rationale of the speed advisory. Not doing so would again lead to customer dissatisfaction, because – from the subjective driver perspective – in this case the speed advisory is inaccurate. What's more, while technically such a system behaviour might reach an optimum in a given scenario, it will embody negative, probably unintended effects.

Consider the strategy to minimise the overall fuel consumption. In this case, a vehicle that consumes more fuel than another would be treated preferentially. This, however, appears unfair to the driver of the more fuel-efficient vehicle. Even worse, it would generate an incentive to buy vehicles that consume more fuel in order to gain an advantage at traffic lights.

Speed advisory could also take into account tailbacks on different lanes and may give advice in advance about which lane should be taken. Obviously, this requires knowledge about the amount of tailback on different lanes, which could for instance be detected and broadcasted by means of roadside infrastructure. Another approach is to exploit the stopping points of vehicles ahead reported by them, leading to a lower bound of the tailback. In fact the tailback could be longer in the case that non-AutoNet-enabled vehicles queue behind. Using the latter approach does not require any standardisation efforts. Instead, it is a matter of in-vehicle software complexity how much effort is put into enhanced speed advice. Using infrastructure assistance requires standardisation of the respective messages and data formats. It is unlikely that such roadside detection systems will be deployed in the initial phase. Therefore, such protocols are currently not being discussed in standardisation, but foreseen for future enhancements.

4.3.2   Example: TRAVOLUTION

The main purpose of the German project TRAVOLUTION[9] was to reduce waiting times in the
city, to optimise the city's traffic flows, and to reduce fuel consumption and air pollution.
Therefore, the city of Ingolstadt, Germany, equipped 46 traffic lights with communication technology. In order to optimise the traffic flows within the city, the traffic control was based on
genetic algorithms in order to find the best traffic light control for the entire test site and for all
road users. This optimisation also included a progressive signal system (called 'green wave'),
which reacted to the current traffic.

Evaluations showed that TRAVOLUTION had significant economic and ecological potential: according to the results published on the website, fuel consumption could be reduced by
700,000 litres (or an equivalent of 1,600 tons of CO_2, respectively) in the city of Ingolstadt,
a city of 125.000 inhabitants, each year. In addition, fewer (and shorter) traffic jams have
the potential to reduce economic costs significantly: according to several measurements, the
waiting times could be reduced by 21% on a daily average.

Besides traffic efficiency, the 'informed driver' was a second aim of TRAVOLUTION. The
traffic lights in Ingolstadt communicated the time of the next green phase to the on-board
computer in an Audi test vehicle. The driver was advised with an appropriate driving speed,
so that they could pass the intersection without having to stop.

4.4   Business and Convenience Case Study: Insurance and Financial Services

Insurance and Financial Services (IFS) are another use for AutoNets which we will explore
in more detail. Respective usages address financial aspects of owning (and driving) a vehicle,
especially with respect to the insurance services required in almost every country. IFS uses
typically address the driver and their vehicle – the services are not related to driving itself.
Hence, IFS applications are something of a combination of vehicle-related applications and
passenger-related applications according to the classification introduced in Section 2.1.

In general, there are different insurance and financial services that can be applied in typical
AutoNet-based scenarios. The following examples give an impression of the variety of potential
services in this area:

- *Insurance services per use.* In Germany, insurance fees for vehicles are typically calculated
  based on some 'characteristics' of both driver and vehicle. Besides brand, model and mo-
  torisation of the vehicle, insurance companies also consider personal aspects such as the
  expected amount of kilometres per year, job and age (and sex) of the driver, the personal
  situation of the driver (e.g, whether or not he or she owns a parking or a flat) and even the
  other persons who will drive the vehicle. For the example of the amount of kilometres per
  year, the owner of a vehicle has to deal with the insurance company every time the expected
  number of kilometres does not match the kilometres actually driven. A useful IFS application
  would be an automatic adaptation of the fees according to the number of kilometres driven
  in a respective period of time.
- *Personalised services.* Today, fees for insurance services are basically assigned to a vehicle
  and, thus, its owner. Thus the services do not reflect the real usage of a vehicle. For example,

---

[9] Website of TRAVOLUTION: http://www.travolution-ingolstadt.de/index.php?id=2&L=1.

it is up to the vehicle owner to deal with the administrative impacts of an accident – even if another person drove the vehicle. If it would be possible to introduce a personal 'pay per use' model for drivers independent of the vehicle they drive, insurance services could be customised individually to each person. Such an insurance service would increase transparency as well as fairness, because the insurance fees could be calculated based on the vehicles a driver uses as well as their usage profiles. Such services are even more interesting for car sharing scenarios, which are considered as an important aspect of future mobility in mega cities [7].

- *Accident management services.* Besides material or even personal damage, an accident always requires a lot of administrative effort: in the case of a serious accident without personal damage, both the driver and the opponent have to claim the accident to their insurance companies, the vehicles often need to be towed to respective repair shops, the damage of vehicles needs to be estimated, the vehicles must be repaired, eventually the drivers must be provided with rental cars, and finally the financial aspects for the repair must be regulated by the respective insurance companies. Such efforts can be easily handled by accident management services. Such services efficiently offer the drivers the complete claim settlement services – right from the beginning of the accident. AutoNets support these services by offering wireless access to the vehicle data and systems.
- *Emergency call.* In the event of a serious accident with personal damage, an automatic emergency call can provide immediate help to the driver. In the case that the vehicle sensors detect a serious accident, this information will be transferred immediately to an emergency service centre. An employee of this emergency service centre then tries to contact the driver by a cellular phone call. If the driver does not accept the call, the emergency service centre immediately organises help for the driver, that is it sends an emergency vehicle from a nearby hospital to the accident. First implementations of such a system are already available in today's premium vehicles. For example, BMW launched an emergency call system named 'e-call' in 2007 in several markets around the world.

These examples are suggestive of the variety of potential uses dealing with the insurance and financial aspects of driving. In the following, we will emphasise the *accident management services* as an example. From a technological perspective, accident management services are very challenging because they require both complex interactions with many stakeholders and integration into many different system components.

4.4.1   Accident Management Services

An accident management service supports the driver in the event of an accident by reducing the administrative effors while improving the processing efficiency of incident management. This is due to the fact that an accident is always an exceptional situation for a driver: a driver is nervous, sometimes stressed and often does not know what to do in this situation. In the following, we assume for the sake of simplicity the management of an accident without personal damage. Here, the steps necessary for the incident management are outlined in Figure 4.16. After the accident occurred (first step), the regulation process begins in the second step with a survey of the accident by the persons involved. Such persons could be either the drivers themselves, the passengers of the vehicles involved in the accident, or

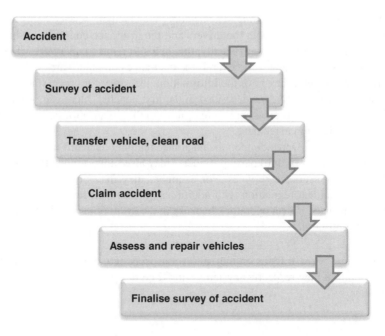

Figure 4.16   Incident management process.

witnesses. Depending on the damage, it may be necessary to call the police to support the survey. In the third step, the vehicles have to be transported to suitable repair shops in order to fix the damage. In the case of serious damage, the vehicles must be towed, and rental cars for the persons involved optionally need to be organised. Possibly, the road must be cleaned and damage to the road infrastructure (e.g. broken crash barriers) has to be repaired. Afterwards, both the driver and the opponent have to inform their insurance companies in the fourth step, which will handle the further process of the accident management in the fifth step. Therefore, the damage of the vehicles will be estimated by independent assessors, and the damage will be repaired. In the final step, both the driver and the opponent will get their vehicles back, the insurance companies regulate the payment and inform the drivers about the impact of the accident on their fees.

This complex process can be alleviated by an AutoNet-based accident managementsService, which supports both the drivers and the insurance companies in some of the process steps. Such a service has three important benefits:

1. *Immediate and specific help for drivers*. In the event of an accident, the drivers immediately get help through their insurance company. Professional support guides the drivers on site, that is they can tell them what to do in this situation. This help includes information that needs to be identified and documented, the professional support may call the police for the driver, or it may organise a tow truck and/or a rental car for the driver. Moreover, this system can be coupled with the emergency call described above. This helps to organise immediate and suitable help in the event of personal damage.

2. *Efficient administrative processing.* The accident management service helps to reduce the administrative overhead for both the drivers and the insurance companies. Since most information will be available electronically, this helps to reduce paperwork while speeding up the accident management. Moreover, if the insurance companies are contacted immediately after the accident, they have the information necessary for processing in their hands (since they guided the driver as described in the first benefit). Moreover, such a system is transparent since the insurance companies can keep the driver informed about the current status of the accident management.
3. *New business models.* Depending on the insurance regulations in different countries, insurance companies (or even third party providers) may also think of potentially new business models for their insurance services. For example, they may offer drivers an additional and improved incident regulation with additional value-added services. This can be very attractive for drivers or fleet operators, because in the event of an accident uncertainty will be kept to a minimum.

In the following, we will focus on the AutoNet aspects of accident management services. We will also outline how respective systems should be designed and how they will interact with each other.

4.4.1.1   Domains, Stakeholders and Requirements

Compared to typical AutoNet-based applications like the hazard warnings described above, insurance and financial services depend on business processes provided by respective stakeholders. From a communication perspective, the focus here is on connecting the business processes of insurance companies to the vehicles or to the mobile devices of the drivers via cellular communication systems. According to our classification introduced in Section 3.1, from the six steps illustrated in Figure 4.16, we can see that the following two domains are participating in a respective AutoNet-based system for accident management systems:

- *AutoNet mobile domain.* In the AutoNet mobile domain, both the vehicle subdomain and the mobile device subdomain are involved. The vehicle may be able to detect the damage. Detection can be realised in two ways: (i) Directly by on-board sensors that indicate an accident. Examples include a fired airbag or a high deceleration of the vehicle that cannot be achieved by the braking system itself. (ii) Indirectly, for example through using information from the parking sensors that may indicate a bump with another vehicle in front or behind. For this situation, it is highly recommended to advise the driver that a crash has been detected and find out whether or not they want to generate a claim.

  The mobile device subdomain plays an important role for insurance and financial services in general, and accident management services in particular. This is due to the fact that not every vehicle will be equipped with integrated applications for such services. However, the services can be outsourced to the mobile devices of the drivers. One may also think of a combined solution with mobile devices of the drivers connected to their vehicles. In such a set up, the vehicle may detect a crash and push this information to the mobile device for further handling. Of course, such a solution would not be as comfortable as the a fully integrated one. But, it can be introduced in the market more easily since nowadays vehicles already provide features to connect mobile devices to their headunits.

- *Generic domain.* Management accident services always rely on respective business processes provided by third parties. This could be either an insurance company providing their processes for claim management, it could be a vehicle manufacturer providing processes and respective services for their customers' relation management, or it could be third party stakeholders offering value-added services. Such processes are hosted in private infrastructure of the respective companies within the generic domain.

Likewise, accident management services have different requirements of AutoNets compared to typical vehicle-to-vehicle-based applications. It is apparent that accident management services do not require hard real-time communication guarantees as an intersection crossing assistant does. Instead, such a system requires reliable and encrypted communication to the infrastructure: in the event of a serious accident, help must be organised immediately and without much delay. It must therefore be prioritised in the wireless system used, typically a cellular system. Moreover, the party offering the service has to provide the overall service in a reliable fashion. This includes a respective service centre with a sufficient number of employees, which guarantees that in case of an accident drivers will be contacted immediately. A second important requirement is the reliable detection of an accident using the vehicles' on-board sensors. Of course, the driver will always have the ability to trigger an accident claim. However, in the event that a vehicle predicts an accident, this prediction has to be robust and reliable to avoid unnecessary driver distraction.

A third important requirement addresses the business processes of the service providers. These processes must be flexible and scalable enough to handle the different and unforseeable situations of accidents. This suggests that such processes will be realised using services-oriented architectures (SOA), which provide this flexibility and scalability. This is of particular importance since these services should be interoperable among themselves to allow for easy data exchange between them. Moreover, these systems must be flexible enough to support cooperation with external service providers, such as hospitals or towing companies. Keeping this flexibility in mind, data protection plays an important role in such a system. Therefore, respective mechanisms and methods have to be implemented to ensure that the data protection regulations of the respective countries are fulfilled.

4.4.1.2 Accident Management Service: Design Example

From a non-technical point of view, the realisation of an accident management service distinguishes three roles: The *driver* (or more precisely, the software component that is used by the driver), who owns a vehicle that is involved in an accident, a *service employee* (named employee in the following) who deals with the insurance claim and a *service system* (named system in the following), which implements the business process for the claim management. Figure 4.17 shows these three roles together with the interaction they use for data exchange with each other.

For the realisation of the business process, it is clear that a ticket-based system should be used. Here, a ticket reflects an accident from the administrative point of view, which can

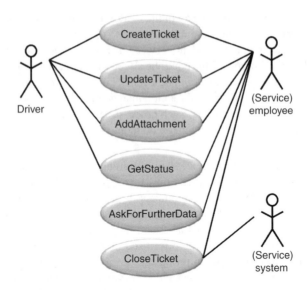

Figure 4.17   Roles and interaction interfaces for an accident management service.

be modified and processed accordingly. As illustrated in Figure 4.17, there are six different interaction interfaces for the example accident management service:

- **CreateTicket.** CreateTicket generates a new insurance claim, without any details about the accident itself. Such a claim can be generated either by the driver or by the service employee.
- **UpdateTicket.** UpdateTicket allows the addition or modification of additional information about the accident to the ticket. This information is based on formalised data, which needs to be specified accordingly.
- **AddAttachment.** Using AddAttachment, a driver or service employee can add additional non-formalised data to the ticket. Examples include pictures or video clips taken by the driver from the accident.
- **GetStatus.** GetStatus allows both the driver and the service employee to be updated about the current status of the claim. This is important for transparency since the driver can get up-to-date information at any time.[10]
- **AskForFurtherData.** This interaction interface is needed to request additional data from the driver.
- **CloseTicket.** CloseTicket completes the insurance claim by closing the respective ticket.

From a communication perspective, Figure 4.18 depicts the interaction between the driver and service system. In the event of an accident, the driver will trigger the creation of a new ticket in the system by claiming the accident. Afterwards, the ticket will be updated with respective information by the driver using either *UpdateTicket* or *AddAttachment*. Moreover, the driver always has the possibility of getting the current status of the claim by using the *GetStatus* interface.

---

[10] An extension could be to inform the driver about changes of the status using, for example, email.

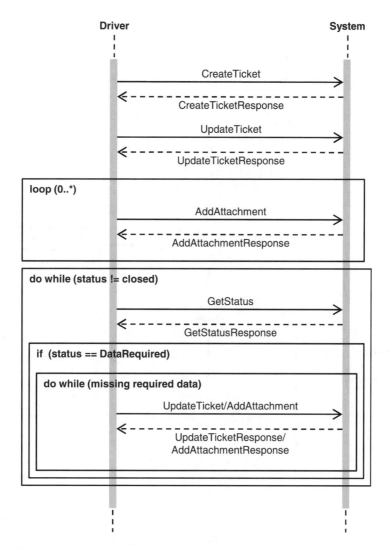

Figure 4.18   Communication sequence diagram for the example accident management service.

The information exchange follows a typical request/response scheme. The driver therefore initiates the data exchange using the respective interaction interface, whereas the response is passed from the system to the driver in a respective response message. This is a typical communication scheme used by the Internet, which enables easy implementation using, for example, web services technology.

4.4.1.3   Accident Management Service: Vehicle Process

The process of the claim management within the vehicle is illustrated in Figure 4.19. A claim can be triggered in two ways: either by an automatic detection of a crash, or manually by the

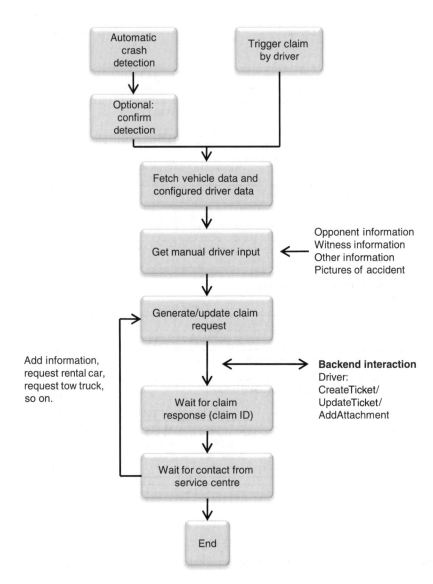

Figure 4.19   Sequence diagram of driver software component.

driver. It is very hard to detect accidents in a reliable way. Hence, an optional step is required for the automatic detection of an accident, where the drivers have to confirm that an accident occurred and that they want to claim this accident. If so, the claim application will continue by collecting relevant data. This could be either vehicle-related data or user-related information:

- *Vehicle data*. If available, the vehicle may provide respective data such as the current mileage, the current position, time and date, an estimate of the accident's severity derived from the vehicle sensors, and so on.

- *Configured driver information.* In the event that the driver configured some information in advance, this information can be also used for the claim. Examples include the driver's name and affiliation, the insurance ID or the vehicle type and number plate.

Afterwards, the driver may add additional information that cannot be pre-configured. Examples include information about the opponent, potential witnesses, information about the crash or pictures of the accident. Based on this information, the claim is triggered as described in the previous section. This means that the driver interacts with the service system in order to claim the accident using the respective interfaces, namely *CreateTicket*, *UpdateTicket* and *AddAttachment*. After a respective claim ticket is generated by the service system (or the service employee), the ticket ID will be transferred to the driver. This ID will be used for the following steps in order to identify the respective claim ticket. Usually, a service employee will then contact the driver for assistance of the accident. This way, the service employee may ask the driver for additional photos or information about the accident, or the employee may tell the driver what to do (or not to do) in this siutation. The driver can also request additional support electronically, such as requesting a tow truck in the event that the vehicle is seriously damaged, or a rental car in order to continue the travel. After all this, the process is finished, the regulation can be started accordingly and the driver can continue with his or her travel.

4.4.1.4   Accident Management Service: Backend (Business) Process

As we mentioned above, the business process located in the generic domain may be realised by a ticket-based system. Therefore, each insurance claim is represented by a ticket, which can be updated by both the driver and the service employee. Tickets are usually assigned to the respective person in charge, who conducts the claim.

Figure 4.20 illustrates the business logic necessary for our exemplary accident management service. An instance of this process can be initiated by the driver as well as the service employee by calling *CreateTicket*. Let us assume that the driver creates a ticket. Then, the process enters the state *New*, where the driver can add additional information and attachments using *UpdateTicket* and *AddAttachment* respectively. By activating the variable *noFurtherAction* in *UpdateTicket* or *AddAttachment*, the driver signals to the service system that no additional information will be transmitted. This way, the state changes to *Created* in order to be assigned to the next free system employee. If a system employee is available, the ticket will be assigned to him or her, and the process enters the state *Assigned*. This would be the system state where the service employee will try to contact the driver in order to assist in handling the accident locally. The *Assigned* state is a composed state: the driver as well as the service employee can add additional information to the ticket using *UpdateTicket* or *AddAttachment*, which is not shown in Figure 4.20. If the service employee requires additional information, he or she may actively request the driver using *AskForFurtherData*. Therefore, the process enters the state *DataRequired*, where the driver can add required information using either *UpdateTicket* or *AddAttachment* to the ticket. When the required information is complete, the service employee finalises the claim using *CloseTicket*, which closes the ticket. In the event that the driver does not provide the required information, the ticket will be closed automatically by the service system after a timeout.

It is worth mentioning that the UML diagram in Figure 4.20 does not contain the *GetStatus* interface for a greater clarity. This interface can be triggered by the driver (or a service

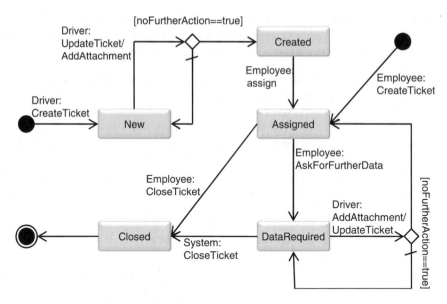

Figure 4.20 UML diagram of the service system.

employee) at any time in order to get the current status of the system itself. After this process, the ticket together with its information is passed to the respective person in charge of the insurance company, who will regulate the claim according to the accident management process of the respective insurance company.

4.4.2 Examples for Insurance and Financial Services (IFS)

From the driver's perspective, IFS can be realised in two different ways. The services can be integrated into the vehicle, for example by a separate application running on a vehicle's headunit. Alternatively, the services may be provided by external consumer electronic devices – this could be, for instance, an IFS app for the iPhone or for Android devices. Optionally, these consumer electronic devices can be connected to a vehicle's infrastructure using, for example, Bluetooth or wireless communication based on IEEE 802.11 a, b or g. In this way, vehicle information can be used by the consumer's electronic device in the event that the vehicle manufacturers provide the respective data.

In the following, we will present two solutions: the vehicle-based IFS application developed in the European research project PRE-DRIVE C2X, and the iPhone app 'Sternhelfer' developed by Daimler AG.

4.4.2.1 PRE-DRIVE C2X IFS Application

One of the text cases developed in the European research project PRE-DRIVE C2X[11] was an accident management system as an example for infotainment and third party applications.

---

[11] Homepage of PRE-DRIVE C2X: http://www.pre-drive-c2x.eu.

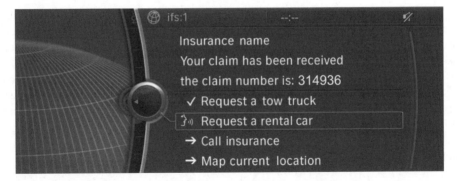

Figure 4.21 Snapshot of PRE-DRIVE C2X IFS application. Reproduced by permission of BMW Group.

Therefore, a back-end system was implemented, which enabled a ticket-based management of the accident. In order to improve reliability, PRE-DRIVE C2X additionally implemented a back-end integration approach, in which a so-called *back-end integration manager* was used for data exchange between vehicles and back-end systems [8]. We will not detail this concept here, because we believe that the benefit of the improved reliability is not worth the effort (including the respective standardisation activities) to bring such a component into the market. Instead, we think that common SOA-based solutions using the Simple Object Access Protocol SOAP with interoperable interfaces will be a more practical approach.

Within the vehicle, the application could be fully integrated into the vehicle's on-board infrastructure. This very simple application collected different vehicle information as well as pre-configured information about the vehicle and the driver. Moreover, it was possible to add additional (pre-defined) information via a vehicle's human–machine interface (HMI). Afterwards, the claim was transmitted to the back-end system, and a claim ID generated and transferred back to the vehicle. The driver then had the possibility of requesting a tow truck or a rental car, calling the insurance company or showing their current position on a map. Figure 4.21 shows a glimpse of what this application might look like.

4.4.2.2 Sternhelfer

In 2010, Mercedes-Benz Bank released an iPhone app called 'Sternhelfer'.[12] Sternhelfer is an IFS application, which supports a driver in the event of an accident. Originally, the application was developed for customers of insurance services provided by Mercedes-Benz Bank. In the meantime, the app can be used for other insurance companies too. The procedure for claiming an accident is the following:

- After the download and installation of the iPhone app, the user first has to personalise the app by adding personal information such as name, address and phone number. Moreover, information about the insurance contract (insurance company, contract number) and the license plate number is required.

---

[12] Sternhelfer homepage: http://www.sternhelfer.de.

- In the event of an accident, the damage can be documented in the iPhone app. This includes the personal information of the opponent and potential witnesses, police support, location and time of the accident, a description of the course of events, photos of the accident and a sketch of damaged parts of the vehicle.
- After the information about the accident is collected, the user can claim the accident to the insurance company. Therefore, the iPhone app generates a structured email that is transferred to the insurance company. After the claim the insurance company will likely contact the driver about the further processing of the claim.

The HMI of the app features the following screens:

- An introduction screen, where the user can choose between a checklist with useful actions in the event of an accident.
- A personal settings screen that needs to be filled by the users after first using the app.
- A claim screen, where the user has to specify whether or not the police was called, if persons were hurt, if there were additional people involved in the accident and if witnesses observed the accident.
- Screens supporting the driver by transmitting an emergency call or by calling the police, the driver can also get additional information about emergency numbers, first aid action or the already mentioned check list for accidents and the accident claim. The driver can specify the time of the accident (pre-initialised with the current system time), the location of the accident (pre-initialised with the current position of the device) and the information about the accident, add photos taken by the iPhone about the accident and mark the parts of the vehicle that are damaged.

References

[1] Dinger, J. and Hartenstein, H. (2006) Defending the Sybil attack in P2P networks: Taxonomy, challenges, and a proposal for self-registration. *Proc. International Conference on Availability, Reliability and Security (ARES 2006)*.

[2] Ostermaier, B., Dötzer, F. and Strassberger, M. (2007) Enhancing the security of local danger warnings in VANETs – A simulative analysis of voting schemes. *Proc. International Conference on Availability, Reliablility and Security (ARES 2007)*.

[3] Weyl, B. (2008) *On Interdomain Security: Trust Establishment in Loosely Coupled Federated Environments*. Verlag Dr. Hut.

[4] Kosch, T. (2004) Local danger warning based on vehicle ad-hoc networks: Prototype and simulation. *Proc. International Workshop on Intelligent Transportation (WIT 2004)*.

[5] Green, P. (1995) A driver interface for a road hazard warning system: Development and preliminary evaluation. *Proc. World Congress on Intelligent Transportation Systems*, 1795–1800.

[6] Harder, K. A. and Bloomfield, J. (2003) The effectiveness of auditory side and forward-collision warnings in winter driving conditions. Report from Minnesota Department of Transportation.

[7] The Economist (2010) Car-sharing: Wheels when you need them: Renting cars by the hour is becoming big business. Appeared September 2nd 2010, also available at http://www.economist.com/node/16945232?story_id=16945232.

[8] Kulp, I. et al. (2010) 2nd Update of PRE-DRIVE-C2X/COMeSafety Architecture Framework. PRE-DRIVE C2X Deliverable D1.5, available at http://www.pre-drive-c2x.eu.

[9] Klanner, F., Ehmanns, D. and Winner, H. (2006) ConnectedDrive: Vorausschauende Kreuzungsassistenz. *Proc. 15. Aachener Kolloquium Fahrzeug- und Motorentechnik*. [German]

[10] Klanner, F. (2008) Entwicklung eines kommunikationsbasierten Querverkehrsassistenten im Fahrzeug. Dissertation. [German]
[11] Hannawald, L. (2005) Unfallanalyse zur Entwicklung und Bewertung von Fahrerassistenzsystemen im Kreuzungsbereich. Research Report. [German]
[12] SAE J2735 (2009) *Dedicated Short Range Communications (DSRC) Message Set Dictionary*. SAE International, U.S. Department of Transportation.
[13] Schulze, M., Noecker, G. and Boehm, K. (2005) PReVENT: A European program to improve active safety. *Proc. 5th International Conference on Intelligent Transportation Systems Telecommunications 2005*.
[14] Lasowski, R. and Strassberger, M. (2006) A new approach for obstacle detection based on dynamic vehicle behaviour. *Advanced Microsystems for Automotive Applications*.
[15] Adler, C. and Strassberger, M. (2006) Putting together the pieces: A comprehensive view on cooperative local danger warning. *Proc. ITS World Congress and Exhibition on Intelligent Transport Systems and Services 2006*.
[16] Meitinger, K., Ehmanns, D. and Heiying, B. (2004) Systematische Top-Down-Entwicklung von Kreuzungsassistenzsystemen: Beispiel Stop-Schild-Warnung. *VDI/VW Gemeinschaftstagung Integrierte Sicherheit und Fahrerassistenz, VDI-Bericht Nr. 1864*. [German]
[17] ETSI TR 102 638 Technical Report (2009) *Intelligent Transport System (ITS); Vehicular Communication; Basic Set of Applications; Definition*.
[18] ETSI TS 101 539-1 Technical Specification (2011) *Intelligent Transport Systems (ITS); V2V Application; Part 1: Co-operative Awareness Application (CAA)*.
[19] Huber, W., Laedke, M. and Ogger, R. (1999) Extended floating-car data for the acquisition of traffic information. *Proc. 6th World Congress on Intelligent Transportation Systems (ITS 1999)*.
[20] Kroes, E., Hagemeier, F. and Linssen, J. (1999) A new, probe vehicle-based floating car data system: Concept, implementation and pilot study. *Traffic Engineering and Control*, Vol. 40, No. 4.
[21] Schulz, W. (1996) Traffic management improvement by integrating modern communication systems. *IEEE Communications Magazine*, Vol. 34, No. 10.
[22] Van der Zipp, N. J. (1997) Dynamic origin-destination matrix estimation from traffic counts and automated vehicle identification data. *Journal Transportation Research Record*, Vol. 1607.
[23] Eichler, S. (2006) Anonymous and authenticated data provisioning for floating car data systems. *Proc. 10th IEEE International Conference on Communication systems (ICCS 2006)*.
[24] Breitenberger, S., Grueber, B., Neuherz, M. and Kates, R. (2004) Traffic information potential and necessary penetration rates. *Traffic Engineering and Control*, Vol. 45, No. 11.
[25] Scheider, T. and Boehm, M. (2010) Extended floating car data in co-operative traffic management. *Traffic Data Collection and its Standardization, International Series in Operations Research and Management Science*, Vol. 144.
[26] Tabuchi, T., Yamagata, S. and Tamura, T. (2003) Distinguishing the road conditions of dry, aquaplane, and frozen by using a three-color infrared camera. *Proc. SPIE Thermosense*, Vol. 5073.
[27] Breuer, B., Eichhorn, U. and Roth, J. (1992) Measurement of tyre/road-friction ahead of the car and inside the tyre. *Proc. 24th International Federation of Automotive Engineering Society Congress*.
[28] Tuononen, A. J. and Hartikainen, L. (2008) Optical position detection sensor to measure tyre carcass deflections in aquaplaning. *Proc. International Journal Vehicle Systems Modelling and Testing*, Vol. 3, No. 3.
[29] Krumm, J. (2006) Real time destination prediction based on efficient routes. *SAE 2006 Transactions Journal of Passenger Cars – Electronic and Electrical Systems*.
[30] Krumm, J. and Horvitz, E. (2006) Predestination: Inferring destinations from partial trajectories. *Proc. 8th International Conference on Ubiquitous Computing (UbiComp 2006)*, Vol. 4206.
[31] Terada, T., Miyamae, M., Kishino, Y., Tanaka, K., Nishio, S., Nakagawa, T. and Yamaguchi, Y. (2006) Design of a car navigation system that predicts user destination. *Proc. 7th IEEE International Conference on Mobile Data Management (MDM'06)*.
[32] Simmons, R., Browning, B., Zhang, Y., Sadekar, V. (2006) Learning to predict driver route and destination intent. *Proc. IEEE Intelligent Transportation Systems Conference (ITSC'06)*, Toronto, Canada.
[33] Berndt, H., Wender, S. and Dietmayer, K. (2007) Driver braking behavior during intersection approaches and implications for warning strategies for driver assistant systems. *Proc. IEEE Intelligent Vehicles Symposium (IV 2007)*.
[34] Tiemann, N., Branz, W. and Schramm, D. (2009) Predictive pedestrian protection - sensor requirements and risk assessment. *Proc. 21st International Technical Conference on the Enhanced Safety of Vehicle (ESV 2009)*.

5

Application Support

Since AutoNet applications are a fundamental part of a cooperative, intelligent transportation system, it is essential to provide functionality for rapid application development. Furthermore, it is important to ensure that each application running on the same AutoNet station is using the same uniform data and information to guarantee a consistent quality of service. Therefore, the AutoNet Generic Reference Protocol Stack introduces a middleware and repository concept called the application support layer. It consists of various facilities, which are not fixed to one particular application; it is typically used by several applications running on the AutoNet station. Hence, the application support layer bridges the gap between the transport layer and the application layer. Figure 5.1 highlights the application support layer in the AutoNet Generic Reference Protocol Stack.

This chapter begins with the integration of the application support layer into the AutoNet communication architecture and introduces a classification of application support functionality called *facilities*. Next, we describe the different planes within the application support layer, followed by some implementation aspects for the facilities located in this layer. Finally, a short summary concludes this chapter.

5.1 Application Support in the AutoNet Generic Reference Protocol Stack

The application support layer is located between the the application layer and the transport layer. AutoNet applications running on top of the application support layer can access functionality of this layer via the AS-SAP. The application support layer itself is located on top of the transport layer and thus exchanges relevant data with this layer using the T-SAP. Moreover, the application support layer interacts with the vertical management layer using the M-SAP. The facilities within the application support layer also rely on security functionality provided by the vertical security layer via the S-SAP (all service access points are illustrated in Fig. 5.2).

Facilities are typically relevant for and used by several applications (or other facilities) of an AutoNet station. Facilities are an important precondition for rapid system deployment and thus for short time-to-market cycles. This is for several reasons: First, a facility is typically developed once and used by several applications. This means applications can be developed

Automotive Internetworking, First Edition. Timo Kosch, Christoph Schroth, Markus Strassberger and Marc Bechler.
© 2012 John Wiley & Sons, Ltd. Published 2012 by John Wiley & Sons, Ltd.

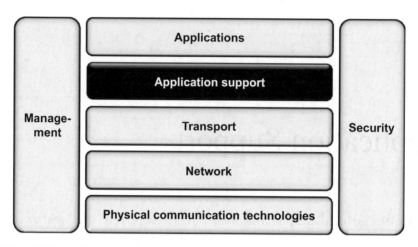

Figure 5.1 Application support layer in the AutoNet Generic Reference Protocol Stack.

faster since several software components for 'typical' tasks are already available and need not be developed, tested and integrated for each application. Moreover, facilities can be tested more easily with respect to interoperability among applications and even across multiple developers and vendors. Hence, the extensive testing of application interoperability is reduced significantly while compatibility and robustness implicitly increases since only few development teams will develop a facility. From a vehicle manufacturer's point of view, facilities are also an important factor to potentially reduce hardware costs. Since basic functionality needs to be stored only once, this helps to reduce storage space within electronic control units, resulting in lower hardware costs.

As illustrated in Figure 5.5, facilities can generally be classified in the following three categories based on the context in which the facility will be used:

- *Management facilities* are facilities providing management information and functionality for the following purposes:
  – overall operation of the AutoNet station and AutoNet station management usages;
  – cross-layer functionalities to optimise behaviour and efficiency of an AutoNet station.

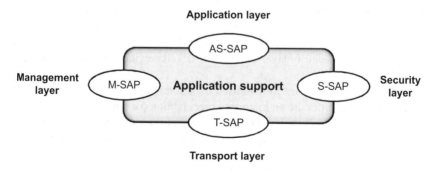

Figure 5.2 Service access points of the application support layer.

- *Domain-specific facilities* are facilities which provide domain-specific functionality to a subset of applications with similar requirements. The AutoNet Generic Reference Protocol Stack therefore distinguishes two different domains:
  - AutoNet domain: this domain defines the functionality related to the exchange of data locally among vehicles via the AutoNet.
  - IP-based domain: Internet-based applications using IP-based communication protocols for data exchange. Such applications typically address interaction between an application on an AutoNet station and third party service providers.
- *Common facilities* are facilities providing common functionalities for all applications running on the AutoNet station, that is for AutoNet applications as well as for IP-based applications.

There are a number of facilities defined for the AutoNet Generic Reference Protocol Stack, which are explained in the following sections. It is worth mentioning that the concept of profiles can also be adopted to the application support layer. This means, that – depending on the applications defined in the respective profile – there will be a subset of the application support tailored for this profile.

5.2   Communication Aspects in the Application Support

The exchange of information is an elementary feature of cooperative transportation systems. This is especially true for entities participating in cooperative systems; according to Section 1.3, this would be in particular vehicle-to-vehicle (V2V) communication and vehicle-to-roadside (V2R) communication. In cooperative systems, the entities as well as the different applications also rely on a basic set of information that needs to be exchanged. This basic set of information needs to be subject to some kind of standardisation in order to provide efficient information dissemination as well as to support future cooperative applications in AutoNets. Hence, the definition of this basic set of information is an important feature of the application support layer, which also ensures a consistent and interoperable information exchange in AutoNets, especially with respect to message formats and information representation.

This means the application support has to provide the basic functionality for facilitating message handling. For AutoNet scenarios, two types of messages are introduced and standardised: *cooperative awareness messages* and *decentralised environmental notification messages*.

5.2.1   CAM: Cooperative Awareness Messages

According to the ETSI specification document for CAMs [5]: 'Cooperative Awareness Messages (CAMs) are distributed within the ITS-G5 (IEEE 802.11p) network and provide information of presence, positions as well as basic status of communicating ITS stations to neighbouring ITS stations that are located within a single hop distance. All ITS stations shall be able to generate, send and receive CAMs, as long as they participate in V2X networks. By receiving CAMs, the ITS station is aware of other stations in its neighbourhood area as well as their positions, movement, basic attributes and basic sensor information. At receiver side, reasonable efforts can be taken to evaluate the relevance of the messages and the information. This allows ITS stations to get information about its situation and act accordingly.'

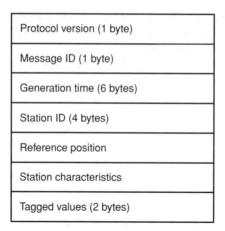

Figure 5.3   CAM message format.

This definition describes the nature of CAMs very precisely: CAMs are a 'general purpose' message format that is used by many cooperative AutoNet applications. CAMs are typically fixed to respective entities in vehicular environments, that is vehicles or roadside stations, which 'announce' their respective general purpose information. As a result CAMs are transmitted periodically with a varying frequency, where the frequency depends on the currently active applications. According to [5], frequencies range from 1Hz to 10Hz in the following way:

- 10Hz: emergency vehicle warning; intersection collision warning; collision risk warning.
- 2Hz: slow vehicle indication; motorcycle approaching indication; traffic light speed optimal advisory.
- 1Hz to 10Hz: speed limits notification.

Figure 5.3 specifies the message format for a cooperative awareness message, which consists of the following information:

- *Protocol version.* Protocol version of the message, will be set to 0.
- *Message ID.* Type of the message, will be set to 0 for CAMs.
- *Generation time.* Time stamp set by the ego vehicle.
- *Station ID.* Identifier of the ego vehicle sending the message.
- *Reference position.* Position of the vehicle at the generation time of the vehicle.
- *Station characteristics.* Specifies the characteristics, could be for example fixed or mobile, private or public authority, and so on.
- *Tagged values.* This parameter may comprise up to 32 attributes to describe the context. For vehicles, this could be the type of a vehicle, vehicle speed and acceleration, further status information about,for example, doors, lights, occupancy and so on according to SAE J2735.

CAMs are also subject to profiling. Currently, three profiles are specified: (i) a basic profile for vehicles, (ii) a profile for roadside station, and (iii) a profile for emergency vehicles. Profiling is not discussed further as it basically specifies pre-defined values for the parameter settings.

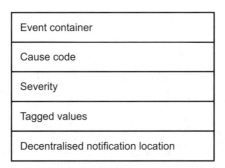

Figure 5.4   DENM message format.

5.2.2   DENM: Decentralised Environmental Notification Messages

Decentralised environmental notification messages are standardised at ETSI in the document 102637-3 [6]. In contrast to CAMs, DENMs are so-called *event messages,* which are not generated periodically but only sent when a specific event occurs. For example, in the event of hard braking a vehicle will notify the following vehicles using a DENM. The following vehicles receiving this DEN will be warned very quickly and are thus able to react accordingly. Hence, DENMs are relevant for the area they were generated and will be kept in (and near) this area. DENMS are also in general highly correlated with respective applications, whereas CAMs are relevant for many different AutoNet applications.

Standardisation of DENMs is still an ongoing activity at ETSI. Figure 5.4 illustrates the basic message format of a DENM. Starting with a management header, DENMs contain the following information:

- *Event container.* The Event container specifies the respective (coded) information following in the rest of the DENM. This could be, for instance, traffic flow, area of a danger warning, delays with respect to the traffic situation on the road and so on.
- *Cause code.* This parameter is a TPEG-TEC-encoded information about the cause of a warning.
- *Severity.* Severity defines the level/importance/danger of a situation, which is also encoded using TPEG-TEC.
- *Tagged values.* This parameter may comprise up to 32 attributes to describe the situation, as introduced for CAMs.
- *Decentralised notification location.* This part of the message specifies the area in which the message is of relevance. This could be a location reference, a destination area and so on in which the message has its relevance.

In order to provide facilitating services for applications, CAMs and DENMs are handled by the message management facility located in the application plane and described in Section 5.3.1.

5.3   AutoNet Facilities

Besides the domain-based classification in management facilities, common facilities and domain facilities, a facility can handle different aspects with respect to their functionality. In

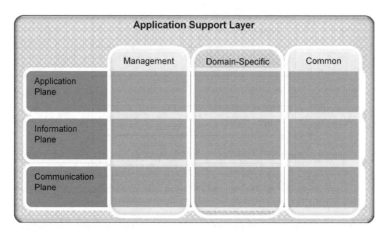

Figure 5.5 Facilities and their classification in the application support layer.

the application support layer, these different aspects are logically clustered into units called *planes*. The AutoNet Generic Reference Protocol Stack defines three planes: the application plane, information plane and communication plane. Similarly to a matrix, the facilities defined in the application support layer can be classified as shown in Figure 5.5. The following sections describe the three planes together with the respecive facilities defined for each plane.

5.3.1   Application Plane

The application plane provides facilities for supporting the applications running on an AutoNet station. Hence, this plane reflects the 'core' functionality of the application support layer, which is typically used by a wide range of both AutoNet and IP-based applications.

The application plane defines two management facilities: service management, and positioning and time. The *service management* facility provides the basic services to manage the AutoNet station. It therefore provides functionality to download and initialise new applications, or to update already installed software in the AutoNet station. It also provides functionality to ensure a safe and stable operation of the AutoNet station. Stability is of particular importance for the software download and initialisation since this procedure should never cause an instable system to avoid any risk or harm for the driver. Finally, the *positioning and time* facility provides information on the AutoNet station's geographical position (longitude, latitude and altitude) and the current time – independent of the positioning technology being deployed in the AutoNet station. A possible source for the information is a global navigation satellite system GNSS or the global positioning system GPS. However, this facility may also use additional information to improve position and time accuracy, such as information provided by the vehicle via the vehicle data provider facility (see below). It is important to mention that an accurate positioning of the AutoNet station is an essential precondition for many AutoNet applications. Due to the dynamic character of an AutoNet system, most information includes a time stamp in order to estimate the actuality of the information.

The application plane supports three domain-specific facilities: message management, local dynamic map and SOA application protocol support.

Message management. The message management facility provides services for the AutoNet domain to handle messaging aspects. This includes, for example, the automatic generation and encoding (and transmission) of CAMs without any intervention from the applications. Message management also includes priority handling of the messages to assign priorities to information (messages) gathered by AutoNet stations. This enables an AutoNet application to classify information based on its importance. This is of particular importance for messages that will be sent by an AutoNet application, since the priority of messages defines the order in which messages are delivered by the lower layers. Another important feature of the message management is the support of service announcement. The use of service announcements depends on the applications an AutoNet station offers. Currently, AutoNet roadside stations will use this facility to announce the (additional) services they offer, which can be used by the passing vehicles. For future applications, it may also be possible that vehicles announce additional services, which can be used by other vehicles in their vicinity. Therefore, the message management facility collects the information about the services running on the local AutoNet station and – if necessary – generates respective messages for the anouncements. It also provides the functionality to transmit the announcements periodically without any interaction from the applications.

Local dynamic map (LDM). LDM is a database to store and maintain information collected about the environment of AutoNet station in its vicinity. In this way, it supports applications of the AutoNet domain. The LDM comprises dynamic information (e.g. the status of traffic lights or the position of vehicles) as well as information about static, non-motorised traffic situations (e.g. the geometry of an intersection or landmarks for referencing). This information is referenced by the LDM to a geographic position (or a geographic area). In this way, the information can be matched to geographic locations. The information itself is represented by the messages an AutoNet station receives. These could be either the messages received from other AutoNet stations via the AutoNet, or they could be traffic-related information received from, for example, traffic management centres. Optionally, the LDM can be enriched by additional vendor-specific information. The different types of information are stored in different layers, as illustrated in Figure 5.6 as an example. The lowest layer represents the plain digital maps from the respective map provider. The upper landmark layercontains several landmarks such as traffic lights, as illustrated in the figure. Such landmarks can be referenced respectively; they can also be mapped on the underlying digital maps. Whereas landmarks are of a static nature, the layer above contains temporary regional information (or events), such as a current traffic accident or congestion on the road. Finally, in the top layer are the messages that are received from neighbouring vehicles or roadside stations. These messages allow for a current 'snapshot' of the traffic situation in the vicinity of a vehicle (called *ego vehicle*). Both the temporary regional information and the communicating nodes can also be mapped onto the digital

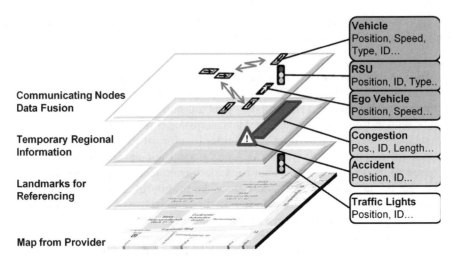

Communicating Nodes
Data Fusion

Temporary Regional
Information

Landmarks for
Referencing

Map from Provider

Vehicle
Position, Speed,
Type, ID...

RSU
Position, ID, Type..

Ego Vehicle
Position, Speed...

Congestion
Pos., ID, Length...

Accident
Position, ID...

Traffic Lights
Position, ID...

Figure 5.6   Local dynamic map facility.

map. Note that we also include same example parameters of the information in
the different layers in Figure 5.6.

SOA application protocol support. Finally, the application plane defines a facility for SOA (service-oriented architecture) application protocol support for
IP-based applications. This facility supports the operation of loosely coupled,
business-aligned and networked services. Thus, this facility provides the basic
functionality for realising a service-oriented information exchange between vehicles and back-end systems, for example by using SOAP-based Web Services.

Finally, the application plane defines the *human–machine interface (HMI) support* facility, which can be used by all applications. This facility supports a bidirectional, programming language-independent information exchange between vehicle applications and in-vehicle
HMIs based on well-defined XML files. This mechanism decouples the application logic from
the HMI properties of the vehicles. Moreover, the HMI support facility enables vehicle manufacturers to optimise their user interface with respect to their display capabilities (e.g. display
size, resolution, colour depth) and input modalities. A first definition of the XML schema for
the HMI support facility is described in [4].

5.3.2   Information Plane

The information plane combines facilities supporting the handling of available information.
In this context, the term 'information' does not refer to a specific type. Instead, the type of
information has to be considered from an abstract point of view. Depending on the respective
domain, information can be vehicle data, messages received from other vehicles, available
services or information about an AutoNet station. Like the application plane, the information
plane provides management facilities, domain-specific facilities as well as common facilities.

In order to monitor the AutoNet station itself, the information plane provides the *station
capabilities access* facility. This facility comprises information about the AutoNet station,

which may be used to monitor the (proper) operation of the AutoNet station. Therefore, the following information is provided:

- Information about the AutoNet station, for example whether an AutoNet station is an AutoNet vehicle station or an AutoNet roadside station.
- Information about the AutoNet station's capabilities, for example the access technologies supported and the current configuration.
- Identity information of the AutoNet station for referencing purposes. This could be either a static identity for AutoNet roadside stations, or dynamically changed pseudonyms for vehicles in order to protect privacy.

Unlike the application plane, the information plane does not define facilities for the IP-based domain; it provides the following facilities only for the AutoNet domain: location referencing, relevance checking and traffic message support. The *location referencing* facility provides mechanisms to map reference information or events onto geographic coordinates. This mechanism is typically used together with the LDM to match the available information from the different layers of the LDM onto road topologies and road traffic (called map matching). This way, the information is referenced accordingly and can be queried based on position. The *relevance checking* facility is also closely linked to the local dynamic map. It provides mechanisms to determine if information – stored in the LDM – is in the requested context. Therefore, an application can define a set of relevance rules to 'query' the LDM for relevant information. A simple task for relevance checking would be the calculation of a distance between two points (e.g. two vehicles) or a query for a specific event; more complex tasks would be whether the position in a received message is located in a specific sector in front of the own vehicle or behind one's own vehicle. In addition to this query mechanism ('pull' principle), the relevance checking facility also provides an active notification ('push' principle) in case a relevant message is received by the vehicle. This way, an application can be notified about a specific event. From a deployment point of view, an application specifies the set of relevance rules together with a callback routine. This routine will be called by the relevance checking facility in the event that a newly received message passes the relevance rules. It is important to mention that the relevance checking facility will neither filter received messages nor give priorities to events. Hence, the relevance checking facility can be seen as an interface to access information stored in the LDM. The *traffic message support* facility supports the functionality to handle different types of messages containing traffic-related information. Today, transportation systems deploy several standards for the representation of data, which are specific for this domain. Examples include the TPEG message format for multi-modal traffic and travel information [1], or AGORA-C for georeferencing defined in ISO 17572-3 [2]. These coding conventions are typically used in traffic management centres to represent their data. The traffic management message support facility provides functions for generation and extraction of the respective messages. Applications will benefit from this facility in two ways: they do not have do deal with the encoding of the messages, and the interoperability is ensured since encoding differences are minimised due to the common use of this functionality.

The information plane also defines common facilities, which are not related to any domain: vehicle data provider and monitoring. The *vehicle data provider* facility enables (controlled) access to the information available in the vehicle. This information is provided by the vehicle itself, that is by electronic control units connected to, for example, a CAN bus. Figure 5.7

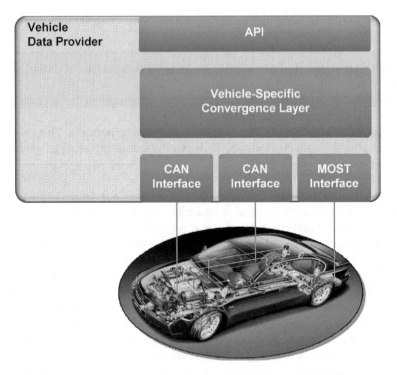

Figure 5.7 Information flow for the vehicle data provider facility.

illustrates the information flow. Examples include the current speed or acceleration of a vehicle or the status of exterior lights. Therefore, the vehicle data provider facility hides the technical realisation within the vehicle from the AutoNet application. This guarantees that vehicle information can be accessed uniformly, independently of whether the information was transmitted on a CAN bus or a MOST ring. The *monitoring* facility supports AutoNet applications to deploy particular monitor functionality in an AutoNet station. Hence, this facility provides mechanisms to control parameters and, in the event that the parameter violates respective specifications, to notify monitoring applications accordingly. Monitoring includes the endpoint (i.e. the address) of the AutoNet station and applications running on the AutoNet station respectively. In addition, dynamic data from a vehicle (provided by the vehicle data provider) can also be monitored. In this way, information such as the current oil level, mileage, or up-to-date information about the different embedded systems csn be monitored by the monitoring facility.

5.3.3 Communication Plane

The communication plane provides facilities to support message transmission and connection management between different AutoNet stations. Unlike the application and information plane, the communication plane only provides management facilities and domain-specific facilities. This is due to the fact that the separation between AutoNet domain and IP-based domain basically relies on the communication aspects used in each domain, resulting in either AutoNet-based communication facilities or IP-based communication facilities.

The communication plane provides one important management facility, the *access technology selector*. In general, an AutoNet station offers several radio access technologies. There are some applications that are highly tailored (and thus fixed) to a particular communication system in order to fulfil their required functionality. However, several applications are not assigned to particular communication systems used for transmitting the information. For example, safety critical applications like brake-light warning will use ITS 5GA technology for transmission, whereas a traffic efficiency application may alternatively use ITS 5GC to transmit its information. For such an application, the access technology selector chooses the appropriate communication system to transmit information. The decision algorithm itself is controlled by both the management layer and the application layer in order to optimise the system's behaviour. Besides the requirements of the AutoNet applications, the decision strategy is basically determined by environmental conditions (such as properties and link status). However, user preferences can also be considered for the selection decision. Examples include the user expectations of costs and performance of the communication systems.

The communication plane defines two important domain-specific facilities: messaging support and SOA session support. AutoNet applications as well as the message management facility may benefit from the *messaging support* facility. It supports compilation, generation, extraction and management of the messages exchanged between AutoNet applications. The messaging support facility therefore provides respective services for the different types of communication possible in AutoNets, namely broadcast, geocast, local unicast and global unicast. *SOA session support* addresses the tailoring of using SOA-based principles in intelligent transportation systems. This facility therefore supports IP-based applications using vehicle-to-business communication with session-related features. Hence, it allows the establishment of a connection to a backend system (i.e. an AutoNet central station) as well as the handling of unexpected connection losses due to the mobility of AutoNet stations. Moreover, the facility is able to manage and maintain sessions even in the event of changes to the used access technology, for example for always realising the best connected concepts. Another important feature of the SOA session support facility is the concept of asynchronous messages. The SOA-based principle typically relies on synchronous communication between the entities, which is not appropriate in vehicular environments. Hence, the SOA session support facility provides functionality for an asynchronous exchange of information to account for the intermittent connectivity of AutoNet stations and to allow for a scalable system architecture. This covers both the pure exchange of information and the invocation of services provided by AutoNet stations.

5.4 Implementation Issues for the Application Support Layer

The facilities described above are separated and classified from a conceptual point of view. From the perspective of implementation, this strict classification would result in a highly complex system with many interfaces, interactions and inefficient system performance. For this reason, it is important for the implementation to mix up the strict classification to some extent by 'clustering' co-related facilities into one functional block. This will minimise the resource-intensive system overhead by avoiding unnecessary exchange of data between the different facilities. The following two examples illustrate this benefit of clustering facilities:

Trace chain matching for relevance checking. In general, reliable relevance checking requires map information in order to determine whether or not an event is of relevance for the vehicle or the driver. This is of particular importance in determining if a vehicle travels towards the event and if the event is located along the route of the vehicle. However, digital maps will not be available in all vehicles – particularly cars in the low pricing segment will not feature digital maps due to the high cost of the licensing. Therefore, an alternative methodology is *trace chain matching*, which does not require navigation maps. For trace chain matching, vehicles 'track' their movements by continuously storing a discrete set of their last positions. Although this methodology apparently provides a significantly reduced quality of checking the relevance of an event, it provides at least the possibility of determining whether or not a vehicle moves towards an event or away from an event.

Clustering LDM, location referencing and relevance checking. The LDM will frequently use the location referencing facility to map messages and traffic information onto its map. Moreover, it will typically rely on the relevance checking to provide access to its data. Hence, a combination of the three facilities will greatly reduce system overheads since the location referencing can be done natively, whereas the relevance checking facility may act as an API (application programming interface) for the LDM to query relevant messages.

Clustering station capabilities access, vehicle data provider and monitoring. Similarly, both vehicle data provider facility and station capabilities access facility only provide access to their respective information. This access can be naturally combined with the monitoring facility, which may act as an API for the two facilities. Hence, applications need not cope with different interfaces to vehicle data provider and station capabilities access; instead they can conveniently use both facilities in the same way using the monitoring facility.

Of course, the clustering of facilities for implementation purposes depends on the requirements of the applications on the AutoNet station, as well as the technology being used for implementation.

In addition, it is worth mentioning that there is no clear separation between the application support layer and the application layer in an implemented system. This is due to the fact that facilities are components that support the applications running on the AutoNet station. As a result, facilities inherently implement parts of the applications' functionality, which is a typical pattern in software architecture design principles. Consider the following example: if the implementation of the AutoNet station will be based on the OSGi framework [3], it would be a likely approach to encapsulate the functionality of both facilities and applications into OSGi bundles. This way, application bundles can immediately use the functionality of the facilities bundles using the respective OSGi interfaces. This software design results in a distributed set of OSGi bundles without any layers and single service access points. But it improves performance and efficiency of the overall system because the message flows between the bundles are minimised accordingly. Note that this difference is similar to the comparison of the upper layers of the ISO/OSI reference model versus the TCP/IP model: whereas ISO/OSI

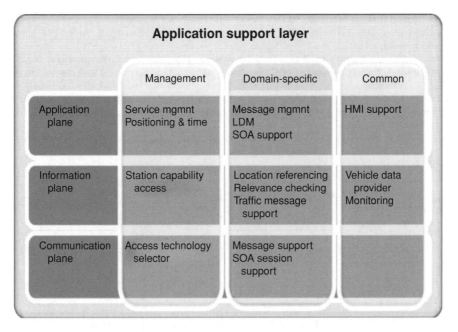

Figure 5.8 Facilities in the application support layer.

suggests three layers (session layer, presentation layer and application layer), the TCP/IP model only defines one layer – the application layer – which basically addresses the same issues.

5.5 Summary

The concept of facilities provided by the application support layer is based on typical software development issues. Here, the first step of the software design is to split the overall subsystem into different components in order to identify functionality that is used by several other modules. Facilities likewise provide the functionality needed for several applications in order to avoid unnecessary overheads for development, testing and integration. Hence, the application support layer provides a significant contribution to reduce time-to-market for the introduction of new functionality. This chapter introduced the very basic characteristics of facilities, their classification along two axes *domains* and *functionality* and described the facilities currently under discussion in the ITS community. Figure 5.8 summarises the facilities introduced in this chapter in each of the different domains and planes. We also outlined some practical issues for the development of facilities, which is based on experiences from related projects to develop respective subsystems. These issues also help to reduce the overall complexity of the system by combining respective facilities from an implementation point of view.

References

[1] Transport Protocol Experts Group (TPEG) (2010) http://www.tpeg.org.
[2] AGORA-C (2008) Intelligent Transport System (ITS) – Location Referencing for Geographic Databases – Part 3: Dynamic Location References. ISO Standard 17572-3.

[3] Open Services Gateway initiative (OSGi) Alliance (2010) http://www.osgi.org.

[4] Kosch et al. (2010) European ITS Communication Architecture Version 3.0, Annex 1.4 http://www
.comesafety.org.

[5] ETSI TS 102 637-2 (2010) Intelligent Transport Systems (ITS); Vehicular Communications; Basic Set of
Applications – Part 2: Specification of Cooperative Awareness Basic Service. *ETSI Technical Specification
V1.1.1*.

[6] ETSI TS 102 637-3 (2010) Intelligent Transport Systems (ITS); Vehicular Communications; Basic Set of
Applications – Part 3: Specification of Decentralized Environmental Notification Basic Service. *ETSI Technical
Specification*.

6

Transport Layer

In the AutoNet Generic Reference Protocol Stack, the transport layer is located between the application support layer and the network layer, as illustrated in Figure 6.1. Therefore, it provides communication services to the application support layer (which may also pass the functionality directly to the application layer). Hence, applications may use the transport layer functionality either directly or via facilities located in the application support layer.

In this chapter, we introduce the transport layer functionality used in AutoNets. Therefore, we will first describe the very basic tasks of the transport layer and the protocols needed for the different application domains. In particular, we will discuss the use of the well-known TCP protocol in AutoNets, which is the most important transport protocol used in the Internet. Based on this discussion, the next section proposes solutions to improve communication in AutoNet scenarios, which is an important issue in order to optimise communication efficiency. We will also illustrate the potential of improvement by some measurements.

6.1 Transport Layer Integration in the AutoNet Generic Reference Protocol Stack

The task of the transport layer is to provide communication services to the upper layers. In terms of the ISO/OSI reference model, this means that the application layer has to ensure the communication between two applications (called *end-to-end*), whereas the networking layer is responsible for transferring the information between the communicating entities (i.e. *endsystem-to-endsystem*). It resides between the network layer and the application support layer and is flanked by the management layer and the security layer. Figure 6.2 illustrates the integration of the transport layer in the AutoNet Generic Reference Protocol Stack together with the service access points to the neighbouring layers. As already described in Chapter 5, the transport layer provides functionality to the upper layers (namely, application support layer) via the T-SAP. Interaction with the management layer and the security layer is performed via the M-SAP and S-SAP, respectively. Using the M-SAP, the transport layer functionality can take full advantage of cross-layer improvements, since it may take into consideration information from other layers in order to improve communication efficiency. We will see in

Automotive Internetworking, First Edition. Timo Kosch, Christoph Schroth, Markus Strassberger and Marc Bechler.
© 2012 John Wiley & Sons, Ltd. Published 2012 by John Wiley & Sons, Ltd.

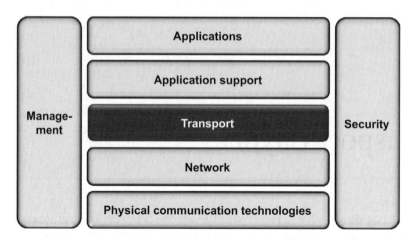

Figure 6.1 Transport layer in the AutoNet Generic Reference Protocol Stack.

Section 5.2.5 that the performance of a transport layer protocol can be improved significantly by using information from other layers.

Similarly, the transport layer functionality may include security functionality via S-SAP in order to secure communication. Finally, the transport layer interacts with the network layer via the N-SAP.

From the transport layer's perspective, there are two application domains in AutoNets: (i) communication using specific AutoNet communication protocols (e.g. for communication among vehicles using ITS G5A) and (ii) IP-based communication, in particular for communication with third party services located in the Internet. These two domains are also reflected in the transport layer of the AutoNet Generic Reference Protocol Stack. Here, the functionality can be separated into two parts as shown in Figure 6.2:

> **AutoNet Transport.** AutoNet transport provides the communication services necessary to communicate between applications, basically among vehicles, and between vehicles and roadside stations.

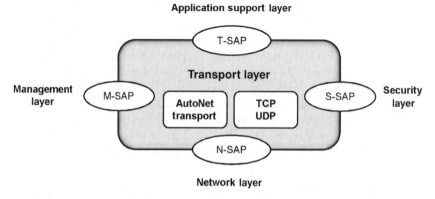

Figure 6.2 Transport layer in the AutoNet Generic Reference Protocol Stack.

TCP/UDP. Both TCP and UDP are the core transport protocols used for communication in the Internet.

Both types are described in more detail in the following sections.

6.1.1   AutoNet Transport

As mentioned above, AutoNettTransport protocols are basically used for communication between vehicles, and between vehicles and roadside stations. Addressing the specific requirements of auto-networked safety applications (cf. Chapter 4), the AutoNet transport protocols have to be highly tailored to vehicular environments, making optimal use of the scarce resources and particularities of vehicular communication environments. Hence, AutoNet transport protocols also should be as highly efficient as the underlying data exchange protocols. As a result, the designed AutoNet protocols may not be compatible with the protocols used in the Internet.

From a standardisation perspective, the AutoNet transport protocol will be standardised as 'Transport protocols over GeoNetworking' at ETSI TC ITS Working Group 3 in the specification TS 102 636-5, which comprises a *basic transport protocol* (TS 102 636-5-1) and *other geonetworking transport protocols*. But right now there is still no significant specification for an AutoNet transport protocol available. Moreover, ongoing large-scale field tests for vehicle-to-vehicle and vehicle-to-roadside-station applications do not deploy any AutoNet transport functionality: respective system architectures (such as COMeSafety as described in Section A.2.3) even aggregate transport layer functionality and network layer functionality into one layer called 'ITS network and transport layer'. Hence, it seems that there is no need for a particular transport layer functionality – in fact, an explanation can be found in the fact that the functionality of AutoNet transport is shifted to upper and lower layers, as rationalised by the following four arguments:

• As we have seen, there are two main principles of AutoNet communication. First, there is a unicast communication between two specific nodes, that is two vehicles or a vehicle and a roadside station, respectively. This is the most common communication principle in well-known computer networks, in particular the Internet. The second principle is a local dissemination of relevant information, which is usually addressed to all nodes in a specific area. Because of their local impacts, typically all safety related uses rely primarily on this type of communication. The overall system is designed in a different manner to standard computer networking. Information is of relevance in a geographic area only, and the receiving vehicle can individually decide what features are supported by the received information. Therefore, for the kind of information dissemination that is designed right now, there is no specific addressing scheme for AutoNet (safety) applications. The messages are stored in the local dynamic map (LDM) of the application support layer. Any application interested in specific information then registers itself for respective messages, or queries the LDM. Hence, an application sending a message does not know which application will receive (and use) the message. In this way, the original application addressing defined in ISO/OSI is shifted to the LDM.
• Likewise, the LDM ensures some kind of reliable data exchange between applications, which is a typical functionality of a transport layer. The LDM stores the messages and

an application can query them within certain time boundaries. This results in a limited reliability, which is a typical feature of, for example, TCP.

- Another typical feature of a transport layer is the potential prioritisation of particular connections between applications. In AutoNets, this functionality is already provided by the networking layer, which itself provides different traffic classes, resulting in a different treatment for data transmission. Here, the applications are 'pre-coordinated', where a defined priority order exists among the different applications running on a system.
- Finally, congestion control mechanisms are also performed by the networking layer in AutoNets, which correlates with the prioritisation of messages described in the previous bullet.

Although it seems that there is no need for an AutoNet transport functionality, we believe that it will play an important role in the future. This is due to the fact that current AutoNet applications basically address vehicle safety and traffic efficiency issues, and therefore it seems that there is no need for an AutoNet transport functionality. However, we may see applications among vehicles using unicast communication in the future. Examples include personalised video-streaming between two vehicles, or even the download of rich media files from a neighbouring vehicle. For such uses, a respective AutoNet transport protocol is necessary, which enables the direct addressing of applications as well as an efficient and fair information exchange. Respective transport protocols may be highly adapted to vehicular environments, or they may be an optimised version of, for example, TCP, which enables a rapid deployment with a short time-to-market for respective applications.[1] In Section 6.2.4 we will introduce one such protocol, which could be a suitable candidate to serve such scenarios.

6.1.2   TCP, UDP

Besides specific protocols for AutoNet tansport, the Internet transport protocols *transmission control protocol* (*TCP*) and *user datagram protocol* (*UDP*) also play an important role in AutoNets. Apparently, both TCP and UDP are used to communicate between vehicles (as well as roadside stations) and communication hosts located in the Internet. This way, TCP and UDP will be used in the following prominent AutoNet communication scenarios:

- *Vehicle-to-central-infrastructure.* Since traffic management centres are typically accessible via IP-based communication, vehicles and roadside stations will use TCP for accessing their services.
- *Vehicle-to-Internet, vehicle-to-private.* Both communication scenarios are – as the name already states – traditional Internet-based communication scenarios. Hence, TCP and UDP will be used by vehicles and roadside stations in order to access the services provided by such hosts.

---

[1] Our expectations are fleshed out by ongoing activities such as the European Research Project GeoNet (http://www.geonet-project.eu), which is currently developing a solution to transmit IPv6 packets over geocast-based networks. Such an approach would enable the deployment of IP-based transport protocols like TCP or UDP in vehicular environments.

Moreover, either TCP or UDP will be used in the event of vehicles exchanging respective AutoNet messages via cellular communication technology (such as UMTS) as described for AutoNet cellular networking in Chapter 7. Here, respective AutoNet messages will be encapsulated in TCP or UDP packets and thus tunnelled via an IP-based service provider to the respective vehicles.

UDP realises an unreliable and connectionless service, whereas TCP provides a reliable and connection-oriented service for data transport. In the Internet, both protocols operate on top of the unreliable Internet protocol IP. Both UDP and TCP are end-to-end protocols; the endpoints are identified by *ports*, which uniquely identify a connection within an end system. From a development viewpoint, the T-SAP for TCP and UDP are *sockets* [19], which act as an API to use TCP or UDP functionality. Since UDP plays a minor role in the Internet and basically provides functionality to address applications, we will focus on TCP in the following.

Originally, TCP was standardised in RFC 793 [2]. It provides a connection-oriented and reliable data delivery service for the application layers. After establishing a connection between two applications, TCP structures the byte stream into segments and ensures that the segments are delivered to their destination in a reliable manner. Therefore, TCP supports several protocol mechanisms [3]:

- Segmentation and reassembly to structure the byte stream into segments and vice versa.
- Timers and retransmissions for lost segments.
- Reordering of segments arriving out of order.
- Transmission error detection.
- Appropriate segment size discovery.
- Flow control and congestion control mechanisms.

As we outlined in the previous section, the use of a transport protocol 'similar' to TCP for communication between vehicles is an interesting option. Such a solution would enable the deployment of traditional Internet applications in vehicular environments – for example directly between vehicles in the multi-hop ad-hoc network. Moreover, such a solution would enable vehicles to communicate via a roadside station directly with Internet hosts without the need for a cellular communication system or a dedicated network bridge.

In the following sections, we will address this issue by detailing the problems of TCP in mobile (and wireless) communication environments in general, and AutoNets in particular. We also suggest a solution to significantly improve communication efficiency, while maintaining compatibility with standard TCP. This aspect will be focused on TCP only – independent of any integration aspect. In Section 10.3.4, we will seize this aspect by describing the integration of such performance-enhancing issues into the overall system architecture and their interaction with the cross-layer functionality provided by the management layer of the AutoNet Generic Reference Protocol Stack.

6.2   TCP in AutoNets

TCP was developed and optimised for communication networks with a fixed topology. In this way, it works well in wired networks and provides an acceptable performance in terms of data throughput. However, TCP provides a poor throughput in both wireless cellular networks and multi-hop ad-hoc networks used for mobile AutoNet applications, although a higher throughput might be possible in theory [4, 5, 6, 9]. This performance degradation mainly results from the flow and congestion control mechanisms deployed in TCP, which ultimately determine the amount of data on the fly between a sender and a receiver and therefore the throughput of TCP.

6.2.1 Congestion Control in TCP

TCP implements flow control based on a sliding window technique. The *window size* specifies the maximum number of unacknowledged bytes that can be sent into the network. TCP transmits segments into the network as long as they do not exceed the window size. Then, it has to wait until the communication peer acknowledges successfully received segments. TCP acknowledges segments cumulatively, that is an acknowledgement admits all segments transmitted previously. The window size is determined by the currently computed congestion window from the sender and the advertised window from the receiver. The TCP flow control algorithm tries to mitigate congestion by controlling the window size. The congestion control consists of two dominant algorithms: *slow start* and *congestion avoidance*. Figure 6.3 illustrates the behaviour of both algorithms.

The slow start algorithm is specified in RFC 2001 [7]. It was developed to avoid congestion collapses in the network. The basic idea of the slow start is to control the data segments at the beginning of a data transfer as well as in the case of segment losses. The slow start algorithm defines a *congestion window* (cwnd) to determine the number of segments that can

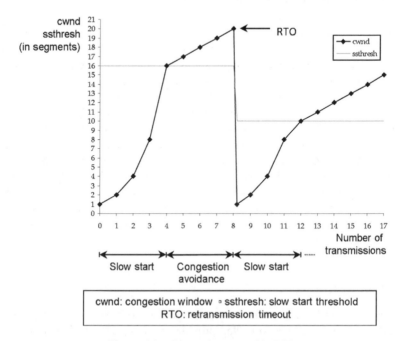

Figure 6.3   Congestion control in TCP.

be transmitted without waiting for an acknowledgement from the communication peer. In the beginning of the slow start phase, the congestion window is initialised with one segment. The slow start algorithm increases the value of cwnd at an exponential rate: if the communication peer acknowledges a segment before a retransmission timeout (RTO) expires, the size of the congestion window is doubled (see Figure 6.3). The current value for the RTO is derived from the round trip time (RTT)^2 permanently measured while exchanging data [10]. The exponential growth of cwnd continues with each acknowledged segment until the value of the congestion window reaches the *slow start threshold size* (ssthresh), which is at transmission number 4 in Figure 6.3.

At this time, the slow start phase is followed by the congestion avoidance phase, which realises an 'additive increase, multiplicative decrease' algorithm [8]. In the beginning, the value of cwnd increases linearly in order to probe for any spare capacity. The linear increase occurs approximately every RTT and will be continued until an RTO occurs (at transmission number 8 in Figure 6.3). The congestion avoidance algorithm then sets the value of ssthresh to half the current cwnd, and the cwnd itself is reset to one segment. Afterwards, TCP enters the slow start phase and communication continues as described previously. This behaviour causes 'sawtooth-like' behaviour of the window size.

The original flow control in TCP was improved over time. An example is the *fast retransmit/ fast recovery* algorithm specified in RFC 2581 [11]. This algorithm is used when segments are dropped in the network, for example due to buffer overflows in intermediate routers. In this situation, the receiver of the missing segment transmits duplicate acknowledgements to inform the sender that the connection is still alive but that segments were not delivered to the receiver. If the sender receives three or more duplicate acknowledgements for one segment, it will retransmit the outstanding segments to the receiver without waiting for the expiry of the RTO. This mechanism is called fast retransmit. After the fast retransmit, the sender performs a fast recovery in the following way:

1. If three duplicate acknowledgements arrive at the sender, the value of ssthresh will be set to the minimum of cwnd and the window size.
2. For each acknowledgement received, the value of cwnd is increased by the segment size, and the sender transmits a new segment.
3. If an acknowledgement for the transmitted segment arrives at the sender, it also acknowledges the previous segments cumulatively. The value of cwnd will be set to the value of ssthresh and the congestion avoidance algorithm restarts as described previously.

Fast retransmit/fast recovery was first introduced in TCP Reno, the implementation currently integrated in common operating systems.

6.2.2   Impact of AutoNets

Originally, TCP was developed for wired networks with low bit error rates and low variances concerning bandwidth and delay. However, communication in multi-hop ad-hoc networks in general and AutoNets in particular is different compared to communication in such stable

---

^{2} Actually, the RTO is equal to the sum of the smoothed RTT and four times its mean deviation.

networks: on the one hand, vehicles are highly mobile and therefore the topology of the AutoNet is subject of a permanent reconfiguration or partitioning. On the other hand, communication is based on wireless radio technology resulting in variations of the transmission quality. Several studies investigated the impact of these aspects on the performance of TCP (see for instance [12, 13, 14]). Based on these investigations, the following observations can be concluded with respect to the characteristics of AutoNets:

- *Bandwidth:* compared to wired networks, the available bandwidth in an AutoNet is rather low. Since radio propagation provides a shared medium, communicating vehicles have to compete for the available bandwidth. This means the available data rate can vary heavily over time. The permanently changing number of communicating vehicles further aggravates this effect.
- *Delays and jitter:* in general, AutoNets exhibit higher transmission delays due to the wireless communication. Moreover, the mobility of the vehicles introduces a potential jitter if the topology changes or if the underlying ad-hoc routing protocol decides to use an alternative route for the IP packets.
- *Segment losses:* wireless transmission links are error-prone resulting in a higher rate of packet losses compared to fixed network technologies. Moreover, multi-hop communication further degrades the packet loss probability with each additional wireless hop between sender and receiver.
- *Temporary disconnections:* the reconfiguration of the AutoNet topology may also partition the AutoNet resulting in a broken communication path between sender and receiver. Hence, communication may be interrupted until the ad-hoc routing protocol finds an alternative path or until the separated network partitions are connected again. Moreover, vehicles may have only a temporary connection to the Internet, either there is no roadside unit available or a connection via cellular communication is not possible. Hence, vehicles may become disconnected for even longer periods of time.

The conservative flow control used in TCP was not developed for such network characteristics, resulting in performance degradations of TCP in mobile and wireless environments [1, 15]. As described in the previous section, TCP approximates itself to the available bandwidth using slow start and congestion avoidance. However, the varying bandwidth in AutoNets requires a permanent adaptation, and TCP therefore has to perform the slow start frequently. Jitter has a similar impact on the congestion control since it causes high variations in RTT. Yet, RTT is one of the fundamental parameters for the calculation of the RTO. The example given in Figure 6.4 illustrates this impact where segment 1 is acknowledged in time. If RTT between sender S and receiver R increases, the RTO for segment 2 will expire at $t = 4$, segment 2 will be retransmitted, and TCP will enter the slow start phase. In this way, a too small value for the RTO burdens the network with needless retransmissions and limits the number of simultaneous TCP segments in the network too restrictively. Vice versa, an overestimated RTO results in large latencies for lost segments, which also decreases TCP throughput.

Another negative effect is that standard TCP is not able to distinguish between congestion in the network and transmission errors. In general, TCP interprets transmission errors as congestion in the network: the RTO expires and thus the slow start phase is activated. Even with the use of automatic request and repeat (ARQ) and forward error correction (FEC), the bit error rate in wireless networks is orders of magnitude higher compared to wired

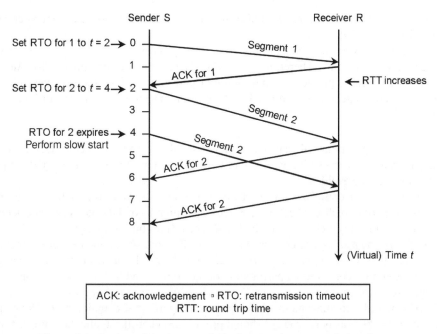

Set RTO for 1 to *t* = 2→ 0

Set RTO for 2 to *t* = 4→ 2

RTO for 2 expires →
Perform slow start

Figure 6.4   Impact of varying round trip times on the retransmission timeout.

networks. Hence, TCP will barely leave the slow start phase, which significantly reduces the data throughput although enough bandwidth is available. This effect is amplified by the long RTTs in AutoNets. The rate at which a TCP sender increases its cwnd is directly proportional to the rate of the acknowledgements received. This way, the cwnd increases at a much lower rate due to the longer RTTs.

TCP also has problems with temporary disconnections, which might be caused by partitioned ad-hoc networks or by the temporary Internet disconnections. In order to reduce the overhead in the case of congestions, TCP increases the RTO for a segment exponentially with each expiry. This exponential back-off is also known as 'Karn's Algorithm'. According to RFC 2988 [10], the retransmission timer may be limited at an upper bound of at least 60 s; in the Linux operating system, the maximum retransmission timer may increase up to 120 s. This way, it can take up to two minutes to detect a reconnection after a longer period of disconnection. During this time, TCP does not transmit any segments, although the sender and receiver are able to communicate with each other.

6.2.3   *Enhancements of TCP and Technical Requirements for AutoNet Scenarios*

Over the years, TCP evolved and was enhanced by several new protocol features. For example, TCP Reno introduced fast retransmit/fast recovery, which was further improved in TCP New Reno according to RFC 2582 [16]. Furthermore, TCP was enhanced by selective acknowl-edgements (TCP SACK, RFC 2018 [17]) and forward acknowledgements (TCP FACK, [18]).

These extensions are already integrated in current TCP implementations of common operating systems like Linux [19].

However, such extensions do not solve the basic problems of TCP running in mobile and wireless environments. Therefore, TCP still provides a poor performance for AutoNet applications. In order to provide an optimised transport protocol for AutoNets, the following requirements have to be addressed for both cellular communication as well as multi-hop ad-hoc communication:

- *Lost segments:* an optimised transport protocol should be able to react efficiently to lost segments caused by transmission errors in the wireless link. In the case of TCP, frequent slow starts must be avoided.
- *Disconnections:* an optimised transport protocol should be able to handle short-term or longer-term periods of disconnection caused by a partitioning of the AutoNet ad-hoc network or in the case that a vehicle is currently disconnected from Internet access.
- *Notifications:* the routing protocol used in the mobile AutoNet ad-hoc network is typically able to notify the communicating peers about network or link characteristics. For example, if the routing protocol detects a partitioning of the AutoNet ad hoc network, the optimised transport protocol has to consider such notifications in order to react quickly.
- *TCP API:* in order to alleviate the development and porting of common TCP applications for AutoNets, the transport protocol should provide a socket-like application programming interface (API) for the applications. This is also an important requirement for short time-to-market cycles of new services for mobile AutoNet ad-hoc networks.

Hence, to improve the end-to-end communication efficiency at the transport layer, AutoNet environments require an optimised transport protocol based on TCP. Such a protocol must be able to distinguish between error-prone links and network congestion in order to handle packet losses appropriately. Moreover, it must be able to utilise information from both intermediate systems and from other protocols. This is necessary for an efficient treatment of both short-term network partitions and longer-term periods of disconnection from the Internet.

An interesting solution for this challenge is called MCTP (MOCCA transport protocol) [20]. MCTP provides a core TCP functionality that can be extended depending on the performance and characteristics of mobile AutoNet ad-hoc networks used in a real-world deployment. The core functionality of MCTP is detailed in the following section, followed by performance evaluations compared to traditional TCP.

6.2.4   The MOCCA Transport Protocol

The MOCCA transport protocol, or MCTP, combines several of the standardised TCP enhancements proposed above. Its core functionality belongs to the category of utilising information from intermediate systems, which is extended by modifications of the congestion control mechanisms used in TCP. In general, MCTP is based on the principles of ad-hoc TCP (ATCP [21]), which relies on information on pending congestion in the network. This idea is combined with an approach similar to TCP feedback [22] and TCP stop-and-go proposed by Ritter [25]. MCTP implements a sublayer between TCP and IP, that is it interacts with both IP and TCP. The basic principle of MCTP is that it observes the IP packet flow between sender and

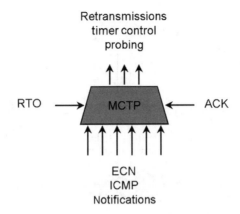

ACK: acknowledgement ▫ ECN: explicit congestion notification
ICMP: Internet control message protocol ▫ RTO: retransmission timeout

Figure 6.5   MCTP information processing.

receiver in order to react appropriately. Therefore, MCTP considers additional notifications from protocols in the underlying communication layers as well as from intermediate systems:

- Pending congestions detected by intermediate systems are indicated using, for example, explicit congestion notification (ECN) according to RFC 3168 [23].
- Intermediate systems indicate a partitioned network using ICMP (Internet control message protocol) messages of type 'destination unreachable' according to RFC 792 [26]. This information is relevant for local communication between vehicles only, that is for communication without accessing the Internet. This can be also handled by the management layer described in Chapter 10.
- The management layer (cf. Chapter 10) notifies MCTP in the event of of disconnection from the Internet.

This available information enables MCTP to distinguish between link errors, congestion, network partition and disconnection from the Internet. Figure 6.5 summarises the information processing in MCTP. Besides the available information from underlying communication layers and from intermediate systems, MCTP also takes into account events caused by TCP itself. Such events are the retransmission timeouts for segments and the arrival of (duplicate) acknowledgements for successfully transmitted segments. Based on this knowledge, MCTP controls the transmission procedure of TCP in different situations by controlling retransmissions and timeouts, and by probing for the network characteristics. MCTP therefore implements its own protocol state machine, which comes into operation after TCP establishes a connection between the end systems.[3]

---

[3] Applied to the TCP protocol state machine [27], MCTP comes into operation after TCP enters the 'ESTAB-LISHED' state.

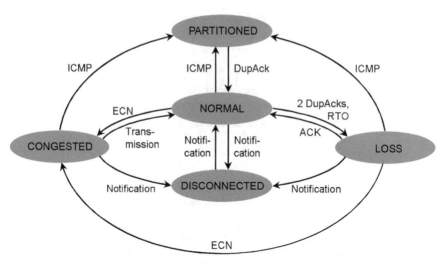

Figure 6.6   MCTP protocol state machine.

6.2.4.1   MCTP Protocol State Machine

In order to process the different information, MCTP implements a protocol state machine that comprises the following five states:

- *NORMAL:* the normal operation mode.
- *LOSS:* the operation mode in case of unreliable, that is subject to loss, communication links.
- *CONGESTED:* MCTP enters this state in case of a congested network.
- *PARTITIONED:* the operation mode for a partitioned network.
- *DISCONNECTED:* this state represents a disconnection from the Internet.

Figure 6.6 shows the transition diagram between these states in the MCTP protocol state machine. A basic feature of this state machine is that it explicitly differentiates between segment losses caused by congestion and segment losses caused by single transmission errors for ongoing connections. Moreover, MCTP distinguishes between a partitioned network and a disconnection from the Internet in the event of a temporary communication breakdown. The first situation only occurs if a vehicle communicates with another vehicle via the AutoNet multi-hop ad-hoc network. It therefore does not include the Internet for its communication. In the second situation, it is assumed that the vehicle communicates with an Internet host. The states NORMAL, LOSS and CONGESTED are the common operation modes of MCTP in the case that a data flow is possible. PARTITIONED and DISCONNECTED are only entered when a communication flow is not possible. The protocol functionality works in the following way:

> **Normal operation, congestions and losses.** An important goal of MCTP is to minimise the number of slow starts in an ongoing connection. As shown above, segment losses activate the slow start algorithm, which degrades the TCP performance in terms of throughput. A TCP sender considers a segment as being lost

in the following two situations: (i) the sender receives three consecutive duplicate acknowledgements for one segment (DupAck) and (ii) a retransmission timeout occurred for one segment at the sender side.

An important feature is that MCTP performs a situation-based handling of lost segments. MCTP therefore classifies losses caused by congestions in the network and losses caused by transmission errors. The MCTP protocol state machine therefore implements the two states CONGESTED and LOSS where segment retransmissions are handled differently. After TCP establishes a connection successfully, MCTP initially operates in the NORMAL state and observes the IP packet flow in the end system. This way, TCP initially starts communication with a slow start phase to determine the initially available bandwidth. If ECN indicates a pending congestion in an intermediate system, MCTP switches into CONGESTED state and does nothing: it ignores DupAcks as well as the expiry of RTOs. This way, MCTP leaves the congestion control completely to TCP and does not interfere with the normal congestion behaviour of TCP. After the TCP sender transmits a new segment, MCTP returns to the NORMAL state.

In the NORMAL state, MCTP counts the number of DupAcks received for a segment. If ECN does not indicate a pending congestion in the network, the segment loss was likely caused by a transmission error. If MCTP receives two DupAcks for a segment in this situation, it enters the LOSS state in order to handle the segment loss appropriately. Since the TCP congestion control reacts only after the third DupAck, it is not affected by the second DupAck and does not interfere with MCTP in this situation. Similarly, MCTP enters the LOSS state if an RTO expires for a segment and ECN does not indicate a pending congestion in the network. In the LOSS state, MCTP forces TCP to enter the persist mode by shrinking the congestion window (cwnd) to zero and by freezing the retransmission timers. This means TCP does not invoke the congestion control, which would be the wrong thing to do in this situation. Instead, MCTP itself retransmits the unacknowledged TCP segment. It therefore controls the retransmission timers accordingly to retransmit the segment in the case that the acknowledgements are not forthcoming. If an acknowledgement for the segment arrives from the peer, MCTP forwards the acknowledgement to TCP, which also removes TCP from the persist mode. MCTP then returns to the NORMAL operation mode.

Partitioned and disconnected mode. The mobility of the vehicles may stall ongoing connections in the VANET for a temporary period of time. These communication disruptions are typically caused in the following two situations: (i) The AutoNet ad-hoc network may be partitioned if communication links between vehicles break and (ii) a disconnection from the Internet.

MCTP considers these two situations and controls TCP appropriately in order to improve the recovery after a connection breakdown. Therefore, MCTP comprises the two states PARTITIONED and DISCONNECTED. The PARTITIONED state represents a network partitioning that is relevant for inter-vehicle communication. This state is entered into in the evenet that an ICMP message indicates a partitioning of the AutoNet ad-hoc network. In contrast, the DISCONNECTED state is entered when the vehicle becomes disconnected from the Internet. In the

PARTITIONED state, MCTP forces TCP to enter the persist mode and freezes its state. In addition, MCTP performs a window probing mechanism similar to the zero window probing used in TCP [19]. Thereby, MTCP probes the connection periodically with acknowledgements. MTCP therefore uses a constant period for sending the probe packets using the last value of the RTO before MCTP switched to the PARTITIONED state. This is in contrast to the zero window probing implemented in TCP, which exponentially backs-off the probing period. If MCTP receives a DupAck from the receiver, the connection is apparently re-established and communication can be continued. In this case, MCTP removes TCP from the persist mode, activates the slow start phase of TCP without reducing ssthresh, and moves itself back to the NORMAL operation mode. The PARTITIONED state is also entered from the LOSS state and the CONGESTED state upon receiving an ICMP destination unreachable message as illustrated in Figure 6.6. If the connection cannot be restored after a pre-defined period of time, MCTP assumes that the connection cannot be re-established and advices TCP to reset the connection.

The PARTITIONED mode is of relevance for AutoNet ad-hoc communication only. MCTP therefore interacts with the management layer to access the relevant information of the lower layers. When MCTP receives a notification for a disconnection while communicating with Internet hosts, it switches into the DISCON-NECTED state. The disconnected mode will also be activated if MCTP operates in the CONGESTED state or the LOSS state. In this mode, MCTP completely stops TCP transmissions and freezes TCP. Both TCP and the MCTP sublayer remain in this state until MCTP is notified about a reconnection to the Internet. MCTP then restores TCP and moves itself back to the NORMAL operation mode. In addition, MCTP activates the slow start phase of TCP without modifying the threshold for the slow start. This allows TCP to converge its data rate to the new situation. Finally, MCTP triggers TCP to retransmit queued segments immediately. If such segments are not available, MCTP sends two acknowledgements in order to generate a duplicate acknowledgement.

6.2.5   Evaluation Results

In this section we will highlight one typical communication scenario in AutoNets: travelling at night on an empty motorway. The overall evaluation of MCTP can be found in [20]. We assume that four roadside stations provide Internet access for the passing ego vehicle. The ego vehicle therefore uses a roadside station for accessing a host in the Internet. In this simple scenario, it is assumed that no multi-hop communication occurs. Furthermore, the following characteristics and assumptions were made:

- The ego vehicle is traveling at high speeds. Hence, the duration of a connection with a roadside station is rather short.
- The distance between the first and the second roadside station is larger than the diameter of a roadside station's radio transmission range. For this reason, communication is not possible

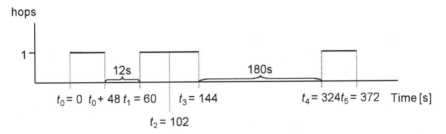

Figure 6.7 MCTP protocol state machine.

in the small area of between the first and the second roadside station, resulting in a short-term disconnection from the Internet.

- The radio transmission areas of the second and the third roadside station overlap each other. Hence, the ego vehicle has to perform a handover to make continuous communication possible while passing the second and the third roadside station.
- The distance between the third and fourth roadside station is assumed to be rather large, which causes a longer period of disconnection. This is justified by the assumption that, especially at an early deployment phase, roadside stations will not have an area-wide coverage.

For this scenario, Figure 6.7 shows the connection times to the Internet (in terms of hops between vehicle and roadside station) for the ego vehicle against the time. The first road-side station becomes accessible at time t_0. The ego vehicle can communicate with the first roadside station for 48 seconds. Afterwards, Internet access becomes temporarily unavailable for 12 seconds. At $t_1 = t_0 + 60$ s, the ego vehicle enters the radio transmission range of the the second roadside station and thus again gets access to the Internet. At $t_2 = t_0 + 120$ s, a handover occurs at to the third roadside station. The latency caused by the processing of the handover is assumed with 24 ms, which corresponds to the measurements in [24]. During this time, all packets are assumed as being lost. Afterwards, the ego vehicle is able to continue its communication until t_3, where the ego vehicle leaves the radio transmission range of the third roadside station, resulting in another disconnection from the Internet for 180 s. At t_4, the ego vehicle enters the radio transmission range of the fourth roadside station and thus has again access to the Internet for 48 s.

As mentioned above, the scenario assumes Internet connectivity for vehicles via roadside stations. This way, the end-to-end connection between the ego vehicle and the Internet host comprises two basic segments: (i) a mobile (and wireless) segment between vehicle and roadside station and (ii) a wired segment between roadside station and the Internet host. To compare the optimised transport protocol with existing approaches, we assume a performance-enhancing proxy in the Internet as described in Section 10.3.4. We compare the following three alternative realisation strategies:

No proxy. Here, a continuous TCP connection was assumed between the ego vehicle and the Internet host.

TCP split. In order to separate the communication characteristics of the wireless and mobile AutoNet, the continuous TCP connection was split up into the two

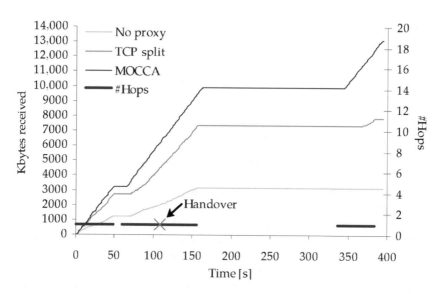

Figure 6.8   MCTP protocol state machine.

partions mentioned above: a mobile and wireless segment[4] and the fixed networking segment. On both segments, TCP was used for communication.

MOCCA. This scenario is basically the same compared to TCP split. The only difference is that the optimised transport protocol MCTP was used to communicate on the mobile segment.

For the measurements, the maximum data rate between ego vehicle and Internet host was determined. Measurements were performed with real end-system implementations for ego vehicle, performance-enhancing proxy and Internet host. The network characteristics for the mobile and the fixed networking segment were emulated. The overall result for each realisation strategy was determined by the average of three measurements each. An intensive overview of the parameters assumed for this emulation is given in [20]. Figure 6.8 depicts the comparison of the averaged measurements for the three realisation strategies. This chart shows clearly the advantages of split connection approaches in this scenario. Although a rather high bandwidth is available for the vehicle, the no proxy test has obvious problems transfering data during the whole emulation time. The measurements also show the impact of the different periods of disconnection from the Internet. Whereas the handover at about 110 s does not affect the TCP throughput significantly, it takes a longer time for TCP to realise the reconnection to the second roadside station after the disconnection at about 50 s. The long disconnection of three minutes (starting at about 160 s) between the third and the fourth roadside station cannot be handled sufficiently by the no proxy realisation: TCP was not able to realise the reconnection when the vehicle travelled through the service area of the fourth roadside station. In contrast,

---

[4] To be more precise, the mobile segment also comprises the segment from the roadside station the the performance-enhancing proxy located in the Internet.

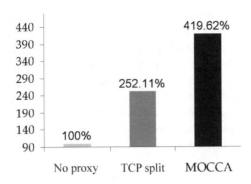

Figure 6.9   MCTP protocol state machine.

the split connection approaches exhibit different performance characteristics: the throughput increases while connectivity is available from the first three roadside stations. However, the TCP split realisation also had problems with the longer period of disconnection. It takes on average more than 70 s until the reconnection to the fourth roadside station is realised and TCP continues its transmission. Figure 6.8 demonstrates the advantage of the optimised transport protocol MCTP. While connected to a roadside station, the throughput increases continuously and steeply. The chart also shows that MOCCA is able to handle even longer periods of disconnections very efficiently compared to the other realisations.

In fact, MCTP outperforms the other two realisations. Figure 6.9 summarises the percentage improvement of TCP split and MOCCA over the end-to-end TCP connection without a performance-enhancing proxy. After the emulation runs, the following averaged amount of data was transmitted in the different tests:

- No proxy transmitted 3.132 MB.
- TCP split transmitted 7.897 MB, which is more than twice the throughput of the no proxy measurements.
- MOCCA transmitted 13.144 MB, which is more than four times as much as achieved in the no proxy test.

MCTP is further described in [20], comprising the modelling of the environmental conditions and the scenarios behind the models. The measurements were taken with simplified model assumptions, still clearly showing the potential of improving transport layer efficiency in AutoNets.

6.3   Summary

For application and supporting services, the transport layer is the interface to the communication system. In this chapter, we discussed the need for both an AutoNet transport protocol as well as the well-known protocols from the Internet, TCP and UDP. Although the development of a transport protocol for AutoNet transport is currently not the focus of standardisation, we have no doubt that such a protocol will be an important issue in the future for the introduction of unicast V2V communication in AutoNets. This chapter also addressed the deployment of TCP as a transport protocol for V2V communication. The profound description and discussion of TCP showed that it is very unsuitable for vehicular environments. In order to provide a suitable transport protocol for V2V communication – which also improves V2Central, V2Internet and V2Private communication – we introduced a highly optimised transport protocol called MCTP. MCTP provides mechanisms to cope with the characteristics of vehicular environments while maintaining compatibility with standard TCP. Combined with a performance-enhancing proxy architecture, which will be taken up in Section 10.3.4, we showed that MCTP is able to improve throughput and, thus, communication efficiency significantly in AutoNet scenarios.

References

[1]   Schiller, J. (2003) *Mobile Communications*, 2nd Edition. *Addison Wesley.*
[2]   Information Science Institute 1981 Transmission Control Protocol. RFC 793, Internet Engineering Task Force (IETF).
[3]   Stevens, W. (1994) *TCP/IP Illustrated, Volume 1: The Protocols. Addison Wesley.*
[4]   Caceres, R. and Iftode, L. (1994) The effects of mobility on reliable transport protocols. *Proceedings of the 14th International Conference on Distributed Computing Systems (ICDCS).*
[5]   Holland, G. and Vaidya, N. (1999) Analysis of TCP performance over mobile ad hoc networks. *Proceedings of the 5th ACM/IEEE International Conference on Mobile Computing and Networking (MOBICOM).*
[6]   Bae, S., Xu, K., Lee, S. and Gerla, M. (2002) TCP behavior across multihop nwireless and wired Networks. *Proceedings of the 2002 IEEE Global Telecommunications Conference (GLOBECOM).*
[7]   Stevens, W. (1997) TCP Slow Start, Congestion Avoidance, Fast Retransmit, and Fast Recovery Algorithms. RFC 2001, Internet Engineering Task Force (IETF).
[8]   Chiu, D.M. and Jain, R. (1989) Analysis of the Increase and Decrease Algorithms for Congestion Avoidance in Computer Networks. Computer Networks and ISDN Systems.
[9]   Lim, H., Xu, K. and Gerla, M. (2003) TCP performance over multipath routing in mobile ad hoc networks. *Proceedings of the 2003 IEEE International Conference on Communications (ICC).*
[10] Paxson, V. and Allman, M. (2000) Computing TCP's Retransmission Timer. RFC 2988, Internet Engineering Task Force (IETF).
[11] Allman, M., Paxson, V. and Stevens, W. (1999) TCP Congestion Control. RFC 2581, Internet Engineering Task Force (IETF).
[12] Dyer, T.D. and Boppana, R.V. (2001) ATP: a reliable transport protocol for ad-hoc networks. *Proceedings of the 2th ACM International Symposium on Mobile Ad Hoc Networking and Computing (MobiHoc).*
[13] Fu, Z., Zerfos, P., Lu, S., Zhang, L. and Gerla, M. (2003) The impact of multihop wireless channel on TCP throughput and loss. *Proceedings of the 22nd IEEE Conference on Computer Communications (Infocom).*
[14] Ott, J. and Kutscher, D. (2004) Drive-thru Internet: IEEE 802.11b for automobile users. *Proceedings of the 23rd IEEE Conference on Computer Communications (Infocom).*
[15] Caceres, R. and Iftode, L. (1995) Improving the performance of reliable transport protocols in mobile computing environments. *IEEE Journal on Selected Areas in Communications.*
[16] Floyd, S. and Henderson, T. (1999) The NewReno Modification to TCP's Fast Recovery Algorithm. RFC 2582, Internet Engineering Task Force (IETF).
[17] Mathis, M., Mahdavi, J., Floyd, S. and Romanow, A. (1996) TCP Selective Acknowledgment Options. *RFC 2018, Internet Engineering Task Force (IETF).*
[18] Mathis, M. and Mahdavi, J. (1996) Forward acknowledgement: Refining TCP congestion control. *Proceedings of the 1996 ACM SIGCOMM Conference.*
[19] Wehrle, K., Pählke, F., Ritter, H., Müller, D. and Bechler, M. (2004) *The Linux Networking Architecture.* Pearson Education.
[20] Bechler, M. (2004) *Internet Integration of Vehicular Ad Hoc Networks.* Logos Verlag.
[21] Li, J. and Singh, S. (2001) ATCP: TCP for mobile ad hoc networks. *IEEE Journal on Selected Areas in Communications.*

[22] Chandran, K., Raghunathan, S., Venkatesan, S. and Prakash, R. (1998) A feedback based scheme for improving TCP performance in ad-hoc wireless networks. *Proceedings of the 18th International Conference on Distributed Computing Systems (ICDCS)*.

[23] Ramakrishnan, K., Floyd, S. and Black, D. (2001) The Addition of Explicit Congestion Notification (ECN) to IP. RFC 3168, Internet Engineering Task Force (IETF).

[24] Forsberg, D., Malinen, J.T., Malinen, J.K., Weckström, T. and Tiusanan, M. (1999) Distributing mobility agents hierarchically under frequent location updates. *Proceedings of the 6th International Workshop on Mobile Multimedia Communications (MoMuC)*.

[25] Ritter, H. (2001) *Bedarfsorientierte Dienstgüteunterstützung durch adaptive Endsysteme*. VDI Verlag.

[26] Postel, J. (1981) TCP Internet Control Message Protocol. RFC 792, Internet Engineering Task Force (IETF).

[27] Tanenbaum, A. (2003) *Computer Networks*, 4th Edition. Prentice Hall.

7

Networking

To support the application types classified in Section 2, it is necessary to be able to deliver data from any type of data source to any type of data sink, that is between any two AutoNet entities. The network layer in the AutoNet Generic Reference Protocol Stack provides services for the layers above and utilises the capabilities of the underlying wireless technologies, as illustrated in Figure 7.1. These technologies, presented in Chapter 8, provide a physical link between certain nodes at certain times. In order to deliver data to other nodes not participating in the physical link, it is necessary to transport the data via a certain number of wireless links or via intermediate wired communication. This issue is addressed by the network layer.

In this chapter, we will first take a closer look at the networking principles. Based on this, we describe both ad-hoc and cellular network characteristics, as well as respective address and routing schemes inf more detail. Subsequently, we discuss the suitability (of dedicated mobility extensions) of the evolving IPv6 standard with respect to AutoNet requirements.

7.1 Networking Principles in the AutoNet Generic Reference Protocol Stack

The integration of the network layer into the AutoNet Generic Reference Protocol Stack is shown in Figure 7.2, along with the functional core components of this layer. Like the transport layer, the network layer interacts with the management layer and the security layer via the service access points M-SAP and S-SAP, respectively. Furthermore, it accesses the services from the lower physical communication technologies via the P-SAP. The network layer itself provides its services by the N-SAP, which is basically used by the transport layer – but also by the management layer and the security layer.

7.1.1 Network Layer Functionality in AutoNets

As we already discussed in Chapter 4, the different types of applications have different requirements with respect to the transfer of information among communicating nodes. Driver-related applications as well as vehicle-related applications are often based on Internet services

Automotive Internetworking, First Edition. Timo Kosch, Christoph Schroth, Markus Strassberger and Marc Bechler.
© 2012 John Wiley & Sons, Ltd. Published 2012 by John Wiley & Sons, Ltd.

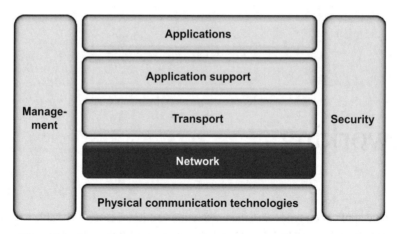

Figure 7.1 Network layer in the AutoNet Generic Reference Protocol Stack.

and, thus, rely on IP-based communication. Note that we assume interacting applications here. Broadcast services are not considered here, since they are typically uni-directional data transfer from a broadcast station to the vehicles. For the driving-related applications, non-IP cellular and ad-hoc networking technologies provide different addressing mechanisms and lower latencies for the applications in the manoeuvering and the stabilisation domain (of course it can also be an option for some applications in the navigation domain). This is reflected in the AutoNet Generic Reference Protocol Stack by the fact that driving-related applications mostly use the AutoNet transport at the transport layer, whereas driver-related applications and vehicle-related applications are typically based on TCP or UDP.

In the network layer, we distinguish three types of 'networking' from an AutoNet station's perspective. The three networking types correspond to the respective functional blocks inside the networking layer, as illustrated in Figure 7.2:

- Communication with any other node in the generic domain (cf. Section 3.1). This communication is typically based on the Internet Protocol IP and is naturally implemented by 'IPv6 and mobility extensions'.

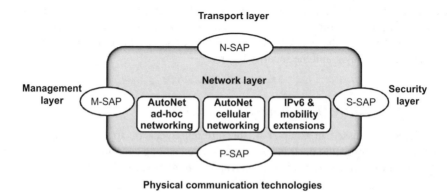

Figure 7.2 Service access points and functionality of the network layer.

- Communication with any other AutoNet station within the AutoNet ad-hoc network. This may be either directly between two AutoNet stations, or using multi-hop communication, which also may include roadside stations. This kind of networking is most often referred to as a *vehicular ad-hoc network*, or *VANET* for short. Following our naming conventions, this functionality is provided by the component 'AutoNet ad-hoc networking'.
- Communication with any other AutoNet station within the AutoNet ad-hoc network, but using communication via the generic domain instead of the AutoNet ad-hoc network protocols. The network layer functionality for this type of communication is realised by the 'AutoNet cellular networking component'.

Figure 7.3 illustrates the correlation between the components at both transport layer and network layer. Most often, both TCP and UDP top-layer IP and thus are using IPv6 and mobility extensions, whereas AutoNet transport may use either AutoNet ad-hoc networking or AutoNet cellular networking. Besides these apparent mappings, we also have to consider a third scenario, delineated by the dotted arrow in Figure 7.3: a communication path using TCP or UDP over IPv6 and mobility extensions, operated via one of the two options of AutoNet specific networking. This networking functionality enables the transmission of common IP-based traffic via the respective AutoNet networking components. It is motivated by the fact that we already introduced in Section 6.1.1: future applications among vehicles may also be based on unicast connections using IP-based networks. The dotted arrow in Figure 7.3 thus allows the transfer of IP-based traffic, for example, among two vehicles using the AutoNet ad-hoc network. It is worth mentioning that this mechanism will likely depend on management issues. It may also be considered as one conceptual building block of the mobility management described in Section 10.3.3. Therefore, the IP-based applications send their data via a common socket, which is internally mapped onto respective AutoNet networking protocols, that is either AutoNet ad-hoc networking or AutoNet cellular networking, both having the same interface. Therefore, the IPv6 address of the targeted vehicle needs to be mapped onto the

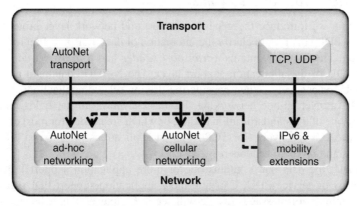

Figure 7.3 Mapping of transport layer components onto network layer components.

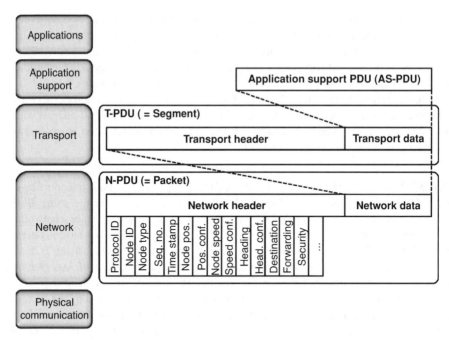

Figure 7.4   Transport and network layer PDUs.

actual geographical position of the vehicle (and vice versa for the target vehicle). Respective approaches for this mapping are under development in, for example, the GeoNet project.[1]

Before focusing on the three components of the network layer, we will first explain the cooperation between transport layer and network layer. For this, we use as an example a driving-related application that uses AutoNet transport together with AutoNet ad-hoc networking.

7.1.2   Network Protocol Data Units

In the following, we introduce the generic transport and network layer protocol data units (T-PDU and N-PDU) in order to clarify the interaction of data transport. Figure 7.4 illustrates this interaction. The T-PDU contains a *transport header* and a *transport data* field, which contains the PDU of the application support layer. The N-PDU contains a network header and the T-PDU, respectively. When considering driving-related applications using AutoNet transport and AutoNet ad-hoc networking, Figure 7.4 shows all the fields that have been defined by any one of the standardisation bodies, or which are needed for most of the proposed research protocols, respectively (for TCP/IP-based applications, we refer to [46] describing the interactions and protocol headers respectively).

In the T-PDU, the data part contains one or more application support PDUs, that is a safety, efficiency or service PDU. Carrying more application support PDUs allows a redu-cuction in the number of messages (as we will see in Chapter 8, this is beneficial in ad-hoc

---

[1] GeoNet Homepage: http://www.geonet-project.eu.

communication scenarios with a large number of network participants, because it minimises transmission collisions on the wireless channel). Every group communication message (e.g. geocast message) contains one or more message identifier (MsgID). The message data is always connected to a message identifier. It may contain message specific sub-addressing. Thus, one T-PDU may contain information from different types of applications in different messages or application support PDUs, respectively. Messages that have to be forwarded according to a multi-hop routing approach may also be part of the T-PDU, concatenated with ones orginated by the transmitting vehicle. Recursively, the N-PDU consists of a *network header* and a *network data* field, whereas the latter carries the T-PDU. Figure 7.4 also shows the network header fields that are currently under discussion in the different standardisation bodies. These parameters are not yet complete, but they give an impression of the aspects of the AutoNet-based components of the network layer:

- Protocol ID describes the identification of the protocol being used.
- Node ID and node type specify the identification of the node and its type, that is the vehicle type or a roadside station.
- Sequence number is used in case a message is too long for the network layer and thus needs to be segmented.
- Time stamp specifies the time a message was sent.
- Node position and node configuration define the current position of the sending node.
- Node speed and speed configuration define the current speed of the vehicle node ID.
- Heading and heading configuration define the direction a vehicle is currently driving.
- Destination and forwarding define the destination (area) for this message and how to reach this area using multi-hop ad-hoc communication.
- Security defines specific security issues for the message.

Driving-task related applications will mainly be based on group communications, or, to be more specific, by a local dissemination of messages. Packets are usually not addressed to a certain group of individuals defined by their identifiers (as, e.g., in case of IP multi-cast) but rather by their contextual situation, in particular their current position. Two types of context-addressing are typically supported by AutoNets: explicit and implicit context addressing. With explicit addressing, the sending application of a packet defines these contextual address parameters. With implicit addressing, the network layer is handling the packets without any addressing at all, purely based on individual relevance assessment. In this case, the network layer needs to be supplied with rules on relevance assessment. One particular instance of explicit context parameter addressing is geocasting, where all nodes residing in a well defined geographical area become addressees of the message.

In the following, we will describe the information flow in more detail. Specific AutoNet addressing and routing mechanisms will be presented in Section 7.2.2. Ad-hoc networks for these types of applications have received considerable research effort and are usually referred to as *VANETS (vehicular ad-hoc networks)*. These usually comprise both V2V and V2R communication. These communication types can be concatenated, that is vehicles exchanging data within a VANET through a roadside station. VANET network properties are detailed in Section 7.2.1. Addressing and routing in VANETS is laid out in Section 7.2.2. AutoNet group communication mechanisms for cellular networks are presented in Section 7.3.

In the following sections, we presume that nodes in an AutoNet have at least one unique network address. As we will see in Chapter 9, for privacy reasons this might also be some kind of pseudonym disguising the drivers' real identities.

7.2 AutoNet Ad-Hoc Networking

In this section, we describe the network characteristics of the AutoNet ad-hoc network and explain the most efficient known protocols to route, forward and disseminate messages. We also present results from theoretical analyses and simulations on how certain AutoNet ad-hoc network properties change with a varying number of communicating vehicles, namely with respect to connectivity, message reception probability and message dissemination delay. Additionally, we have a quick look at traffic effects with respect to the penetration rate.

7.2.1 AutoNet Ad-Hoc Network Characteristics

The most prominent feature of AutoNet ad-hoc networks is that they are self-organised networks. This means that no communication infrastructure is necessary for communication among the network nodes. However, AutoNet ad-hoc networks are possibly supported by network infrastructure services to facilitate network management and security. For generic (and usually less) dynamic ad-hoc networks, point-to-point packet routing is an important feature. However, AutoNet ad-hoc networks have different requirements. In AutoNet ad-hoc networks, mainly applications from the entertainment domain require the establishment and maintainance of a route in the network between two vehicles for unicast communication. Applications from the driving-related domain need quick and reliable distribution of vehicle parameters and sensor data to those recipients that can make the best use of this information. Since this type of information regularly features both spatial and temporal relations, it is usually used by applications running in vehicles in the vicinity of the georeferenced origin within a limited time after the information is generated. The latter is due to the fact that such information typically becomes quickly outdated as the vehicles are moving.

With mainly vehicles and possibly roadside stations representing the communicating nodes in AutoNet ad-hoc networks, this special type of ad-hoc network features special characteristics. Vehicles can move fast, which results in a very dynamic network topology. Since vehicle movements are regularly bound to roads, the network nodes travel along the road network leading to specific topology patterns. Road traffic can be anything from very sparse to completely jammed. For network protocols, this means they have to deal with rapidly changing network structures, short connectivity, network partitioning and scalability challenges.

To describe and analyse the characteristics and behaviour of AutoNet ad-hoc networks, we need a mobility model, a network model and a communication model. The mobility model formalises the movements of the vehicles. The network model describes the network topology at a given point in time, based on the specific channel and link characteristics of AutoNet ad-hoc networks. In the following, we use graph theory to model the network. This model simplifies the real world situation by making the assumption that the possibility that two nodes can communicate is a binary characteristic, depending only on their distance and represented by the communication range parameter r. It does not represent the fact that there is a probability distribution for the correct reception of a packet send from one node to another and that this relation is usually an asymmetric one. The communication model describes the behaviour of AutoNet ad-hoc networks with respect to the sending, receiving and forwarding of data packets.

7.2.1.1   The Penetration Rate Challenge

Aside from the technological challenges required for AutoNets to take off, one fundamental problem needs to be solved: system introduction. Customers buying an AutoNet-enabled vehicle can only reap the benefits if there is a communication partner available with whom to exchange data. This is true for both vehicle-to-vehicle and vehicle-to-roadside communication. With AutoNet ad-hoc networks, the problem is aggravated by additional network connectivity characteristics. In the following, we concentrate purely on network characteristics (for a discussion on the necessary penetration rated from an application perspective, please refer to Chapter 12). In Section 7.3, we introduce an important feature of the AutoNet Generic Reference Protocol Stack, which alleviates this system introduction by significantly reducing the necessary penetration rate in the beginning of a roll-out scenario for AutoNet technology.

What can be observed is that the step from the penetration rate at which an AutoNet application does not work properly to the penetration rate at which the application can be beneficially used, often features a rather quick transition. This quick transition stems from the fact that the ad-hoc network formed by AutoNet equipped vehicles is very similar to the more generic mathematical model of geometric random graphs. Just like these random graphs [37], AutoNet ad-hoc networks feature almost immediate phase transitions for some important properties, foremost connectivity. For some other properties, the phase transition is a little smoother, but since it is typically related to connectivity properties, there is always a phase where

- the system would not work properly; then
- a transition phase where it starts to become useful; and then
- a phase where the penetration rate is sufficient for the system to work in a satisfactory manner.

Phase transition phenomena are presented in the following for one and two dimensional networks with respect to

- connectivity;
- message delivery success rates;
- message dissemination times, that is how long it takes until a certain share of nodes within a message dissemination area have received a message.

In the matrix shown in Figure 7.5, AutoNet applications are organised in a way that reflects the necessary penetration rate of both vehicles equipped with AutoNet capabilities on the x-axis and AutoNet infrastructure on the y-axis, respectively. Some applications solely depend on the availability of infrastructure entities, like a work zone warning. This matrix should be read as providing relative penetration prerequisites rather than exact percentages of penetration rates necessary for each type of application.

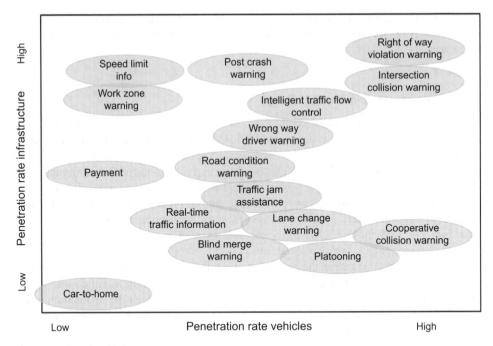

Figure 7.5   Penetration rate dependency of certain application types.

While a car-to-hotspot communication starts to be interesting with a certain number of hotspots available, a work zone warning will only be useful if the majority of work zones actively provide information about their existence, spatial extent, speed allowed and so on. Other applications can be realised by pure V2V communication, like real-time traffic information or cooperative collision warning. They would benefit from and be assisted by infrastructure units but could also work autonomously, given a necessary crucial share of vehicles participating in the information exchange. In addition, there are applications that will only work well in a cooperative fashion when supported by both V2V and V2R communication. Prominent examples are intersection assistance systems. They will require a high penetration rate of both V2V-enabled vehicles and roadside stations.

7.2.1.2   Connectivity Properties

Figure 7.5 shows that many applications already work without every vehicle participating in an AutoNet system or having full AutoNet infrastructure equipment. However, every application type has its own requirement on the quality of service and thus on certain network properties. This is true for the performance of the data transmission, such as data rate and latency, but also for the basic capability to transmit data from one entity to another. This basic capability requires a network connection between the two entities or nodes of some sort. For instance there needs to be a network connectivity between the traffic light and the vehicle to provide the status information of a traffic light. The knowledge of the traffic situation along the route of a vehicle based on probe data depends on the number of vehicles providing probe data and the way this probe data is transferred from the providers to the customers. If this data transmission is to be done within an AutoNet ad-hoc network, this network needs to provide a certain amount of connectivity to ensure the data can be delivered.

In general, a well connected network is helpful for a high quality of service, even if full connectivity may not be necessary for the application in question. Since connectivity only provides the basis to be able to deliver data from some entity to another, it can generally be stated that higher connectivity is always desirable from an application's perspective. The analysis of the connectivity properties provides some valuable insights into the properties of ad-hoc networks in general and AutoNet ad-hoc networks in particular. Existing work on ad-hoc network connectivity mainly considers nodes randomly distributed in a plain rectangular area [29, 30, 31, 32, 33, 34, 35, 37]. This more general work, however, is not directly applicable to AutoNet ad-hoc networks with their specific node distribution, movement patterns and external conditions.

As a metric for the degree of AutoNet ad-hoc network connectivity we propose using the mean number of nodes c that each node is connected to at a certain point in time. If n denotes the total number of nodes in the network and c_i denotes the number of nodes, to which a path exists in the network graph from node i, then $c = \frac{\sum_{i \in V} c_i}{n}$. The closer this value is to the optimum ($= 1$), the better the connectivity.

Let us have a look at the phase transition phenomenon for connectivity in the one dimensional case. Consider the probability of being connected to a communication partner at a certain distance along a straight road. Assume that AutoNet ad-hoc network-enabled vehicles (called *AutoNet vehicles* in the following) are randomly distributed along this road. For highway scenarios with multiple lanes (projected on a single lane), the assumption of a Poisson distribution is justified.

Let's assume AutoNet vehicles in the model would feature a fixed transmission range of 400 m. Using this simplistic disc communication model, we abstract from any physical channel properties where a more realistic model would assume a probability distribution for correct data reception depending on the distance.

If we calculate the probability of a connection to a vehicle at different distances, we find – not surprisingly – that with a growing density of AutoNet vehicles, that is a higher product of traffic density times penetration rate, the probability for connectivity is growing. There exists a lower boundary for the probability $P_{connection}$ that all n nodes randomly placed along a straight line of length l are connected with each other (for details please refer to [34]).

Dousse et al. present a formula for the exact calculation of this probability [32]. Based on this formula, the probabilities in Figure 7.6 have been calculated. It can be easily seen from the figure that there is a relatively quick transition from an almost zero probability to almost one. This zero–one property is known as the phase transition property of random graphs [36, 37]. Thus, in order for the network to feature a high connectivity, a particular critical penetration rate of vehicles is needed. This phase transition phenomenon can also be observed in the two-dimensional case [65]. Since in AutoNet ad-hoc networks vehicles are moving along roads, they are not randomly placed on a plane. Even more than in the one-dimensional case, successful transmission on a two-dimensional road network depends on the physical environment between two nodes. While in the one-dimensional case we have argued that a line of sight indexLine of Sight between any two nodes can be assumed as an acceptable model of the real situation, the simple distance model of reception is not as easily justified for the two-dimensional case. This is because buildings and other obstructions do not allow

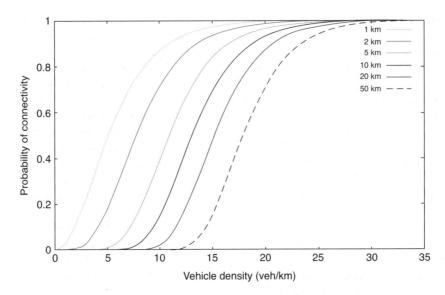

Figure 7.6 Probability for connection to a communication partner at different distances (1 km to 50 km) depending on the density of AutoNet vehicles in a one-dimensional highway scenario (with a communication range of 400 m).

a line of sight between two vehicles placed on different roads. For this case, we assume that communication between two nodes is only possible if there is a line of sight between them and other vehicles do not obstruct this line of sight. We account for obstruction of the direct line of sight using a simple model (cf. Figure 7.7). In this model, communication between two nodes v_1 and v_2 is only possible if there is no point on the direct line connecting the two nodes that is further away from the street network than some threshold value, which we call the obstruction value δ.

The effect of the obstruction value δ becomes evident in particular in high density scenarios, where connectivity is very much restricted by the street network topology. As a result,

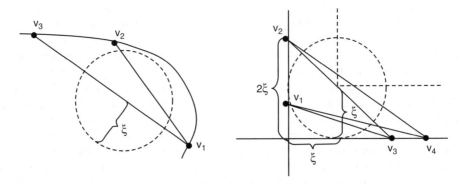

Figure 7.7 Line of sight obstruction.

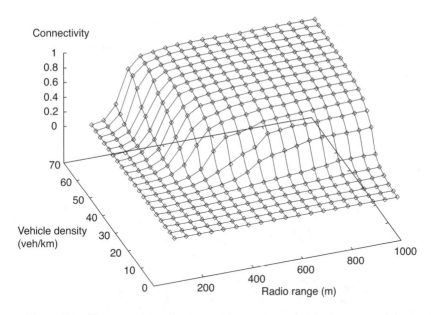

Figure 7.8 Phase transition phenomenon in two-dimensional urban connectivity.

multi-hop forwarding of messages mainly takes place along single roads, neglecting crossing roads in urban scenarios. Thus, even at high node densities, isolated nodes often exist in by-roads when obstruction is taken into account.

In low node-density scenarios, connected network parts consist only of a few nodes and occur in a localised fashion. If connections are obstructed by buildings, the connected network parts grow much more slowly than in the unobstructed case. While this leads to a smoother transition from zero to almost certain full connectivity, the phase transition phenomenon is still noticeable. This becomes evident in Figure 7.8. Connectivity is plotted on this graph in relation to both communication range and penetration rate measured as density of AutoNet vehicles. The plot shows the connectivity measure c depending on the communication range r and the density of equipped vehicles. Figure 7.9 presents some specific combinations of connectivity for fixed penetration rates depending on r. The affinity with the one-dimensional case is evident. At lower values of r, a high node density is necessary for good connectivity. As soon as r exceeds a certain critical value, further increasing the transmission power and thus the communication range only has minor effects.

7.2.2 *AutoNet Ad-Hoc Network Addressing and Routing*

In this section, we consider networking mechanisms incorporating priorities for both point-to-point and point-to-multipoint communication. Both types of communications in an AutoNet ad-hoc network make use of the fact that the positions of the nodes are known to the network layer. This also allows the communication protocols to address destinations based on geographical regions or areas, called *geographical broadcast* (or *geocast*), if all nodes within a specified area shall be addressed, and *geographical anycast* if any one node within such area may serve as the destination.

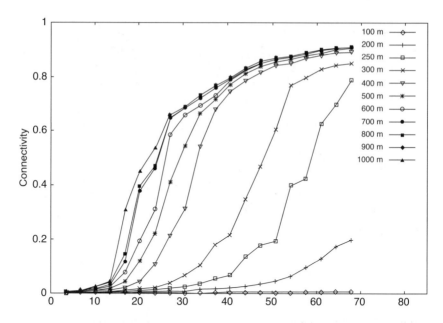

Figure 7.9   Connectivity depending on penetration rate for different communication ranges.

7.2.2.1   Store and Forward Investigations

If all network nodes residing within the dissemination area are connected, they form a graph within which messages are forwarded immediately after reception. If the nodes are not all connected to each other, messages can still reach all nodes within the area if appropriate *store-and-forwarding* mechanisms are used (sometimes also referred to as *carry-and-forward*). In the case where all nodes are connected, the challenge for dissemination schemes is mainly to reduce the overhead generated by redundant messages, that is the optimum scheme would reach each node within the network part addressed with the minimum number of messages without any overhead caused by additional network control messages. Latency is not such a big issue in these network situations. In the event where there are disconnected parts or singular nodes, the challenge for the dissemination scheme is to reach all nodes by intermediately storing the messages in forwarder caches. Metrics for the performance of the schemes in this case are redundancy and latency as well, but this time, minimisation of the latency caused by intermediate message storage is the relevant factor, whereas possible network overhead is usually only caused by few nodes and thus does not cause a problem in sparse scenarios. In the following, we will first consider the sparse network case and then have a more detailed look at the dense network situations.

Within sparse network scenarios, let us first study the store and forward case in one dimension, that is along single streets. In this case, one of the questions to answer is: until which distance does a node have to store a message in order to be able to deliver it to the next node with a very high probability? We assume that the message contains information about a situation at one particular point on the road, such as a dangerous situation. Let us furthermore consider that one node detects this situation on the road when passing by. The goal in this case is to communicate this information to the next following vehicle, such that this vehicle will

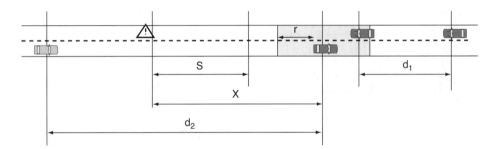

Figure 7.10 Minimum dissemination distance of a hazard warning along one single road.

receive the message with a very high probability before reaching the event. In Figure 7.10, this question is visualised with a hazardous section marked by a warning sign.

To be able to react in time, vehicles need to be informed about the danger at the latest at a (safe) distance s before the hazardous area. In a fragmented AutoNet ad-hoc network, the message can only be forwarded to following traffic by making use of oncoming traffic. The vehicle on the lower lane on the right forwards the message at a distance x from the hazardous area to an oncoming vehicle moving towards the danger. This vehicle receives the message at a distance $x + r$, depending on the communication range r. Let us consider the following question: up to what distance x should a vehicle on the lower lane transport the message before dropping it? Assume for simplicity that all vehicles are moving at the same constant velocity v. Let the vehicle density on the upper lane be ρ_v^1 and ρ_v^2 on the lower lane (in vehicles per kilometre). Let the distance d to the next vehicle follow the Poisson distribution $f(d) = \rho_v e^{-\rho_v d}$ as we had above for connectivity considerations. Then, whenever two vehicles meet on this road, the distances to the respective next vehicles follow the common distribution function

$$f(d_1, d_2) = \rho_v^1 \rho_v^2 e^{-\rho_v^1 d_1 - \rho_v^2 d_2}$$

This function is plotted in Figure 7.11 for $\rho_v^1 = \rho_v^2 = 0.2$, that is only one vehicle every 5 km.

To answer the above question about when to drop a warning message (i.e at what distance x) there is one major constraint: the probability that the following vehicle on the lower lane (which is receiving the message from the leading vehicle on the upper lane) crosses the line at distance s from the hazard earlier than the following vehicle on the upper lane. This is the case if $d_2 - x + s < d_1 + r + x - s$. Solved for x we get $x > \frac{1}{2}(d_2 - d_1 - r) + s$. Solving the problem independently from s and r, we are looking for a value of x for which the probability that $d_1 + x > d_2 - x$ is high. This probability can be calculated on the basis of the density distribution depicted in Figure 7.11 as the integral over the shaded area, which is

$$P_x = \rho_v^1 \rho_v^2 \int_{d_1=0}^{\infty} \int_{d_2=0}^{d_1+2x} e^{-\rho_v^1 d_1 - \rho_v^2 d_2} dd_1 dd_2$$

Simplifying this expression leads to

$$P_x = \frac{\rho_v^2}{\rho_v^1} - \frac{\rho_v^2}{\rho_v^1 + \rho_v^2} e^{-2\rho_v^2 x}$$

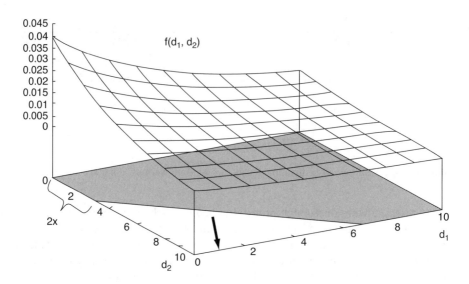

Figure 7.11   Common distribution of the distances between two vehicles on opposing lanes on a street.

and for $\rho_v^1 = \rho_v^2$ to

$$P_x = 1 - \frac{1}{2}e^{-2\rho_v x}$$

This function is plotted in Figure 7.12. It is evident that again we have some sort of phase transition phenomenon where certain critical combinations of penetration rates and message transport distances lead to the targetted high probabilities where there cannot be much more gained by further enhancement of either the penetration rate or the storing distance.

More interesting, however, are the iso probability lines along this curve: how do vehicle density and minimum transportation distances of messages relate to any given probability that a message reaches the following vehicle in time? This relation is plotted in Figure 7.13 for probabilities of 99%, 99.9% and 99.99%. This shows that even at low densities of AutoNet vehicles, it is in general sufficient to store messages for less than 5 km. For any AutoNet that incorporates such functionality, it is sensible to adjust the minimum transportation distance according to the traffic situation observed. For these considerations we assume a straight road topology with no intersections, and a fixed communication range r. However, the findings also match with real world road topologies and real world radio propagation. The reasons are:

- A phase transition phenomenon also appears with a randomly distributed range of successful data transmissions. The findings are valid for both minimum and average communication ranges. In order to achieve a fixed probability of, for instance, 99.9%, the specific distance x may vary in different scenarios according to changing environmental conditions. A sensible distance to store messages is still in the presented magnitude.
- A phase transition phenomenon also appears if vehicle density is not commonly distributed, but vehicles travel in a sort of cluster, that is a couple of vehicles travel in close vicinity. In

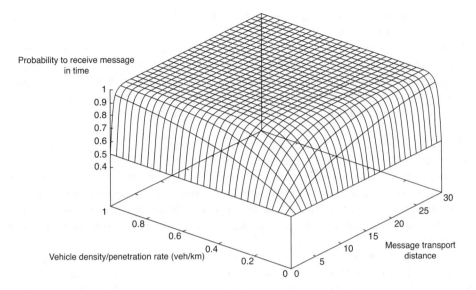

Figure 7.12 Probability that hazard message is received in time depending on density of AutoNet vehicles and transportation distance of hazard warning message.

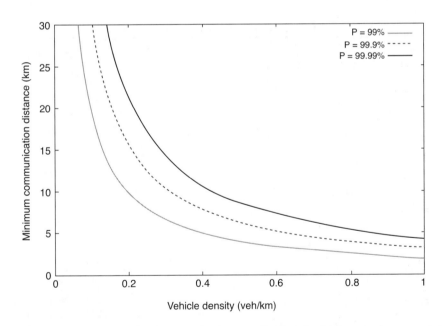

Figure 7.13 Relation between vehicle density and message transport distance for a given probability that the message is received in time.

this case, the model is still valid when applying the distribution of the respective clusters as a common distribution function.

- Every road (segment) with intersections can be modelled by flattening the topology into further straight segments as presented, one for each possible path from a source to a destination. For each flattened straight segment, the store and forward distance along with the distribution probabilities follow the same principle. Assuming different vehicle densities for crossing roads, the segment where the reported event is located is decisive.

Based on these considerations, an adaptive storing distance x can be applied based on the current vehicle density, or, to be more precise, the density of auto-networked vehicles.

Considering the connectivity property, it is hardly surprising that – when disseminating messages – a phase transition behaviour also reappears for the latency of message reception while distributed in a certain area of the road network. Naturally, there is a direct relation between the degree of connectivity and the time a message needs to be disseminated within a certain area.

The percentage of vehicles within a certain area that have received the message after a certain amount of time follows an S-shaped curve and can once more be interpreted as a phase transition behaviour – however, in many cases in a slightly smoother kind of way. Figure 7.14 shows the simulation results for disseminating a single message in an area of 8 km². The y-axis shows the shares of all addressed vehicles reached depending on the time elapsed on the x-axis. Again what we can see is that the necessary penetration rate for a sufficiently quick

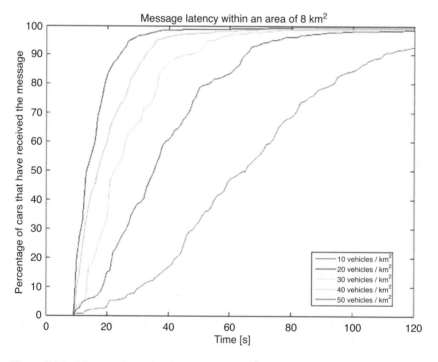

Figure 7.14 Message dissemination delays in 8 km² area, depending on penetration rate.

dissemination of messages within an area of a certain size also needs some critical mass of equipped vehicles.

For the message dissemination delays, there is a direct relation to connectivity. For fully connected nodes within the area, the message delay is negligible. Further increasing the penetration rate does not improve the situation any more. The transition from poor behaviour to good behaviour, however, is noticeably smoother than in the case of pure connectivity. This can be explained by the fact that the nodes' movements lead to connectivity over time which allows for the forwarding of the message. The movement thus helps to bridge the connectivity gaps.

7.2.2.2 Message Dissemination

Message dissemination methods can generally be classified as explicit or implicit dissemination. Explicit dissemination requires the addressing of a group of vehicles to which a packet or message shall be delivered. Implicit dissemination works without addressing. Although it is not exactly the same issue, it is often used synonymously with packet-centric and information-centric forwarding [1, 68]. In the following, we introduce both the packet-centric forwarding (as an example of explicit) mechanisms and the information-centric (as an example of implicit) mechanisms.

Driving-related applications in the manoeuvering and stabilising domain require precise knowledge about the positions and other variables of the surrounding vehicles. Driving-related applications in the navigation domain require data from other vehicles along the driving corridor. Since usually a significant number of vehicles are interested in this same information, this information is locally spread among the AutoNet vehicles in a broadcast fashion. For all kinds of driving-related applications, this is typically via DEN messages, as introduced in Chapter 5. For manoeuvering and stabilising applications, it is often sufficient if each vehicle broadcasts its sensor data without any retransmissions by others. The CAM messages as introduced work this way. In some cases, it may be beneficial to rebroadcast CAMs, too. The benefit-based approach as an information-centric scheme presented in Section 7.2.4 would automatically adapt to the network conditions and decide on this. The information required by applications in the navigation domain is usually further away than that covered by a single broadcast within the AutoNet ad-hoc network domain. In this case, messages need to be forwarded or disseminated.

7.2.2.3 Explicit Message Dissemination: Geocasting

As an explicit dissemination method, *geocasting* defines all nodes within a geographical region as the addressees of a message. Usually, a geocast is a best effort method, that is the message is sent to and forwarded to the destination area. However, it is neither kept within the area for a certain time, nor is the delivery of the message guaranteed. If the message shall reach all vehicles that can be found within this region for a certain time-frame, this is usually referred to as *stored geocast*. Mechanisms for stored geocast can be found in, for example, [39].

The simplest method to achieve a geocast is to let every node receiving the geocast packet retransmit it. This, however, results in flooding the AutoNet ad-hoc network, which is inefficient for dense network scenarios and can quickly lead to what is known as the *broadcast storm problem* [2]. In the flooding approach, many retransmissions are redundant. Thus, they consume wireless bandwidth resources without providing any significant benefit. In the following, we present some more intelligent geocast methods and assess their efficiency with the following criteria:

- Message delivery rate. (Does every node receive the message?)
- Message latency. (How long does it take until a node receives the message?)
- Number of redundant messages. (How many messages are sent above the necessary?)

For any given network topology, there is a minimum number of relaying nodes, which forward a message towards its destination. In order to reduce the number of retransmissions, geocast algorithms use different mechanisms to select nodes that relay a message. Another possibility is to adapt the network topology. Positions of nodes may not be changed for networking reasons. But the networking topology can be influenced by adapting the transmission power to send a message [71].

To choose the best relay nodes, geocast algorithms usually make use of node position knowledge. For these algorithms, relay nodes with specific position properties need to be chosen out of all nodes that receive a data packet. The known algorithms mostly follow a greedy approach. With the absence of global knowledge, they make a locally optimal choice as a heuristic to come close to the global optimum. For instance, they choose one or more of the nodes being the farthest away from the previous relay node. With respect to retransmitting, these nodes will usually cover the largest additional area reaching the biggest number of new nodes.

In order to choose nodes as relay nodes based on position properties, many methods store and maintain network neighbourhood information in each node. This information has to be obtained by all nodes, which is realised by sending beacons, including their positions, usually in a cyclic fashion (e.g. once per second). It can also be obtained on-demand, that is a short request message is sent to which all nodes in the neighbourhood answer with a message containing their positions. Other methods work in a stateless fashion and abstain from using beacons. We will come back to both issues below.

Position-based geocasting algorithms are often an extension of position-based unicasting methods that forward messages or packets in a multi-hop fashion to a particular destination. These methods generally assume that the location of the destination node is known to the sender who will provide this information in the network header of the packet. In the last few years, there have been innumerable publications of geocast improvements, in particular with special focus on AutoNet ad-hoc networks. The following protocols illustrate particular methods of the respective algorithms. Our extraction covers the main principles (which have been reinterpreted in various ways).

GPSR *greedy perimeter stateless routing* simply chooses the node which is closest to the destination based on a Euclidean distance calculation [8].

MFR The MFR algorithm *most forward progress within radius* [5] projects the positions of neighbour nodes onto the virtual straight line drawn between the sender or forwarder and the destination and chooses the farthest node on that line.

TRADE The *track detection* protocol (TRADE) protocol [7] extends these approaches to geocasting in AutoNet ad-hoc networks. It uses vector comparisons to distinguish between neighbouring vehicles on the same road ahead, on the same road behind, and on different roads. For the forwarding strategy, TRADE selects the farthest vehicles on the same road ahead and on the same road behind and all vehicles on different roads. In this fashion, TRADE spreads messages in an efficient way along the road network. In contrast to position-based unicast mechanisms where the destination is defined as a point, the destination for geocasting is defined as an area, such as a polygon or an ellipse.

REAR The forwarding scheme REAR (*reliable and efficient alarm message routing*) [23] uses a method to estimate the correct reception probabilities of all neighbouring nodes. The node with the highest number of neighbours that are believed to correctly receive the message will then rebroadcast the message. This scheme is based on a beaconing mechanism.

GRUV The approach of *geocast routing in urban vehicular ad-hoc networks* enhances the geocast performance in the presence of intersections [72]. Because of line of sight obstructions, radio propagation often only purely covers side rows. As a result, information often is not disseminated to vehicles travelling on such side roads, because the vehicle selected as the next forwarder is unfavourably located with respect to radio propagation at the intersection. To overcome this phenomena, GRUV distinguishes vehicles according their locations in crossroads nodes and in-road nodes, respectively. In the next-hop decision, cross-road nodes are privileged. As a variant – and if present – a vehicle that is located right in the centre of an intersection is the most suitable candidate to forward a message. As a draw back, such an approach requires a digital maps, or some intelligent heuristics to make guesses, for instance based on the headings or trajectories of various vehicles nearby.

UMB Approaches similar to the *urban multi-hop broadcast protocol* [14] address the hidden node problem in wireless ad-hoc networks. Such approaches introduce a request-to-broadcast/clear-to-broadcast mechanism similar to the RTS/CTS scheme used in, for example, IEEE 802.11. However, this mechanism introduces considerable overhead in dense and high packet load scenarios.

Various cluster-based approaches Few publications propose cluster-based solutions to AutoNet ad-hoc network communication [17, 18, 19]. These proposals are not discussed further because the overhead to create and maintain clusters in AutoNet ad-hoc networks is generally too high due to the high mobility of the nodes. Instead of explicit clustering schemes, methods that make use of knowledge about the street network as a different way to structure the network seem to be more efficient and promising for AutoNet ad-hoc networks.

Further alternative solutions realise geocasting in a stateless fashion. Such protocols make use of one particular method: delaying retransmission of packets for a certain time, which is calculated based on position information. When the time elapses, the packet is either sent or suppressed based on information obtained by other packets overheard in the meantime, as proposed by [11]. This approach is sometimes also called *contention-based forwarding* [16]. Contention-based forwarding was intentionally designed to deliver a data packet in a unicast fashion into from anywhere outside into a dedicated target area. Once the message has reach that area, the area is commonly flooded. The overall principle is still valid for many refinements addressing geocast message dissemination. As before, we want to give an illustrative overview of the main principles:

DDT The *distance defer transmission protocol* makes use of waiting times that are inversely proportional to the distance from the source and thus assure that the farthest node is sending first. In addition to the farthest node, other nodes will also rebroadcast the message after their counter has expired, if they reach a minimum additional area size.

ODAM Slightly varying DTT, the ODAM protocol (*optimized dissemination of alarm messages*) lets all nodes rebroadcast that have not received the message in order to increase the message delivery rate [9].

SBA The SBA approach (*smart broadcast algorithm protocol*) makes use of the street layout and divides the street network into segments [10]. The waiting time procedure is the same again. A node rebroadcasts if it has not received the message from another node behind on its street segment.

DWOP The *distributed wireless ordering protocol* [41] uses priorities which are reversely proportional to the packet waiting times. Priorities are calculated for the next head-of-line packet. The priorities for packets of all nodes within one transmission range are exchanged to approximate a 'global' scheduler. The priority exchange is realised by piggybacking the information into the MAC frame.

TRRS One of the challenges of stateless methods is how to calculate appropriate waiting times. Good waiting time function selection would need to be configured according to node density in order to avoid collisions in high density scenarios and in order to avoid unnecessarily high delays in in sparse network scenarios. Such adaptive timing is addressed by the *time reservation-based relay node selecting algorithm* (TRRS) [13].

7.2.2.4   Implicit Message Dissemination

Implicit message dissemination does not rely on specific (geographical) addressing. Instead, each vehicle within the communication range of a transmitting vehicle decides autonomously whether or not to retransmit the received message. For this reason, implicit message dissemination is sometimes also referred to as *receiver driven dissemination* (explicit message dissemination is then called *sender driven dissemination*, accordingly). In the following we present some examples of respective algorithms proposed in the last few years.

NFP As an example of an implicit dissemination mechanism, NFP (nearest with forward progress) adapts the network topology for unicasting. The sender calculates the minimal distance to its surrounding nodes. It then reduces the transmission power to reach only that node [6]. With such an approach, redundant rebroadcasts from different receiving vehicles are minimised, but not completely avoided. This is because of the varying signal propagation in the real world, where each specific constellation determines the current reception characteristics. Refinements of this approach aim to adapt the transmission range in a way that the signal reaches all vehicles that are located at different directions from the sender. This way, the message can be spread in all directions. On the other hand, this again increases the number of redundant retransmissions. Adapting the transmission power in such s sense has different effects on the AutoNet network. On the one hand, it needs more hops to distribute a message within a specific area, which leads to an increased latency. On the other hand, a reduced transmission power minimises the hidden node problem. In addition, the overall channel busy times are reduced, which leads to better utilisation of the available overall capacity (for channel access strategies please see Chapter 8).

DPC Another implicit disseminiation approach is the so-called *distributed power control for reliable broadcast*, where each node rebroadcasts a message until it has overheard a specific number of rebroadcasts that is adapted to the node density [24]. In addition, the transmission power is chosen based on a calculation that uses average path loss information per neighbour obtained by beaconing.

D-PAV *Distributed fair transmit power adjustment for vehicular Networks* is again a power adjustment method aiming at fair channel utilisation [25].

I-BIA In order to improve reliability, some protocols include acknowledgement mechanisms. *Intelligent Broadcast with Implicit Acknowledgement* (I-BIA) basically turns the idea of stateless forwarding around. Again vehicles wait to forward a message depending on their distance, but this time they wait longer the further they are away from the sender [28].

As another example of an implicit dissemination mechanism, dissemination can be controlled by the utility (also referred to as *benefit*) brought by the information carried in messages to their recipients. The generic concept of information utility serves as the basis for adaptive message dissemination schemes when applied to autonomous communication. In contrast to the routing schemes presented above – which are usually packet-centric – using the benefit assessment leads to an information-centric network communication [20–22]. This potential benefit is calculated by the sender or forwarder, respectively, by summing the potential utility values of all adjacent vehicles. By doing so, the global utility of the network is increased. The principle of shaping network traffic in order to optimise the network participants' benefit is a well known field in computer networks and is usually called *network utility maximisation* (NUM) We will introduce the model and mechanisms for such a utility maximisation in AutoNet networks in more detail in a dedicated section below (cf. Section 7.2.4).

While the approaches presented so far all address the dissemination of messages once, information is often valid for vehicles within a certain area for a longer period of time. After a message has been disseminated, vehicles still enter and leave such an area of relevance. In order for these vehicles to also receive this information, mechanisms have been introduced to keep information within an area for a certain period of time. The stored geocast [39] and the cached greedy geocast [4] are methods to keep a message alive within a certain time and within a geographical area. This way, the 'addressing' used for geocasting can be specified in the following two ways:

- Definition of a geographical area and a time frame, in which the message is of relevance.
- Geographical reference together with a time stamp. This information is used to calculate the relevance of the respective message for the vehicle.

7.2.3   Beaconing

In ad-hoc networks, certain data is useful for network management and routing decisions and is used by many proposed network protocols. For AutoNet ad-hoc networks, the most important data elements are node IDs and time stamped positions. Therefore, the network layer comprises its own network managament protocol. The main task of this management protocol is to exchange this data according to the needs of the network protocols. To do so, usually a network heartbeat packet is sent at a fixed frequency, called a network layer beacon.

At first glance, for all nodes that are sending cooperative awareness messages (CAMs), these can be used as substitutes for network layer beacons, because they contain the same information and are sent at least at the same rate. Since CAM-based cooperative safety applications are one of the main drivers for AutoNets, basically all vehicles will perodically transmit such safety messages. If a node does not send CAMs (e.g. a roadside station), it will periodically send network layer beacons. The fixed payload of these beacons depends on the role of the station and contains other node information. Variable payload is foreseen to be inserted if requested by applications and if authorised by congestion control.

There is still an intense discussion going on at standardisation bodies over whether or not it is sensible to combine network layer beacons with driving related safety messages (CAMs for EU and BSM for the US, respectively). The main arguments for combining both beaconing approaches are:

- AutoNets are driven by safety applications, in particular by collision avoidance based on periodic safety messages.
- Such safety messages comprise the necessary information for network management.
- Safety messaging requires higher update frequencies.
- The wireless channel capacity is the most limiting factor for future developments, because once a standard is finished and deployment has started, the radio systems have to be stable. Thus, transmitting redundant data should be avoided.
- A separated management of neighbouring vehicles' locations for safety applications and network also increases resource consumption in terms of local storage, which is still rather limited in embedded systems like vehicles.

In contrast, the main arguments to keep the network layer beacon a dedicated business are the following:

- As is common for computer networks, the network layer should be able to perform its designated task without relying on top-layer information.
- Fusing both beaconing approaches complicates system architecture and breaks with common ISO/OSI communication layering.

- Accuracy requirements for location information and timing are different.
- Safety messages will most likely be sent on a dedicated channel (the control channel). AutoNet networking capabilities should also be available in multi-channel operations, where there is no need to tune to a dedicated channel.
- Network beacons and safety beacons may underlie different congestion control mechanisms.
- The AutoNet network can be organised better the more neighbouring vehicles are known. For this reason it is necessary to either transmit a network beacon with a rather high transmit power, or to forward the respective beacon messages in a limited area. Since the network beacon frequency is comparably low, this will not lead to a significant network overhead. In contrast, safety messages might be transmitted with high frequency in certain situations, but require less communication range. Therefore transmission power could be decreased, or the transmission could be shifted to other channels (for multi-channel beaconing approaches see for instance [73]). In addition, security requirements are different. Hence, beaconing goals are competing.

In short, not combining safety and network beaconing will keep the system simpler, in line with common network architects and will better suit the competing requirements – for the only sake of only a few bytes more to communicate.

7.2.4   Network Utility Maximisation in AutoNets

The goal of communicating messages in AutoNets is to provide all kinds of applications with the information they need. The challenge is that in most cases the nodes where this information is available do not know about the other nodes where this information is needed, or would be beneficial for that matter. With limited network resources, it is not possible to disseminate all information to all network participants under all circumstances. As we have seen in Chapter 4, the quality of foresighted safety applications increase along with the amount of information available. As a consequence, from an application point of view it is always sensible to transmit as much valuable information as possible. As a result, all information dissemination schemes – packet-centric as well as information-centric – aim at the optimum usage of the limited resources in order to provide the applications with the required information.

Such maximisation of the so called global, aggregate utility is usually referred to as *network utility maximisation*, or in short *NUM*. It has become an important principle underlying several rate allocation schemes and congestion control protocols for diverse communication networks. In this section, we consider decisions on message dissemination, that is whether or not a message should be repeated or forwarded by a node. Instead of pre-determined, static propagation limits with respect to either time or space, the benefit-based scheme leverages an autonomic, situation-adaptive content evaluation and dissemination strategy. To do so, the nodes apply a mechanism to evaluate the benefit a message will provide to the set of its potential recipients when transmitted. In order to do so, we want to give a brief introduction to the matter of NUM in common networks.

7.2.4.1   Network Utility Maximisation in Common Networks

The basic approach to realise utility-based data rate control and thereby maximise the aggregate utility of all network nodes works as following (for a more detailed elaboration see for instance [71, 74, 75]):

Consider a system of J links (also referred to as resources) and a set of I users, where C_j is the restricted capacity of link $j \in J$. Each user i utilises a fixed set of J_i links in its path, which is a nonempty subset of J. The authors then define a matrix A consisting of ones and zeros, where $A_{i,j} = 1$ in case link j is part of user i's actual route J_i and $A_{i,j} = 0$ otherwise. The data throughput of user i is denoted as x_i, such that the user receives utility $U_i(x_i)$. $U_i(x_i)$ is an utility function that has to be defined according to the mission-specific goals of the respective application. For example, the mere data rate may work as an adequate means for measuring the utility delivered by a file transfer application. In their mathematical model, increasing, strictly concave and continuously differentiable functions of $x - i$, where $x_i \geq 0$ are assumed [76]. For elastic traffic, concave functions may adequately reflect the utility provided to end users. The higher the data rate, the more useful is the application for the user. In the case of so called inelastic traffic with minimal performance requirements, however, convex or sigmoid functions are more suitable.

Getting back to the utility-based rate control model: utilities are assumed to be additive such that the aggregate utility of rate allocation $x = (x_i, i \in I)$ is $\sum_{i \in I} U_i(x_i)$. The resulting optimisation problem can then be described as follows:

Maximise

$$\sum_{i \in I} U_i(x_i)$$

with

$$A_x^T \leq C$$

where

$$x \geq 0$$

Thereby C is defined as $C = (C_j, j \in J)$. The first constraint says that the total data rate of a link may not be larger than its capacity. Assuming that the specific utility function is known, this problem can be solved mathematically. However, in practical scenarios, a central system coordinator knowing all the individual utility functions is unlikely to exist. For that reason, a fully decentral, utility-oriented rate allocation based on a pricing scheme is proposed (see [77]). Here, the users are to solve the problem of maximising their individual utility, which is the utility received by applications less the amount they (virtually) pay. The equation above is also solved by so so called *rate-based algorithm* [76]. The underlying methodology relies on a so called shadow price charged by links depending on the total traffic load going through them. P_i thereby denotes a user's willingness to pay per unit time, where $x_i(t)$ is the current data rate at time t. Each link $j \in J$ is supposed to charge a price per unit data flow μ_j which is calculated according to the following equation:

$$\mu_j(t) = b_j \left[\sum_{i:j \in J_i} x_i(t) \right] \tag{7.1}$$

$$\frac{d}{dt} x_i(t) = \kappa \left[p_i - x_i(t) \sum_{j \in J_i} \mu_i(t) \right] \tag{7.2}$$

Therefore b_j increases with the total data rate going through it. b_j is the function relating the provided data rate to the price the node has to pay. κ represents a system constant.

The latter differential equation can be motivated as follows: the links (resources) in the network are assumed to continuously give feedback to the users and thereby notify them about the actual price per unit data flow $\mu_j(t)$. If the product of a user's current data rate $x_i(t)$ and the price per unit data flow is lower than the price the user is willing to pay $p_i(t)$, the user may increase its data rate. It can be proved that each user reaches an equilibrium state, where the price he is willing to pay equals its aggregate cost.

This approach assumes the network to be able to provide necessary feedback to the users, and users then adjust their rates based on the feedback information to maximise the global, aggregate utility. An extension of this utility optimisation scheme completely passes on any kind of feedback from the network [74]. Assuming that the data rate through one of a network's links is smaller than its capacity, there is no contention since each user receives its desired rate. As soon as the rate approaches the capacity, however, the increase in the data rate of any of the users leads to a backlog at the link (packet queues emerge). The resulting increased queuing delay at that link can then be regarded as an increase of the implicit cost the users pay due to the larger delay. The network has to account for this change by recovering the increased system cost through a new pricing scheme: q_j, $j \in J$ is denoted as the current backlog at link j. As soon as the link is congested (the total rate through it equals its capacity), the link has to charge the users a price, where the price per unit flow per unit time g_j is determined as the queuing delay at the link: $g_j = q_j/C_j$. J is governed by the following equation. The total price per unit flow per unit time is equal to

$$\sum_{j \in J_i} .$$

The total price per unit a user i pays is determined by

$$x_i \sum_{j \in J_i} g_j$$

The network utility received by a user i is then defined by

$$U_i(x_i) - x_i \sum_{j \in J_i} \frac{q_j}{C_j}$$

To make explicit feedback from the network superfluous, the actual size of the users' packet queues is the basis for the computation of link-backlogs q_i.

7.2.4.2   Network Utility Maximisation in AutoNet Ad-Hoc Networks

Because of the long product life-cycles of vehicles it is of particular importance to use a network that is as optimal as possible. However, there are significant differences when transferring the NUM problem to vehicular ad-hoc networks:

- No fixed links: connectivity in AutoNet ad-hoc networks is highly dynamic. Connections may only last for a few seconds, and there is no per-link capacity. Instead, all nodes within mutual radio range have to share the same medium and have to share that capacity.

- No rate-based utility calculation: since connectivity is only short-lived, occasions to emit a whole flow of data are rare. The calculation of utility must not be based on the rate of transmission but rather on the information that is transmitted.
- Utility functions are manifold: nevertheless, in general the types of utility functions are not known beforehand. Utility calculation must be possible using arbitrary functions. In addition, utility functions may change and are not static. This is in particular true for applications that depend strongly on context information, such as active safety applications.

For most AutoNet applications, utility provided to users cannot be quantified with the help of simple convex or concave functions, which only depend on measures such as packet delay, throughput and jitter. For auto-networked vehicles, the information contained in data packets must be evaluated against the background of the vehicles' respective contexts, as will be explained later.

Apart from that, NUM schemes mostly head for a globally fair allocation of resources to the different network participants. Because of its dedicated purpose to (also) increase safety, fairness in AutoNets must be defined significantly differently when compared to the common understanding of fair usage of available network resources (see also [71]). The common sense of fairness usually guarantees all network participants a certain amount of bandwidth (either the same for all nodes, or staged based on priority considerations), that is no node is sustainably precluded from accessing the communication channel. However, as one major consequence of the safety support of AutoNets, AutoNets are inherent altruistic, that is a single node does not have to gain any advantage when offering an information artefact to others. Each node may only gain utility from participating in the network if all participants are willing to cooperate in an altruistic way. Especially if the network operates at its limit, it must be ensured that:

- Messages comprising safety critical information (stability subtask) can access the channel with very low latency, if necessary (i.e. the message carries unknown and critical information and the intended receivers are close to the critical incident).
- Messages supporting the drivers' task (guidance and navigation subtasks) can (but need not) be prioritised over unspecified messages (deployment applications), if the carried information has significant utility to others.
- If the channel capacity is insufficient to transmit all desired messages, those messages that explicitly provide the least utility may starve. Because of the network objective, this will typically concern (but is not limited to) messages of deployment applications.

Therefore in certain critical situations certain nodes should be temporarily suppressible for the sake of others if they cannot contribute to the altruistic network objective. As a consequence, NUM in AutoNets also requires a modified resource allocation, that is stations are not granted access to the medium on an equal basis, but depending on the utility of their available information. This principle is termed *joint fairness* in order to emphasize the difference to the common sense of fairness [71].

As another difference to common networks, most AutoNet applications are broadcast-based and do not feature stable unicast links. As a consequence, the constraint of the optimisation problem as argued before is not valid in AutoNet environments. Since individual data flows, which can be assigned specific data rates, do not exist in this case, the utility or utility functions leveraged cannot be a function of data rate, jitter or delay, but of a single data packet and the utility provided to its potential recipients.

Thus, in the context of AutoNet ad-hoc networking, the constraint to the maximisation problem is the per node available channel capacity (for a detailed elaboration on the capacity of wireless ad-hoc networks, please see [78, 79, 71]) The dominant factors influencing the network's capacity are network size, traffic patterns applied and detailed local radio interactions. The total one-hop capacity of an ad-hoc network is governed by the amount of spatial reuse. Assuming a constant radio range, this spatial reuse is proportional to the network's physical area. Further assuming that the node density δ is uniformly distributed, the physical area of the network (called A) is related to the total number of nodes by $A = n/\delta$. Therefore, the total one-hop capacity (called C) is also proportional to the area with $C = kA = kn/\delta$ for some constant k.

If now each node transmits packets at a rate λ, and the expected physical length of a communication path from a source to a destination vehicle is L (that is the number of hops), the minimum number of hops required to deliver data packets is L/r, where r is the fixed radio transmission range. Thus, the total one-hop capacity in the network required to send and forward data packets can be described by:

$$C \succ n\lambda \frac{L}{r}$$

Combining this with $C = k\frac{n}{\delta}$, the capacity available at each node, λ is

$$\lambda \prec \frac{kr}{\delta}\frac{1}{L} = \frac{C/n}{L/r}$$

Due to the decentrally organised contention for the shared, wireless medium, nodes waste a lot of time while backing off, deferring their medium access and resolving channel collisions (cf. Chapter 8). In the case of unicast communication data flows, where one node communicates with exactly one other node over one or several hops, the per-node capacity is decreased even further due to the requirement that nodes forward each others' packets. For a unicast-based ad-hoc network, the expected path length is governed by

$$L = \frac{2\sqrt{A}}{3}$$

In order to optimise the overall utility of the vehicular ad-hoc network, two fundamental methodologies form the basis of our approach: first, the utility data packets provided to potential recipients in the local neighbourhood must be quantified. Second, the message transmission must be prioritised according to the resulting utility values to maximise the utility received by all the vehicles participating in the network.

To describe the basic characteristics of optimising the overall network utility in AutoNet ad-hoc networks, the following idealised and simplified scenario can be studied. Suppose there is a set NB (e.g. $|NB| = 6$) of wireless stations within mutual communication range. At one specific point of time t, each of them has at least one packet ready for broadcast. Since all network nodes have to share the wireless medium, only one of them is able to transmit at the same time. Each of the nodes $n_i \in NB$ has a dedicated interest in certain information (in the form of packets $p_{n_i,j}$) that can be quantified with the help of a utility function $ur_i(p,t)$ that varies with time. To maximise the utility provided to the whole network, a global scheduler that has available detailed knowledge about all the nodes' packets and their respective utility functions would conduct a two-step process. First, a node-internal packet utility ranking is established: for each of the data packets $p_{n_i,j} \in P_{n_i}$ that are queued at a specific node n_i, the utility provided to all the adjacent nodes is thereby computed. This normalised network utility ($\overline{U}$) is computed as the sum of all the $|NB|$ nodes' individual utility values provided by packet $p_{n_i,j}$. The utility for the transmitting node n_i is obviously zero, since the information is known. Therefore,

$$\overline{U}(n_i, p_j, t) = \frac{1}{1 - |NB|} \left(\sum_{k=1}^{|NB|} ur_k(p_{n_i,j}, t) \right)$$

The packet $p_{n_i}^{max}(t)$ of node n_i providing the maximum utility to all of the reachable neighbours is considered to be the packet the node should broadcast as the next one. Thereby:

$$p_{n_i}^{max}(t) = \left\{ p \in P_{n_i} \mid \overline{U}(n_i, p_j, t) = \overset{max}{\underset{p \in P_{n_i}}{}} [\overline{U}(n_i, p_j, t)] \right\}$$

However, the internal packet ranking is not sufficient to maximise the global, aggregate utility provided to the whole network. In fact, the six packets which have won the node-internal ranking process probably provide different utility to the rest of the network. As a second step, an external ranking is to be established that ensures that the node n_k whose highest ranked packet $p_{n_k}^{max}$ provides the highest utility to the network (maximum normalised network utility $\widehat{U}$) is finally granted access to the shared, wireless medium, transmitting message p_{NB} with

$$p_{NB}(t) = \left\{ p \in \bigcup_{n_k \in NB} p_{n_k}^{max}(t) \mid ur(p) = \overset{max}{\underset{n_i \in NB}{}} \left(ur \left(p_{n_i}^{max}(t) \right) \right) \right\}$$

The resulting network utility $\widehat{U}(n_i, p_j, t)$ when transmitting message p_{NB} is accordingly

$$\widehat{U}(n_i, p_j, t) = \overset{max}{\underset{n_i \in NB}{}} \left\{ \left[\frac{1}{1 - |NB|} \left(\sum_{k=1}^{|NB|} ur_k(p_{n_i,j}, t) \right) \right] \right\}$$

For the application in a vehicular ad-hoc environment, however, a global, central packet scheduler that is able to see the network from a bird's eye view does not exist. AutoNet ad-hoc networks are operated decentrally. Thus, packet scheduling must be realised without the help of any central entity. In fact, each node will have to approximate the utility each of its packets provides to its adjacent nodes and thus establish a node-internal packet broadcast sequence. The necessarily limited knowledge of the nodes also prevents the realisation of a perfect schedule. None of the network nodes exactly knows about its neighbours' individual interest in information as represented by the respective ur-functions.

Following the above principle, the utility provided by a certain message must be computed prior to message transmission. To be accurate, there exists a difference between the actual receiver utility ur and the utility that is estimated by the transmitter, referred to as ut with $ut \in [0; 1]$. ut represents the normalised utility of a message to be sent for other network participants, which the sender estimates. Thereby obviously the sender only has limited knowledge to do so.

In this context, the sum of all individual utility accounts of all vehicles at a certain point in time is called the *global benefit*. It represents the utility provided to all auto-networked vehicles within a scenario and is the measure to be maximised. By integrating over all the receiver utility values ur_{n_i} computed by vehicle n_i, one ends up with the collective benefit provided to the applications of vehicle n_i. The global benefit $GB(t)$ is then the sum of all the respective values for all $|N|$ AutoNet vehicles at time t. Therefore

$$GB(t) = \sum_{i=1}^{|N|} \int_0^t ur_{n_i}(p, t)dt.$$

Another important assumption underlying the differentiating traffic according to its respective utility for other nodes is that adjacent nodes evaluate certain messages in a similar way, that is the benefit values estimated by a sender and provided to a receiver of the same message must be similar. This rationalises that the expected utility can be properly estimated prior to message transmission.

Two major assumptions must be fulfilled to allow for a sound utility approximation: first, functions for determining ut and ur must be similar in all the different vehicles produced by different car manufacturers. If vehicles estimate information utility completely differently, the intrinsic paradigm of altruism is obviously violated. Second, we can assume that the majority of vehicles in a neighbourhood (within mutual communication range) are within a similar context too. Interest in certain information can thus be regarded as similar, too. Provided that these postulates hold, a different prioritisation of data according to ut helps meet the requirements of a scalable and efficient communication protocol for broadcast-based vehicular ad-hoc networks.

Straightforward dissemination limitation strategies such as geocasting can also be interpreted as heuristic approaches to optimise the benefit for the participating nodes as just described. However, they neither account for the individual nodes' benefit gain nor are they able to optimally leverage network resources in both sparse and dense networks. NUM can be applied to any decision that is necessary within the AutoNet communication system based on a cost-benefit evaluation of this decision. In AutoNet ad-hoc networks, the network can basically be used for free up to its capacity limits. Thus, the cost borne by a transmission can be interpreted as the opportunity cost for not being able to transmit another message or for the necessary delay of this other message.

7.2.4.3 Quantification of Traffic Information Utility

Benefit provided by traffic information can be quantified with the help of measurements such as reduced number of accidents, reduced environmental pollution or shorter travel times (see for instance [42]). But measurements like these can hardly be computed or observed in an objective way before drivers or vehicles make their decisions. A first estimate of the benefit provided by a certain message must be possible exactly at the point in time when the message is scheduled for transmission. With a benefit-oriented message dissemination scheme, both the individually varying demand for information is met and available resources can optimally be leveraged in the network.

Therefore, the specific utility of a specific piece of traffic-related information provided to a receiving vehicle or the driver respectively, depends on how much this information influences further decisions. To illustrate this main principle, just imagine the following few example scenarios [71]:

The 'Well, I don't care anyway' scenario: Assuming vehicle A has encountered a critical road condition, such as an icy spot on a curvy road. The information is disseminated through the AutoNet ad-hoc network. Vehicle B receives this information. Obviously, the utility for vehicle B is zero in the event that B will not pass the critical location. This is particularly true if B has already passed the location, or the location is not on its ongoing route. The utility is also zero if B has just approaching the critical location and there is no time left to intervene appropriately. Similarly, the utility of information about a traffic jam a couple of hundred kilometres away is also close to zero, even if it is on the planned route. This is obviously because the travel time to the reported area is comparatively long and therefore it is very likely that the traffic situation will have changed again before arrival.

The 'Yes, I already know' scenario: Again, assuming vehicle A has encountered a critical road condition, such as an icy spot on a curvy road. The existence of the event has been confirmed by various other vehicles before. Vehicle B receives all of those confirmations. Obviously, even if there is still some additional utility in sharing those confirmations, the overall utility approximates again zero, because vehicle B is already sufficiently informed and confident. What makes a difference indeed in this scenario is the information that a vehicle that has just passed the respective location could not observe the reported event characteristics. The utility of such *revocation messages* is very high, but again decreases over time as further revocation reports are communicated.

Those two short examples make it obvious that even the same piece of information may have different utility for different vehicles, depending on the individual's previous knowledge and the current driving task, the latter in particular comprising the planned route. The utility of an information artefact can therefore be quantified as weighted influence upon the individual local view on the vehicle's driving context, where the weight correlates to the impact on further decisions in the system.

The utility depends on a variety of different environmental parameters, leading to a noticeable computational overhead. However, it can be realised that in most cases the utility depends on similar parameters in a similar way. In particular, the utility typically decreases with increasing distance to the origination of the information and its age. It is therefore reasonable to estimate the expected utility based on a variety of different heuristics taking into account specific sets of context parameters, with individual specific configuration. The configuration thereby deals with the usual dynamics with which a specific context parameter changes over time. A spot of aquaplaning for example usually does not change within minutes, whereas the position of the end of a traffic jam does. It is important to note that beside the estimation of the utility for the immediate receiving nodes, the further dissemination progress must also be considered.

This can be realised by taking the following context information into account for the benefit calculation (for a more detailed list please see reference [71]):

- Message context C_m, which is characterised by age, time since last broadcast, time since last reception of a message and so on.

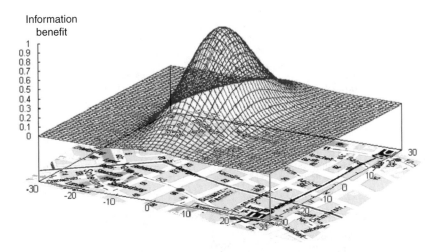

Figure 7.15 Exemplary utility subfunction for computing the *parameter distance to information source* [71].

- Vehicle context C_v, which is described by driving direction, route flexibility, vehicle speed, current road type and so on.
- Information context C_i, which is specified by time of day, distance to the source, information accuracy, information category, change rate of the information and so on.

Figure 7.15 shows for instance a reasonable utility subfunction for computing the parameter *distance to information source* [71].

As argued above, all these parameters now have to be translated into preliminary utility values with the help of so called subfunctions. A model comprising subfunctions for each of these parameters and weights for combining the results of these subfunctions needs to be based on road traffic statistics and other empirical observations. The results are then weighted and summed up to generate one single utility value (ut) per message.

The following formula highlights the way the benefit of a specific piece of information can be quantified. Suppose there are n parameters such as distance to the information source or the age of the message. They all have to be computed with the help of the message context C_m, the vehicle context C_v, the information context C_i, and the already stored messages S_m. The n parameters are evaluated with application-dependent (heuristic) subfunctions, resulting in values b_i. The parameters are then weighted with application-dependent factors a_i, where $\sum_i a_i = 1$. In the context of a crossing collision warning, for example, the message age parameter is more important for the overall message benefit than the parameter evaluating the quality of the information. All weighted parameters are then summed up and divided by n:

$$ut(p,t) = \frac{1}{n} \cdot \sum_{i=1}^{n} a_i \cdot b_i(C_m, C_v, C_i, S_m)$$

Figure 7.16 delineates the general principle of determining one single utility approximation for a certain piece of information.

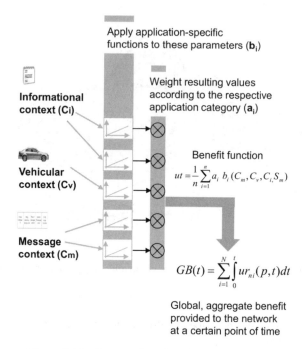

Figure 7.16 Computation of the packet utility value.

All the so called subfunctions b_i and also the weights a_i differ with regard to the respective application. Certain parameters play a major role for determining the utility of messages from one application, where the same parameters are irrelevant in the context of another service. The utility provided by crossing collision warning messages, for example, does not vary with the purpose of travelling of individual vehicles.

Another important insight is that there is not one generally valid and optimal function for determining *ut* or *ur* existing. In fact, car manufacturers will have to agree on a common understanding of utility quantification and thereby set up a comprehensive model featuring all required subfunctions and weights covering all envisioned applications. Car manufacturers will also have to set up similar field test with a significant number of test drivers in order to take into account human factors and find out about the experienced utility (in real world scenarios) of various information artifacts, as for instance breakdown warnings, weather information or traffic jams.

7.2.4.4 Deployment of the Utility-Based Approach

The deployment of benefit-based communication requires respective queueing mechanisms for the packets containing the respective information. Such mechanisms include a benefit-based queueing, a potential reordering of packets within the queue, and even the deletion of packets. Such a feature can be implemented at the network layer very easily. However, this solution may interfere with the scheduling mechanisms implemented at the link-layer for medium access, where a different packet scheduling may be implemented. Moreover, packets leaving the network layer queue cannot be treated further since they are handled autonomously by the link-layer of the respective communication technology. For instance, if the communication technology currently being used is not able to transmit the packet, neither the network layer nor the application layer will be able to remove the packet from the transmission queue in the link-layer in case the benefit decreases. Hence, the packet may be (re-)transmitted afterwards although there is currently no further benefit of the transmission.

In order to realise utility-based communication efficiently, a cross-layer approach is necessary, considering respective mechanisms provided by the link-layer of several communication technologies [64]. This allows one to implement the packet handling mechanisms directly in the link-layer using already available communication protocol mechanisms defined for respective communication technologies. Therefore, the AutoNet ad-hoc networking component of the network layer interacts with the respective link-layer protocols via a cross-layer mechanism provided by the management layer, which is introduced in Section 10. This way, the benefit-based packet handling can be realised very efficiently at the link-layer, whereas it is controllable by the network layer.

Benefit-based communication is considered as an important possibility for message dissemination in AutoNet ad-hoc networks. We will visit this topic again in Section 8. After the introduction of several communication technologies, we will outline potential approaches to realise benefit-based communication using available protocol mechanisms of the respective communication technologies.

7.3 AutoNet Cellular Networking

A second important aspect of the network layer is the support of cellular communication technologies for exchanging AutoNet messages. Figure 7.17 illustrates the basic exchange of AutoNet Messages among vehicles using AutoNet cellular networking. Apparently, AutoNet cellular networking has an increased delay for the exchange of AutoNet messages compared to communication technologies for AutoNet ad-hoc networking. Hence, AutoNet cellular networking does not fit for all applications: it is typically unsuitable for driving-related safety applications since they require very low communication latencies. Depending on the cellular technology being used, it can be an interesting option for driving-related efficiency applications, for vehicle-related applications and for passenger-related applications.

Although the transmission of AutoNet messages via cellular networks sounds somewhat redundant, it has three crucial benefits for the success of AutoNets:

- *Short time-to-market of AutoNets.* AutoNets can be rolled out efficiently since nowadays vehicles are mostly equipped with cellular communication technology – either built-in or via respective cellular phones of the drivers. Hence, the required hardware platform is already available and respective applications can be developed and introduced much faster. This is of particular importance for the market introduction of AutoNets since there will be more communicating vehicles in this phase. Hence, applications requiring a particular penetration rate of communicating vehicles can be used (and rolled out) early.
- *Higher penetration rates.* In general, it is expected that AutoNet ad-hoc networking will be introduced for vehicles of the lower price segment rather later due to the additional cost of this technology. This way, AutoNet cellular networking is an interesting alternative to

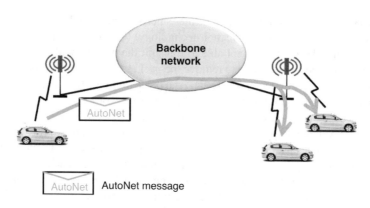

Figure 7.17   Scenario of AutoNet cellular networking.

serve this market with AutoNet applications. This also results in a higher penetration rate
and, thus, enables the market introduction of AutoNet applications demanding a rather high
penetration rate for their proper functioning.

- *Consideration of low-density locations.* AutoNet ad-hoc networking inherently requires
  a sufficient number of vehicles to transfer AutoNet messages to a destination far away.
  Moreover, rural areas are likely have a very low density of vehicles and thus the information
  could not be communicated in such areas. This is somewhat alleviated by roadside stations
  (i.e. by forwarding the AutoNet message via roadside stations very close to the destination),
  but especially in the beginning of the market introduction, there will be a very low coverage
  of roadside stations. In such scenarios, AutoNet cellular networking couls be used to transfer
  relevant AutoNet messages directly to the few vehicles in the respective destination, which
  would not be possible with pure AutoNet ad-hoc networking.

An important requirement for the exchange of AutoNet messages is for them to be transparent for the applications being deployed; from an applications' perspective there is in general no
differentiation of an AutoNet message whether it is received by AutoNet ad-hoc networking
component or by an AutoNet cellular networking component.[2] Consequently, both components have to process the same AutoNet message formats and the same security mechanisms
have to be applied. The transparency requirement also includes the fact that AutoNet cellular
networking has to provide similar communication mechanisms, such as geocasting. In Section 7.3.2, we will briefly describe potential mechanisms by which such geocasting could be
realised in actual cellular networks.

Besides transparency, AutoNet cellular networking has to be completely independent from
a particular network operator or network provider. Moreover, it has to be independent of
the communication technology being used. This independence also includes the fact that
it should be possible to use different communication technologies for uplink and downlink
communication.

---

[2] However, some AutoNet applications need the possibility of differentiating among AutoNet messages received
by AutoNet ad-hoc networking and AutoNet cellular networking. This is an important issue since both messages may
have a different but crucial quality of service, which may result in a different application behaviour.

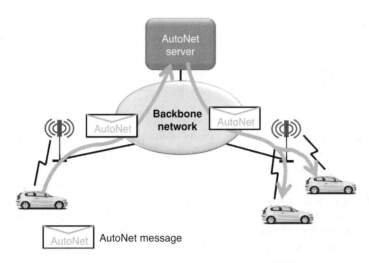

Figure 7.18   Communication architecture for AutoNet cellular networking.

7.3.1   *Communication Architecture for AutoNet Cellular Networking*

In order to address the requirements mentioned in the previous sections, the communication architecture for AutoNet cellular networking requires an additional component in the generic domain, which is called the *AutoNet server*. The AutoNet server is typically a server located 'on top' of the cellular communication technologies infrastructure. The AutoNet server is not necessarily fixed to a particular network operator or communication technology. The AutoNet server acts as a geocast entity: based on the information in the AutoNet message header, the AutoNet server can determine the target area for which the message is of relevance. If a vehicle wants to send an AutoNet message via AutoNet cellular networking, it transmits the AutoNet message to the AutoNet server using the IP over the respective communication technology (uplink), as illustrated in Figure 7.18. The AutoNet server receives this message, determines the target area for the AutoNet message, and forwards the message to the relevant vehicles in the target area using the respective communication technology. Since this mechanism is transparent, the AutoNet message will be treated within the vehicle in the same way as if it was sent by AutoNet ad-hoc networking. Hence, the received messages will be inserted into the local dynamic map (LDM) facility for further processing.

The AutoNet server is not limited to communication with vehicles. The AutoNet server can also be used by other Internet hosts – from the generic domain as well as from the AutoNet (central) infrastructure domain. This enables, for example, traffic management centres to inject respective AutoNet messages in the AutoNet system. Vice versa, this link can be also used to provide respective hosts with the AutoNet messages sent by vehicles. The messages can be used, for example, by traffic management centres for a more precise calculation of the current traffic situation on the roads. This feature also shows the particular importance of AutoNet cellular networking for the market introduction of AutoNets: although the roadside station coverage – necessary for transmitting AutoNet messages of vehicles to traffic management centres – will be sparse in the beginning, traffic managements centres will benefit in this time by receiving the AutoNet messages from the AutoNet server.

Besides the geocast feature, ongoing projects for the development of an AutoNet server such as CoCar/CoCarX or simTD (please consult the Appendix for more information on research projects) introduce additional functionalities for the server: a *reflector* functionality and an *aggregation* functionality. Both functionalities may become relevant in the future. However, we do not consider them for AutoNet cellular networking for the following reasons:

- *Reflector.* As the name suggest, the reflector 'reflects' AutoNet messages received by the AutoNet server to the vehicles in the vicinity of the sending vehicle. In the communication architecture mentioned above, this feature is provided by the geocast feature, which basically sends the message to the relevant vehicles.
- *Aggregation.* This feature aggregates messages received by the AutoNet server. For example, several identical (or similar) AutoNet messages will be aggregated to one AutoNet message, which will be sent to the relevant vehicles. We do not consider this feature in our communication architecture for two reasons. First, the processing is costly and based on the pure data only without considering the requirements of the applications. Second, several applications may also rely on the number of messages. As we saw in Section 2, the number of similar messages from different vehicles may be an important input parameter for estimating the reliability of the information. The aggregation may result in a different estimation and, thus, a different behaviour of the respective application.

7.3.2   Deployment Strategies

An important issue of AutoNet cellular networking is the geocast feature, which has to ensure that an AutoNet message sent by a vehicle is forwarded to the vehicles in the target area. This forwarding can be realised in different ways, depending on the scenario as well as the communication technology being used. Hence, this aspect has to be considered in a deployment scenario in order to ensure that the AutoNet messages are forwarded correctly. Depending on the technology being used, AutoNet cellular networking may require (slightly) modified network headers or even additional protocol extensions in order to realise the respective geocast functionality.

One possibility for performing the geocast functionality is through emulating the geocast with unicast connections to respective vehicles. This approach is an interesting option for IP-based heterogeneous communication networks, including the variety of cellular communication technologies, such as UMTS, LTE or even IEEE 802.11 a/b/g. This approach is used, for example, in the simTD project. In order to map the IP addresses of the vehicles onto a geographic area, the AutoNet server maintains a mapping of the registered IP addresses onto their current position. This mapping is updated periodically by the messaging support facility in the communication plane of the application support layer (cf. Section 5) of the registered vehicles, which periodically sends the current position of the vehicle together with the IP address being used.[3] The distribution of AutoNet messages in this approach is straightforward, as illustrated in Figure 7.19. If a vehicle sends an AutoNet message via AutoNet cellular networking, the AutoNet message will be forwarded by the AutoNet server. There, the geocast functionality

---

[3] Together with the use of mobile IPv6 described in Section 7.4.2, the IP address used for the update will be fixed, independent of the communication technology being used.

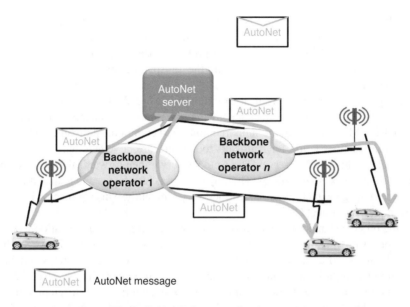

Figure 7.19 Unicast-emulated geocast.

will derive the target area from the address information in the AutoNet message, determine the IP addresses of the vehicles located in the target area, and forward the message via unicast to each of the respective vehicles. This forwarding can be realised either via the connectionless (and unreliable) transport protocol UDP, or via TCP connections between the vehicles and the AutoNet server. The unicast emulation can be improved in some communication scenarios. An alternative optimisation of this processing would be the use of IP broadcast, which can be used in a rather limited communication scenario. For example, if all targeted vehicles are located within the IP broadcast domain of a wireless IEEE 802.11 a/b/g network, the AutoNet messages can be sent to the targeted vehicles via IP broadcast. However, such optimisations require additional functionality in the AutoNet server and do not fulfil the required transparency functionality, since they typically do not support the heterogeneity of the various wireless networking environments.

The second deployment scenario is the use of inherent mechanisms provided in todays' cellular wide-area communication networks, such as UMTS or LTE, which is realised, for example, in the project CoCar. This approach utilises the area-wide coverage of the respective cellular networks, which ensures that communicating vehicles will always be connected to the AutoNet server. The CoCar approach was to utilise the random access channel RACH (for upstream traffic) and forward link access channel FACH (for downstream traffic) for the transmission of respective messages. Both channels can be used for cellular broadcasts, that is to send one message to all mobile devices within the same cell. The idea was to map the target area (coded in the upstream messages) onto the respective cells of the communication system. In this way, a message is sent to a 'geocast server' (which corresponds to the AutoNet server), which determines the target area and reflects the message to the respective cells. Using RACH and FACH channels is in general possible for todays' communication systems, since they are specified for the communication system and thus need to be supported by every network operator. However, this approach needs to be revised and improved significantly, since RACH and FACH only support small packet sizes, which are not sufficient for the transmission of AutoNet messages. This way, additional protocol extensions are required (resulting in higher communication delays) in order to fulfil the transparency requirement mentioned in the previous section.

A third deployment alternative is the use of multimedia broadcast multicast service (MBMS) for downstream traffic, which is also specified for actual communication technologies like UMTS and LTE. MBMS enables the support of multicast communication for cellular networks, which is an interesting option for AutoNets too. Therefore, a vehicle will send an AutoNet message via IP to the AutoNet server. The AutoNet server then forwards the AutoNet message via MBMS to the respective vehicles. Thereby, multi-cast can be used to realise the geocast of the AutoNet message, which can be performed very efficiently. This approach will be realised in, for example, the project CoCarX.

Using the broadcast and multi-cast features of actual cellular communication technologies is an efficient way to realise AutoNet cellular networking functionality. However, both approaches require a close and European-wide cooperation among the network operators and network providers in order to ensure the transparency requirement. The AutoNet server needs to be operated in close cooperation with all network operators, since different vehicles may use different telecommunication providers for Internet access. The close cooperation, thus, has to ensure that an AutoNet message sent by a particular vehicle will always be forwarded from the AutoNet server to the targetted vehicles, independently of the telecommunication providers and the communication systems being used by the targetted vehicles. Another important aspect is the operation of the AutoNet server in a real-world deployment scenario. The operation has to be done by an independent institution, which is able to ensure interoperability and cooperation across the different European-wide telecommunication providers.

7.3.3   Interactions and Cross-Layer Optimisations

Both AutoNet ad-hoc networking and AutoNet cellular networking provide the basic protocol functionality to transfer AutoNet messages. However, they do not address the important aspect of which component needs to be used for the transmission. This is a typical management issue, which is realised by the management layer of the AutoNet Generic Reference Protocol Stack, namely by the access technology selector component in the communication plane. Based on the current status of the available communication technologies within a vehicle station and other important parameters of the network, transport, application support and application layer, the access technology selector is able to determine the most suitable communication technology (and, thus, network layer component) being used for the transmission of an AutoNet message. This feature ensures the optimum communication performance accordung to the current requirements of the applications, the user, and the networking environment of the vehicle. Further details on the access technology selector can be found in Section 5.3.3.

7.4   IPv6 and Mobility Extensions

A crucial feature for the success of AutoNets is Internet access; todays users, as well as drivers, expect a new network architecture or communication technology that can be used to access services on the Internet. This is also true for AutoNets, since Internet access enables the integration of the generic domain into the AutoNet architecture, that is it enables vehicles to access services outside the AutoNet ad-hoc networking domain, such as third party services, access to traffic management centres or the deployment of automotive back-end services provided by vehicle manufacturers. For example, BMW's ConnectedDrive services are completely based on IP since they use common cellular networking technologies for access to the BMW back-end services. Moreover, the support of IP paves the way for the deployment of IP-based applications within the AutoNet ad-hoc network. This aspect is also called *Internet integration*, that is the task is to integrate the AutoNet ad-hoc network domain into the Internet. Here, the term 'integration' means that the AutoNet ad-hoc network has to appear as a transparent extension of the Internet, that is all vehicles and roadside stations have to appear as IP-based nodes (called *hosts*) to any host located on the Internet.

As a result, the support of the Internet Protocol IP is an imminent feature of AutoNets, since all communication in the Internet is based on IP for networking. In the network layer of the AutoNet Generic Reference Protocol Stack, IP is provided by the component *IPv6 and mobility extensions*, which is highlighted in this section. Since there is a good deal of literature on IP-based networking, we only will only describe the basic functionality in this section and refer ambitious readers to respective further readings mentioned in the references.

7.4.1   IPv6

Current communication in the Internet is based on IP version 4 (IPv4) providing an address space of 2^{32} nodes. Since this address space is not sufficient to address all hosts in the Internet uniquely,[4] it is obvious that it will not be possible to address every single vehicle or roadside station with a globally accessible IPv4 address. Consequently, efforts have increased to move the Internet towards IP version 6 (IPv6) in the past few years, which provides an address space of 2^{128} addresses. This is sufficient to uniquely address all Internet hosts in the future as well as all mobile consumer electronics devices and all vehicles.[5]

In order to scale with the number of communicating vehicles, the standardisation activities for the AutoNet Generic Reference Protocol Stack consequently rely on IPv6. This way, each vehicle is fitted with one or more unique IPv6 addresses and thus can generally be accessed from hosts in the Internet. The final configuration of IPv6 addresses of the vehicles greatly depends on the final deployment of AutoNets. In this way, one may think of using static IPv6 addresses pre-shipped with the vehicles, or dynamically assigned addresses using respective mechanisms of IPv6. Moreover, how interoperability with the IPv4-based Internet is maintained in the AutoNet also depends on deployment. This will usually be realised by the standardised mechanisms for IPv4-IPv6 translation protocols.

In the following, we will not describe the functionality of IPv6, the assembly of the IPv6 PDU and respective routing protocols used in the Internet. Additional and detailed descriptions of IPv6 can be found, for example, in [59]. We will describe one deployment solution for AutoNets

---

[4] In fact, todays' Internet is only able to work with IPv4 due to the intensive use of network address translation (NAT) according to RFC 3022 [49] combined with classless inter-domain routing (CIDR) according to RFC 1519 [48].

[5] Indeed, the address space of IPv6 will even be sufficient to address every single atom uniquely: experts assume that the total number of atoms in the universe is between 2^{66} and 2^{80}.

in Section 7.4.3, which includes architectural aspects, IPv6 addressing, IPv6 translation and mobility optimisations.

7.4.2 Mobility Extensions

An important feature for the Internet integration of AutoNets is that both vehicles and roadside stations should be accessible for all or some hosts located in the Internet. This would allow an efficient and tailored service provisioning to the vehicles. In general, cellular communication technologies like UMTS are used by vehicles to access Internet services. Although such communication technologies provide a seamless roaming for instant Internet access, they typically 'hide' mobile devices, and thus vehicles, by using network address translation (NAT) [49]: mobile devices receive private (and temporary) IP addresses for communication, which are mapped dynamically to global IP addresses. This way, it is not possible for an Internet host to connect to a mobile device or vehicle since the mapping is only temporary and not available for the Internet host. Moreover, a vehicle may have temporary connectivity to the Internet in several ways:

- The vehicle may connect to the Internet either via GSM, GPRS, UMTS, EDGE or LTE. It may also connect to the Internet via a mobile telephone located inside the vehicle. For each communication technology, the vehicle will be assigned a different IP address, typically from a different subnet.
- The vehicle may also have temporary high-speed access to the Internet via a WiFi hotspot located in the vicinity. One may even think of connecting the vehicle to the Internet via Ethernet, for example in the home garage of the driver.
- Future scenarios may also allow the accessing of Internet services via (IP-based multi-hop communication to) roadside stations, which themselves are connected to the Internet.

As a result, an Internet host cannot connect to a vehicle since it does not know which address to use. This way, the mobility management of vehicles plays an important role for Internet integration in AutoNets. The mobility management has to ensure that the vehicle is accessible via exactly one globally unique and fixed IP address as long as the vehicle has Internet access – independent of the communication technology and its temporary IP addressing being used. Therefore, the Internet Engineering Task Force (IETF) developed mobility supports for both IPv4 and IPv6, which are called Mobile IPv4 [52] and Mobile IPv6 [51], respectively. In the following, we will introduce the basic principles of mobility support and describe some important peculiarities of Mobile IPv6.

Mobile IP provides protocol enhancements for a transparent routing of IP packets to mobile nodes (MNs) in the Internet. Mobile IP therefore defines a *home network*, which is a typical network located in the (fixed) Internet. Hence, each MN is always identified by a globally unique *home address*, regardless of its current point of attachment to the Internet. From the topological perspective, the home address is located in the home network of the MN. While situated away from its home network, an MN is also associated with a *care-of-address* (CoA) providing information about the MN's current point of attachment to the Internet. Therefore, Mobile IPv4 deploys an agent-based system comprising a *Home Agent* (HA) and a *Foreign Agent* (FA) as illustrated in Figure 7.20. If a correspondent node (CN) in the Internet sends an

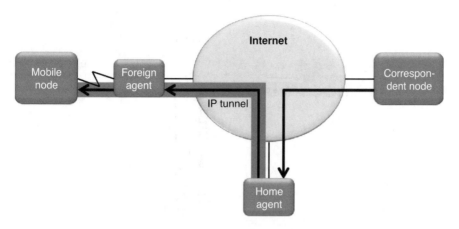

Figure 7.20   Mobile IP.

IP packet to the home address of the MN, the packets will be routed to the home network of the MN. In the event that the MN is currently located in a foreign network, the HA of the MN in the home network accepts the IP packets on behalf of the MN. The HA then encapsulates the IP packet and tunnels it to the current CoA of the MN. In the example depicted in Figure 7.20, the CoA of the MN is located with the FA of the foreign network. The tunnelling is realised by putting a new header in front of the original IP header with the CoA as the new destination address and the HA as the source address. Hence, the packet is forwarded through the Internet to the FA, which unpacks the IP packet and forwards it to the MN. Vice versa, IP packets from the MN to the CN are transmitted either directly to the CN or they are first tunnelled back to the HA ('reverse tunnelling' according to RFC 3024 [50]), which unpacks the IP packets and forwards them through the Internet to the CN.

An alternative variant of Mobile IP is based on co-located CoAs (CCoAs), which is illustrated in Figure 7.21. In this case, the FA functionality is integrated into the mobile nodes since the CoA is co-located with the MN. If an MN enters a foreign network, it receives a temporary CCoA from within the foreign network. This CCoA can be configured either statically by the user or dynamically using, for example, the Dynamic Host Configuration Protocol (DHCP [53]). The MN then registers the CCoA with its HA. Thus, the MN has a topologically correct IP address being used for further communication. IP packets sent by a CN are routed to the home network of the MN. The HA in this network accepts the IP packets on behalf of the MN and tunnels them to the CCoA of the MN. On the MN, the additional tunnel IP header is stripped off and the original packet is delivered to the upper protocol layers. Alternatively, the MN can notify its CNs directly about its CCoA with a *binding update* message. In this case, the CN can send the IP packets directly to the MN since it knows the current CCoA of the MN.

7.4.2.1   Mobile IPv6

Mobile IPv6 [51] adds mobility support for IPv6-based nodes. MNs are thus able to move between wireless IPv6 networks. For this purpose, Mobile IPv6 is based on an agent-based system similar to Mobile IP(v4) using co-located CoAs. An important difference from Mobile

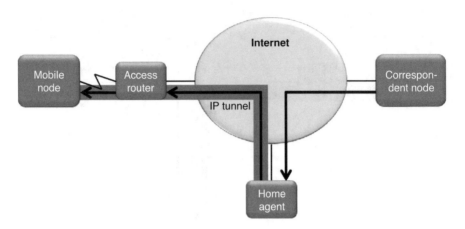

Figure 7.21   Mobile IPv6.

IPv4 is that Mobile IPv6 does not use any foreign agents. If an MN moves to a new point of attachment in another subnet, it has to acquire a new valid IPv6 address from within this foreign network, the CoA. As with Mobile IPv4, the HA of the MN in the home network acts as a representative while the MN is located in a foreign network. In contrast to Mobile IPv4, the MN in Mobile IPv6 has to register its current CoA not only with its HA but also with the CNs it currently communicates with. The association made between the home address and the current CoA of an MN is also called a *mobility binding*.

An MN detects its movement into a new subnet by analysing router advertisements broadcasted periodically by the access routers of the foreign network. An MN can also request an access router of a foreign network with a router solicitation message [54] to transmit a router advertisement. The information contained in the router advertisements allows the MN to create its new CoA autonomously. Therefore, the MN performs the following three steps [55]:

1. The MN performs a duplicate address detection algorithm on its assigned link-local IPv6 address to verify the uniqueness of the link-local address [54].
2. Afterwards, the MN generates a topologically correct IPv6 CoA using either stateless [56] or stateful [57] address configuration.
3. The generated CoA is verified for its uniqueness using duplicate address detection according to [54].

Once the CoA construction is finished, the MN updates the mobility bindings in the HA and its current CNs by sending a *mobility binding update* message with the new CoA. However, it is important to mention that Mobile IPv6 requires some modifications of IPv6 and respective IP protocols. These modifications will not be detailed further on; ambitious readers may refer [51] for additional and detailed information.

7.4.2.2   Network Mobility

Besides Mobile IPv6, current standardisation activities also discuss network mobility as an inherent feature of the IPv6 and mobility extension feature in the network layer of the AutoNet

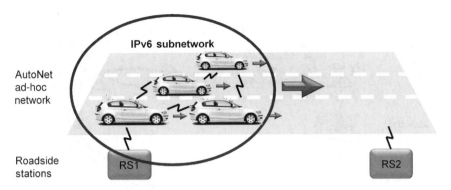

Figure 7.22   Example scenario for network mobility.

Generic Reference Protocol Stack. In general, Mobile IPv6 enables the handover of single nodes or vehicles from one IPv6 subnetwork to another IPv6 subnetwork in order to support the mobility of the MNs. If a number of MNs are organised in a 'mobile network', which moves from one IPv6 subnetwork to another IPv6 subnetwork, the handover procedure of Mobile IPv6 needs to be performed for each MN separately, resulting in a significant signalling overhead. This issue is illustrated in Figure 7.22, where a mobile network moves from access router AR1 to AR2. This is an important issue for vehicular environments, because vehicles often travel in so-called bulks: although a group of vehicles may drive at high speeds, their relative speed may be rather low, enabling the local exchange of data via ad-hoc multi-hop networking within this group of vehicles. In such a scenario, *network mobility* addresses the mobility support of entire mobile networks by reducing the signalling overhead for the handover of a mobile network.

In addition to the mobility support of single nodes, the NEMO (Network Mobility) working group of the Internet Engineering Task Force develops enhancements for Mobile IPv6 to manage mobility of an entire mobile network [47]. In such a mobile network, one or more mobile routers connected to the Internet provide connectivity and reachability for the nodes inside this IPv6 subnetwork. For the mobility support, a bi-directional tunnel is maintained between the mobile router and its HA. This way, IPv6 packets from a CN to the MN inside the subnet of the mobile router are routed to the home network of the mobile router. The HA of the mobile router accepts these packets and tunnels them to the CoA of the mobile router. Finally, the mobile router decapsulates the IPv6 packets and forwards them to the MN. Vice versa, IPv6 packets from the MN are tunnelled back via the mobile router to the HA, which unpacks the original IPv6 packets and forwards them to the targetted CN. As a result, NEMO only requires a handover procedure for the mobile routers within a mobile network instead of all nodes within the mobile network.

7.4.3   Deployment Issues

Both Mobile IPv6 and NEMO provide the protocol support for the mobility of vehicles for IP-based networking. However, the Internet integration of AutoNets also depends on the architecture of such a system, which supports the transparent integration of mobility support protocols described above – and which also has the potential for using highly optimised and efficient communication protocols while maintaining compatibility with the common communication protocols used in the Internet. Bechler therefore suggests a proxy-based communication

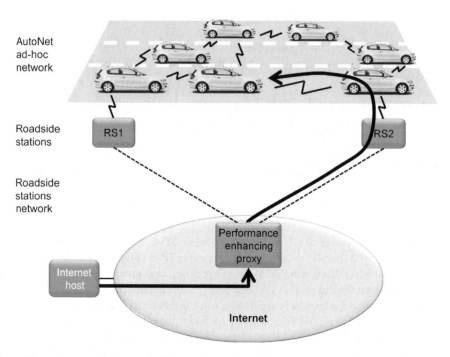

Figure 7.23   Deployment of IP-based communication in AutoNets according to [58].

architecture, in which a performance-enhancing proxy (PEP) 'separates' the characteristics from the AutoNet ad-hoc networking and the communication characteristics of the Internet [58] (cf. Sections 6.2 and 10.3.4), as illustrated in Figure 7.23. Here, the overall AutoNet ad-hoc network is represented by a PEP from a logical point of view. The entire AutoNet ad-hoc networking appears as one large IPv6 subnetwork (with a global and unique IPv6 address prefix), which comprises every vehicle and every roadside station. The PEP is the logical access router to the AutoNet ad-hoc network.

In order to maintain the scalability of such a system, the PEP will likely be realised as a 'cloud' of distributed PEPs, which are loosely coupled among each other. Figure 7.24 illustrates this cloud for two performance-enhancing proxies. Thereby, each vehicle can be accessed from Internet hosts by exactly one PEP. In this scenario, each PEP has the same global IPv6 prefix, so the AutoNet ad-hoc network still appears as one IPv6 subnetwork. However, each PEP forms its own logical IPv6 subnetwork within this global IPv6 subnetwork, as illustrated in Figure 7.24. In this example, the overall AutoNet ad-hoc network is identified by the global IPv6 prefix 001A:BBBB:CC::/40. This IPv6 subnetwork is logically partitioned into two subnetworks:

1. PEP1 with the IPv6 prefix 001A:BBBB:CC11:1111::/64 comprising vehicles V1 and V3,[6]
2. PEP2 with the IPv6 prefix 001A:BBBB:CC11:1112::/64 comprising vehicle V2.

---

[6] See Kurose et al. for an explanaition of IPv6 addressing conventions [59].

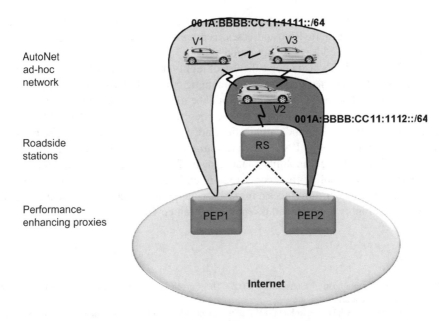

Figure 7.24 IPv6 addressing with multiple performance-enhancing proxies.

In this communication architecture, Mobile IPv6 as well as NEMO can be deployed very easily and transparently: PEPs are predestined for handling the mobility management of vehicles (and roadside stations) as well as the remaining Internet integration functionality. Therefore, the PEPs additionally have to perform two important tasks:

- *Home network for Mobile IPv6.* The PEPs also maintain the home agent functionlity of Mobile IPv6. This way, mobile nodes are logically located at the PEPs. Using reverse tunnelling of Mobile IPv6, it is also guaranteed that IPv6 packets sent by the vehicles also pass the PEPs since reverse tunnelling always tunnels IPv6 packets back to the home agent, which itself forwards the packets to the respective Internet host.
- *IPv4-IPv6 Translation.* Since all traffic from and to the AutoNet ad-hoc network has to pass the PEP, it is the perfect location for performing translations between IPv4 traffic in the Internet and IPv6 traffic in the AutoNet ad-hoc network. Therefore, several standardised protocols can be used, depending on the address configuration being used in the deployment scenario. Alternatives include dual stack solutions acording to RFC 4241 [60], 6-to-4 translation according RFC 3056 [61], stateless IP/ICMP translation according to RFC 2765 [62], or a transport relay translator according to RFC 3142 [63].

It is important to mention that a PEP may deploy several home networks, which are logically separated from each other. Hence, different home networks can be deployed, for example different home networks for vehicles of different manufacturers or for fleet operators. The use of Mobile IPv6 in such a scenario results in a fixed mapping of a vehicle to its home network. Hence, a vehicle can always be accessed via the same IPv6 address and thus via the same PEP – independent of the current communication technology being used, and independent of the IPv6 subnetwork the vehicle is currently registered with. This configuration even allows us to equip vehicles with pre-defined static IPv6 addresses, as proposed in [58]. This configuration further reduces the overhead for address configuration.

The introduction of PEPs also allows the use of highly optimised communication protocols within the AutoNet ad-hoc network while maintaining compatibility with the communication protocols used in the Internet. One example would be to overcome the deficiencies of TCP by using an optimised TCP at the transport layer as described in Section 6.2.4. At the network layer, additional improvements and optimisations are possible. For example, Bechler describes an optimised version of Mobile IPv6, which re-introduces the foreign agent functionality and combines it with hierarchical mobility management. This approach was combined with an efficient discovery protocol of the access routers and thus the foreign agents: besides the efficient discovery, Bechler introduces a fuzzy-based approach to determine the most suitable foreign agent in a highly heterogeneous communication environment.

References

[1] CAR 2 CAR Communication Consortium (2007) The Car-2-Car Communication Consortium Manifesto. http://www.car-to-car.org.

[2] Tseng, Y.C., Ni, S.Y., Chen, Y. and Sheu, J.P. (1999) The broadcast storm problem in a mobile ad hoc network. *Proceedings of the 5th Annual ACM/IEEE International Conference on Mobile Computing and Networking (MobiCom 1999).*

[3] Karp, B. and Kung, H.T. (2000) Greedy perimeter stateless routing for wireless networks. *Proceedings of the 6th Annual ACM/IEEE International Conference on Mobile Computing and Networking (MobiCom 2000).*

[4] Maihöfer, C., Eberhardt, R. and Schoch, E. (2004) CGGC: Cached greedy geocast. *Proceedings of the 2nd International Conference on Wired/Wireless Internet Communications (WWIC 2004).*

[5] Takagi, H. and Kleinrock, L. (1984) Optimal transmission ranges for randomly distributed packet radio terminals. *IEEE Transactions on Communications*, vol. 32, no. 3.

[6] Hou, T.C. and Li, V. (1986) Transmission range control in multihop packet radio networks. *IEEE Transactions on Communications*, vol. 34, no. 1.

[7] Sun, M., Feng, W., Lai, T., Yamada, K., Okada, H. and Fujimura, K. (2000) GPS-based message broadcast for adaptive inter-vehicle communications. *Proceedings of the 52nd IEEE Vehicular Technology Conference (VTC 2000).*

[8] Karp, B.N. and Kung, H.T. (2000) GPSR: Greedy perimeter stateless routing for wireless networks. *Proceedings of the 6th Annual ACM/IEEE International Conference on Mobile Computing and Networking (MobiCom).*

[9] Benslimane, A. (2004) Optimized dissemination of alarm messages in vehicular ad-hoc networks (VANET). *Proceedings of the 7th IEEE International Conference on High Speed Networks and Multimedia Communications (HSNMC 2004).*

[10] Fasolo, E., Furiato, R. and Zanella, A. (2005) Smart broadcast algorithm for inter-vehicular communications. *Proceedings of the Wireless Personal Multimedia Communication (WPMC 2005).*

[11] Briesemeister, L., Schäfers, L. and Hommel, G. (2000) Disseminating messages among highly mobile hosts based on intervehicle communication. *Proceedings of the IEEE Intelligent Vehicles Symposium (IV 2000).*

[12] Torrent-Moreno, M. (2007) Inter-vehicle communications: assessing information dissemination under safety constraints. *Proceedings of the 4th Annual IEEE/IFIP Conference on Wireless On Demand Network Systems and Services (WONS 2007).*

[13] Kim, T., Hong, W. and Kim, H. (2007) An effective multi-hop broadcast in vehicular ad-hoc network. *Proceedings of the Architecture of Computing Systems (ARCS 2007).*

[14] Korkmaz, G. and Ekici, E. (2004) Urban multi-hop broadcast protocol for inter-vehicle communication systems. *Proceedings 1st ACM Workshop on Vehicular Ad-hoc Networks (VANET 2004).*

[15] Plazzi, C., Ferretti, S., Roccetti, M., Pau, G. and Gerla, M. (2007) How do you quickly choreograph inter-vehicular communications? A fast vehicle-to-vehicle multi-hop broadcast algorithm, explained. *Proceedings of the 4th IEEE Consumer Communications and Networking Conference (CCNC 2007).*

[16] Füssler, H., Widmer, J., Käsemann, M., Mauve, M. and Hartenstein, H. (2003) Contention-based forwarding for mobile ad-hoc networks. *Elsevier's Ad Hoc Networks*, Volume 1, Number 4.

[17] Durresi, M., Durresi, A. and Barolli, L. (2005) Emergency broadcast protocol for inter-vehicle communications. *Proceedings of the 11th International Conference on Parallel and Distributed Systems (ICPADS'05)*.

[18] Bononi, L. and Di Felice, M. (2007) A cross layered MAC and clustering scheme for efficient broadcast in VANET. *Proceedings of the IEEE International Conference on Mobile Ad hoc and Sensor Systems (MASS 2007)*.

[19] Chen, W. and Cai, S. (2006) Dynamic local peer group organizations for vehicle communications. *Proceedings of the Vehicle-to-Vehicle Communications Workshop (V2VCOM 2006)*.

[20] Eichler S, Schroth C., Kosch, T. and Strassberger, M. (2006) Strategies for context-adaptive message dissemination in vehicular ad hoc networks. *Proceedings of the 2nd International Workshop on Vehicle-to-Vehicle Communications (V2VCOM)*.

[21] Kosch, T., Strassberge, M., Eichler, S., Schroth, C. and Adler, C. (2006) The scalability problem of vehicular ad-hoc networks and how to solve it. *IEEE Wireless Communications Magazine*, vol. 13, no. 5.

[22] Schroth, C., Eigner, R., Strassberger, M. and Eichler, S. (2006) A framework for network utility maximization in vanets. *Proceedings of the 3rd ACM International Workshop on Vehicular Ad Hoc Networks (VANET 2006)*.

[23] Jiang, H., Guo, H. and Chen, L. (2008) Reliable and efficient alarm message routing in VANET. *Proceedings of the 28th International Conference on Distributed Computing Systems Workshops*.

[24] Reumerman, H.J. and Ruffini, M. (2005) Distributed power control for reliable broadcast in inter-vehicle communication systems. *Proceedings of the 2nd International Workshop on Intelligent transportation (WIT 2005)*.

[25] Torrent-Moreno, M., Santi, P. and Hartenstein, H. (2006) Distributed fair transmit power adjustment for vehicular ad hoc networks. *Proceedings of the 3rd Annual IEEE Communications Society on Sensor and Ad Hoc Communications and Networks (SECON 2006)*.

[26] Mittag, J., Schmidt-Eisenlohr, F., Killat, M., Haerri, J. and Hartenstein, H. (2008) Analysis and design of effective and lowoverhead transmission power control for VANETs. *Proceedings of the 5th ACM International Workshop on Vehicular Inter-NETworking (VANET 2008)*.

[27] Artimy, M. (2007) Local density estimation and dynamic transmission-range assignment in vehicular ad hoc networks. *IEEE Transactions on Intelligent Transportation Systems*.

[28] Biswas, S., Tatchikou, M. and Dion, F. (2006) Vehicle-to-vehicle wireless communication protocols for enhancing highway traffic safety. *IEEE Communications Magazine*, vol. 44, no. 1.

[29] Bettstetter, C. (2002) On the minimum node degree and connectivity of a wireless multihop network. *Proceedings of the ACM/IEEE International Symposium on Mobile Ad Hoc Networking and Computing (MobiHoc)*.

[30] Cheng, Y.C. and Robertazzi, T.G. (1989) Critical connectivity phenomena in multihop radio models. *IEEE Transactions on Communications*.

[31] Clementi, A.E.F., Penna, P. and Silvestri, R. (2000) The power range assignment problem in radio networks on the plane. *Proceedings of the XVII Symposium on Theoretical Aspects of Computer Science (STACS)*.

[32] Dousse, O., Thiran, P. and Hasler, M. (2002) Connectivity in ad-hoc and hybrid networks. *Proceedings of the IEEE Infocom*.

[33] Sànchez, M., Manzoni, P. and Haas, Z. (1999) Determination of critical transmission range in ad-hoc networks. *Proceedings of the International Conference on Multiaccess, Mobility and Teletraffic*.

[34] Santi, P., Blough, D.M. and Vainstein, F. (2001) A probabilistic analysis for the range assignment problem in ad hoc networks. *Proceedings of the ACM/IEEE Symposium on Mobile Ad Hoc Networking and Computing (MobiHoc)*.

[35] Gupta, P. and Kumar, P.R. (1998) Critical power for asymptotic connectivity in wireless networks. In: *Stochastic Analysis, Control, Optimization and Applications: A Volume in Honor of W.H. Fleming*; W. M. McEneaney, G. Yin, and Q. Zhang (Eds.), Birkhäuser, Boston, USA.

[36] Krishnamachari, B., Wicker, S.B. and Béjar, R. (2001) Phase transition phenomena in wireless ad-hoc networks. *Proceedings of the Symposium on Ad-Hoc Wireless Networks*.

[37] Penrose, M.D. (1999) On k-connectivity for a geometric random graph. *Random Structures and Algorithms*, vol. 15, no. 2.

[38] Schroth, C., Eichler, S., Kosch, T. and Ostermeier, B. (2005) Simulation of car-to-car messaging: analyzing the impact on road traffic *Proc. 13th Annual Meeting of the IEEE International Symposium on Modeling, Analysis, and Simulation of Computer and Telecommunication Systems (MASCOTS), Atlanta, USA*.

[39] Maihöfer, C., Franz, W.J. and Eberhardt, R. (2003) Stored geocast. *Proceedings of Workshop on Kommunikation in Verteilten Systemen (KiVS)*.

[40] Chiasserini, C.F., Fasolo, E., Furiato, R., Gaeta, R., Garetto, M., Gribadou, M., Sereno, M., and Zanella, A. (2005) Smart broadcast of warning messages in vehicular ad-hoc aetworks. *Proceedings of the Workshop Interno Progetto NEWCOM (NoE)*.

[41] Kanodia, V., Li, C., Knightly, E., Sabharwal, A. and Sadeghi, B. (2002) Ordered packet scheduling in wireless ad-hoc networks. *Proceedings of the 3rd ACM International Symposium on Mobile Ad Hoc Networking and Computing (MobiHoc)*.

[42] Schwarz, R., Schaufelberger, W., Raymann, L., Merz, H., Zaugg, F., Kloth, T. and Farago, P. (2004) Wirksamkeit und Nutzen von Verkehrsinformation. *Vereinigung Schweizerischer Verkehrsingenieure (SVI), Research Proposal SVI 2000/386*.

[43] University of Southern California (2006) The Network Simulator ns-2. http://www.isi.edu/nsnam/ns/index.html.

[44] University of Southern California (2006) Two-Ray Ground Reflection Model. http://www.isi.edu/nsnam/ns/doc/node217.html.

[45] Carnegie Mellon University (2005) Rice Monarch Project Extensions to ns-2. http://www.monarch.cs.cmu.edu/cmu-ns.html.

[46] Wehrle, K., Pählke, F., Ritter, H., Müller, D. and Bechler, M. (2004) *The Linux Networking Architecture*. Pearson Education.

[47] Devarapalli, V., Wakikawa, R., Petrescu, A. and Thubert, P. (2005) Network Mobility (NEMO) Basic Support Protocol. RFC 3693, Internet Engineering Task Force (IETF).

[48] Fuller, V., Li, T., Yu, J. and Varadhan, K. (1993) Classless Inter-Domain Routing (CIDR): an Address Assignment and Aggregation Strategy. RFC 1519, Internet Engineering Task Force (IETF).

[49] Srisuresh, P. and Egevang, K. (2001) Traditional IP Network Address Translator (Traditional NAT). RFC 3022, Internet Engineering Task Force (IETF).

[50] Montenegro, G. (2001) Reverse Tunneling for Mobile IP, revised. *RFC 3024, Internet Engineering Task Force (IETF)*.

[51] Johnson, D., Perkins, C. and Arkko, J. (2003) Mobility Support in IPv6. *RFC 3775, Internet Engineering Task Force (IETF)*.

[52] Perkins, C. (2002) IP Mobility Support for IPv4. *RFC 3344, Internet Engineering Task Force (IETF)*.

[53] Droms, R. (1997) Dynamic Host Configuration Protocol. *RFC 2131, Internet Engineering Task Force (IETF)*.

[54] Narten, T., Nordmark, E. and Simpson, W. (1998) Neighbor Discovery for IP Version 6 (IPv6). *RFC 2461, Internet Engineering Task Force (IETF)*.

[55] Deering, S. and Hinden, R. (1998) Internet Protocol, Version 6 (IPv6) Specification. *RFC 2460, Internet Engineering Task Force (IETF)*.

[56] Thomson, S. and Narten, T. (1998) IPv6 Stateless Address Autoconfiguration. *RFC 2462, Internet Engineering Task Force (IETF)*.

[57] Droms, R., Bound, J., Volz, B., Lemon, T., Perkins, C. and Carney, M. (2003) Dynamic Host Configuration Protocol for IPv6 (DHCPv6). *RFC 3315, Internet Engineering Task Force (IETF)*.

[58] Bechler, M. (2004) *Internet Integration of Vehicular Ad Hoc Networks*. Logos Verlag.

[59] Kurose, J.F. and Ross, K.W. (2009)*Computer Networking: A Top-Down Approach*. 5th Edition, Prentice Hall International.

[60] Shirasaki, Y., Miyakawa, S., Yamasaki, T. and Takenouchi, A. (2005) A Model of IPv6/IPv4 Dual Stack Internet Access Service. *RFC 4241, Internet Engineering Task Force (IETF)*.

[61] Carpenter, B. and Moore, K. (2001) Connection of IPv6 Domains via IPv4 Clouds. *RFC 3056, Internet Engineering Task Force (IETF)*.

[62] Nordmar, E. (2000) Stateless IP/ICMP Translation Algorithm (SIIT). *RFC 27652, Internet Engineering Task Force (IETF)*.

[63] Hagino, J. and Yamamoto, K. (2001) An IPv6-to-IPv4 Transport Relay Translator. *RFC 3142, Internet Engineering Task Force (IETF)*.

[64] Kosch, T. (2005) Efficient message dissemination in vehicle ad-hoc networks. *Proceedings of the 11th World Congress on Intelligent Transportation Systems (ITS)*.

[65] Kafsi, M., Papadimitratos, P., Dousse, O., Alpcan, T. and Hubaux, J.-P. (2008) VANET connectivity analysis. *Proceedings of the IEEE Workshop on Automotive Networking and Applications*.

[66] Fiore, M. and Harri, J. (2008) The networking shape of vehicular mobility. *Proc. 9th ACM International Symposium on Mobile Ad Hoc Networking and Computing (MobiHoc), May 2008*.

[67] Lochert, C. (2008) Avoiding the Gridlock Information Dissemination in Vehicular Networks. *Dissertation, University of Duesseldorf, 2008.*

[68] Torrent-Moreno, M., Festag, A. and Hartenstein, H. (2006) System design for information dissemination in VANETs. *Proc. 3rd International Workshop on Intelligent Transportation (WIT 2006).*

[69] Mauve, M., Widmer, J. and Hartenstein, H. (2001) A survey on position-based routing in mobile ad-hoc networks. *IEEE Network*, Vol. 15 No. 6.

[70] Maihoefer, C. (2004) A Survey of Geocast Routing Protocols. *IEEE Communications Surveys and Tutorials* 6(1–4).

[71] Moustafa, H. and Zhang, Y. (2009) *Vehicular Networks – Techniques, Standards, and Applications.* Auerbach Publications 2009.

[72] Zhang, G., Chen, W., Xu, Z., Liang, H., Mu, D. and Gao, L. (2009) Geocast routing in urban vehicular ad hoc networks. *Computer and Information Science 2009 – Studies in Computational Intelligence*, Volume 208/2009, Springer.

[73] Lasowski, R. and Strassberger, M. (2011) A multi channel beaconing service for collision avoidance in vehicular ad-hoc networks. *Proc. 74th IEEE Vehicular Technology Conference (VT2011).*

[74] La, R. and Anantharam, V. (2002) Utility-based rate control in the Internet for elastic traffic. *IEEE/ACM Trans. on Networking (TON)*, Vol. 10, No. 2.

[75] Gao, X., Nandagopal, T. and Bharghavan, V. (2001) Achieving application level fairness through utility-based wireless fair scheduling. *Proc. IEEE Global Telecommunications Conference (GLOBECOM01).*

[76] Kelly, F., Maulloo, A. and Tan, D. (1998) Rate control in communication networks: shadow prices, proportional fairness and stability. *Journal of the Operational Research Society* 49.

[77] Kelly, F. (1997) Charging and rate control for elastic traffic. *European Transactions on Telecommunications*, Vol. 8 (1997).

[78] Li, J., Blake, C., De Couto, D., Lee, H. and Morris, R. (2001) Capacity of ad hoc wireless networks. *Proc. 7th ACM International Conference on Mobile Computing and Networking (MobiCom01).*

[79] Gupta, P. and Kumar, P. (2000) The capacity of wireless networks. *IEEE Trans. Information Theory*, Vol. 46, No. 2, 2000.

8

Physical Communication Technologies

The lowest layer in the AutoNet Generic Reference Protocol Stack handles the physical transmission in the physical communication technologies layer, as illustrated in Figure 8.1. The physical communication technologies layer deals with the transmission of data among the different AutoNet entities and AutoNet domains. Hence, this layer not only incorporates wireless communication technologies, but also wired networking technologies to connect the different domains or the different communicating entities within the different domains. Compared to ISO/OSI reference model terminology, the physical communication technologies layer incorporates the physical transmission of bits over media (ISO/OSI layer 1, physical layer) as well as the mechanisms for medium access control and logical link control (ISO/OSI layer 2, data link-layer). The physical communication technologies layer therefore provides the functionality to transfer data between network entities, which also typically provides the means to detect and possibly correct errors depending on the communication technology. Both layers are merged into the physical communication technologies layer, because some wireless technologies, as for instance Bluetooth, come with protocols for the entire communication protocol stack specifically designed for use with the respective physical wireless communication characteristics and/or the respective applications. For AutoNets, this is true to a large extent for the wireless LAN (short range) communication technologies, especially in the 5.9GHz band and the infrared technologies.

In general, every relevant (i.e. existing and being used) communication technology can be deployed in AutoNets, as illustrated in Figure 8.2. All of them are extensively described in literature, so we will not detail all of them in this chapter but give respective references for further reading. Instead, we will focus on the upcoming solution for V2X communications, namely the communication standard IEEE 802.11p. We will also introduce potential and promising optimisations for this layer in order to support and to improve the benefit-based message dissemination described in Section 7.2.4, together with a technology comparison with respect to scalability issues. As a result, this chapter is organised as follows. In the next section, we will emphasise the physical communication technologies layer in AutoNets, followed by a detailed description of the upcoming standards for V2X communication in Section 8.2.

Automotive Internetworking, First Edition. Timo Kosch, Christoph Schroth, Markus Strassberger and Marc Bechler.
© 2012 John Wiley & Sons, Ltd. Published 2012 by John Wiley & Sons, Ltd.

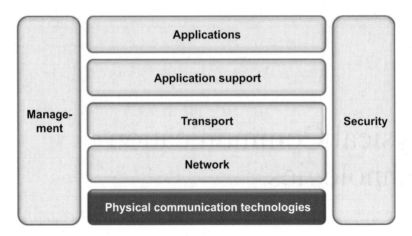

Figure 8.1 Physical communication technologies layer in the AutoNet Generic Reference Protocol Stack.

Section 8.3 introduces optimisations of IEEE 802.11p mechanisms for the benefit-based message dissemination, and Section 8.4 describes the results of comparing simulations in order to show the suitability of different communication systems for being used in AutoNets. Finally, Section 8.5 concludes this chapter with a brief summary.

8.1   Wireless Networks in the AutoNet Generic Reference Protocol Stack

The physical communication technologies provides services to send bits from one communicating node to others using respective communication technologies. As illustrated in

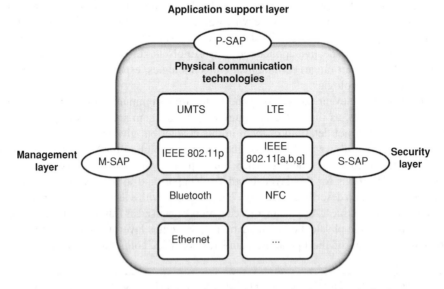

Figure 8.2   Service access points of physical communication technologies layer.

Figure 8.2, it depends on the domain or the entity which communication technology is being used. For vehicles, for example, this could be cellular systems like UMTS or LTE for Internet access or for communication with entities located in, for example, traffic management centres, whereas IEEE 802.11p or IEEE 802.11 a, b or g may be used for data exchange among vehicles in an ad-hoc fashion. For communication with mobile devices – that is vehicle-to-mobile (V2M) – communication technologies like Bluetooth or near field communication (NFC) can be used. In the AutoNet Generic Reference Protocol Stack, this functionality is provided by the physical service access point (P-SAP) for the upper networking layer. Moreover, the two service access points M-SAP and S-SAP are defined in this protocol stack for interaction with the management layer and security layer, respectively.

In this chapter, we take the vehicles' perspective. We are focusing on wireless communication technologies used for communication between vehicles and for the interaction between vehicles and the infrastructure domain. Wireless radio systems have typically been developed for a particular purpose. Thus, they show their best performance if used for this purpose. Cellular systems, for example, were originally developed as telecommunication networks for telephony applications, handling a rather large number of users. Data communication networks such as WiFi according to IEEE 802.11 a, b or g were developed for high data rate Internet-based applications and therefore usually provide high rates and high reliability, but no real-time support. Thus, different types of wireless technologies are suitable for different AutoNet applications. No standard radio system to date can support the very high reliability, low latency real-time requirements of some of the driving-task-related applications, since typically reliability is increased with increased delay (e.g. using retransmissions). Special radio technologies have been developed to address these and other AutoNet specific requirements. To optimally match AutoNet application requirements, various communication systems were proposed, especially in the late 1990s, with a special focus on AutoNet ad-hoc networking (for an overview for see for instance [12]).

For AutoNets, such networks can be classified according to the classical network hierarchy with respect to the typical geographical range and scale. Therefore, we distinguish four different classes:

- *Wide area networks: digital broadcast systems.* Broadcast media have traditionally been used to spread traffic information and hazard warnings to the receivers in a large area. Today, all road vehicles generally feature at least an FM radio receiver. These are in many cases still analogue receivers. Digital broadcasting technologies have recently gained ground, even though there is no global standard for them. In Europe, a variation of digital broadcasting systems like DAB (Digital Audio Broadcast), DMB (Digital Multimedia Broadcast) or DVB (Digital Video Broadcast) have emerged and been installed. In the US, systems like HD Radio have been developed by the company Ubiquity as hybrid analogue/digital broadcast systems on the basis of DAB. While these standards are generic rather than AutoNet specific, a set of AutoNet application messages is sent over these systems today (in particular TMC (Traffic Management Channel) and TPEG (Transport Protocol Experts Group)).
- *Metropolitan area networks: cellular systems.* While originally used for voice communications, cellular networks have been extended to and are now more and more used for digital data communications. Examples include GSM (global system for mobile communications) with GPRS (general packet radio service) and EDGE (enhanced data rates for GSM evolution) extensions, UMTS (universal mobile telecommunications system) with HSPA (high

speed packet access), and as the next evolutionary step LTE (long term evolution) networks. Some recent research has addressed the use of cellular systems for AutoNet safety applications. We summarise some of these results regarding both the analysis of their performance and AutoNet specific extensions and adaptations in Section 8.4.

- *Local area networks: dedicated short range communication (DSRC) and wireless local area networks.* For these types of systems, both off-the-shelf components and AutoNet-specific technologies are used or are in development. This includes for instance Bluetooth and different types of wireless LAN (WLAN) technologies of the IEEE 802.11 family as off-the-shelf components, but also a variety of AutoNet specific technologies. The latter are usually designed to work in the frequency band from 5.8GHz to 5.9GHz. Examples include IEEE 802.11p as draft standard with ETSI ITS G5A as its European profile, the European CEN DSRC (dedicated short range communication), and the Japanese 5.8GHz DSRC.

  DSRC systems are usually classified into *active DSRC* and *passive DSRC* systems. Thereby, passive DSRC systems do not rely on powered on-board units, that is the power for the radio transmission is gained by the reception of the received signal from a DSRC infrastructure node. In contrast, active DSRC systems come with a dedicated power supply for the on-board unit, which enables greater radio propagation and vehicle-triggered communication.

  Additionally, infrared-based systems are popular, mainly for toll-collection applications as standardised in, for example, ISO 21214:2006.

- *Personal area networks: near field communication and RFID.* There are several solutions to connect mobile devices or to identify gadgets, depending on the use. For example, smartphones may be connected to the vehicle infrastructure using Bluetooth, whereas gadgets may feature RFID (radio frequency identification) technology in order to identify themselves within the vehicle system. Identification based on RFID is standardised by ISO. It considers different uses, such as contactless smartcards (ISO/IEC 10536, 14443, 15693, 10373) or item management according to ISO/IEC 18000. In contrast, NFC is an international standard for contactless exchange of information using either passive RFID tags according to ISO/IEC 14443 or ISO/IEC 15693, or active RFID tags according to ISO/IEC 15408.

Since time-critical safety-related applications usually require the exchange of data within the range of a few metres up to at most a few hundred metres, technologies for local area networks are best suited with respect to this basic characteristic. A dedicated communication protocol stack for AutoNets for this purpose has been developed called *wireless access in the vehicle environment* (WAVE) [10], whose physical layers are finally being standardised in the context of IEEE 801.11p. A dedicated frequency is used for this technology to provide the necessary reliability and quality of service. In the following, we will take a closer look at automotive WLAN and DSRC as a representative technology for the AutoNets. An exhaustive description for broadcast systems and cellular communication technologies can be found in [1, 13], wireless LAN according to IEEE 802.11 as well as Bluetooth can be found in [13].

8.2 Automotive WLAN and DSRC

Major challenges for the radio technology on the physical and MAC layer for automotive WLANs and DSRC systems are particular radio channel characteristics, high mobility of the network nodes, congestion stemming from AutoNet-specific data traffic patterns for safety applications (especially CAMs) with the absence of a central managagement entity, and the hidden station problem. At the same time, safety applications require high reliability and low latency. Consider the case of cooperative awareness messages. They require high packet sending rates which leads to a high probability for collisions while accessing the medium. While they would also benefit from a high transmission power, this also leads to a higher collision probability as well as to increased interference.

8.2.1 Spectrum Policies

In the physical medium, the robustness required by AutoNet safety applications can only be provided when transmissions are successful with a very high probability. This requires a somewhat controlled environment. As a fundamental precondition, the physical medium must be available with a very high probability, that is the possibility for non-AutoNet devices to transmit on the same shared medium must be regulated such that it is only allowed in a controlled manner. To ensure this, the frequency spectrum must not be freely accessible but devices using the spectrum must follow certain rules. While it is rather impossible to allocate a private part of the overall spectrum solely to AutoNet applications, this is also not necessary. Instead, some sort of protection is sufficient, which was adopted by European regulation for the 5.9GHz communications. This includes mitigation techniques to avoid interference, for example in Europe between European CEN dedicated short range communication (RTTT DSRC) equipment and intelligent transport systems (ITS) operating in the 5GHz frequency range.

Overlapping and close bands out of the overall radio spectrum in the frequency range of 5.8GHz to 6.0GHz have been allocated in many parts of the world for use by ITS applications. Figure 8.3 shows which parts of this band have been allocated in which parts of the world. In the United States, the Federal Communications Commission (FCC) allocated the spectrum between 5.850GHz and 5.925GHz for DSRC (through CFR 47: Title 47 on Telecommunication). It specified the reservation of seven channels for AutoNet communications together with a certain type of usage for each of the respective channels, as shown in Figure 8.4.

Figure 8.4 also shows the situation of the European spectrum allocation in the 5GHz band, with 30MHz for road safety applications plus an additional 20MHz spectrum for possible future extension. Also, a 20MHz band has been assigned to non-safety ITS applications. This band is not protected. The regulatory authority of radio spectrum allocation in the European Union belongs to the Member States, but is embedded in EU law and international radio spectrum agreements.

8.2.1.1   Standardisation Process

The Radio Spectrum Decision 676/2002/EC adopted by the European Parliament and the Council on 7 March 2002 laid the ground for the development of the EU radio spectrum policy, based on the EU regulatory framework for electronic communications. This framework was defined to ensure that the European market for electronic communications works properly while at the same time being more and more deregulated, guaranteeing competition while market players respect basic user interests. With the Radio Spectrum Decision, the Commissions goal was to facilitate coordination of radio spectrum policy approaches within the EU,

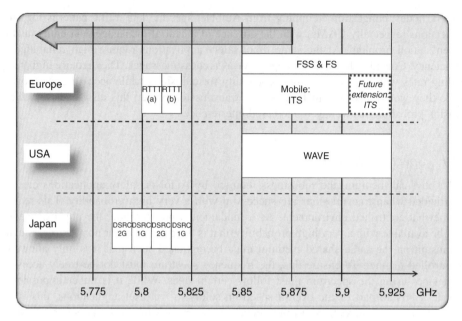

Figure 8.3 International frequency situation for AutoNets at 5.8GHz to 5.9GHz.

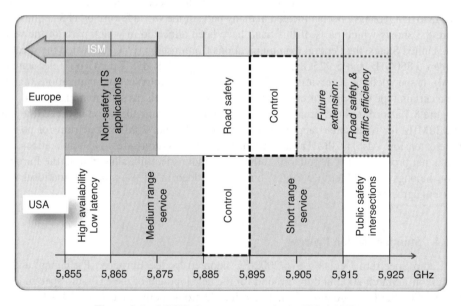

Figure 8.4 5.9GHz frequency usage in the EU and USA.

both to ensure harmonised conditions with respect to spectrum allocation needs stemming from different policy goals internally and to unambiguously represent European interests in international negotiations. While the radio spectrum allocation process is tied to certain organisational institutions, it tries to ensure that all relevant stakeholders are involved in the process, including the Member States, Community institutions, industry, radio spectrum users and other interested parties. The EC receives advice and assistance from an expert group called the Radio Spectrum Policy Group. This group came into existence with the Radio Spectrum Decision in 2002. Whenever a Community policy passes which requires the usage of radio spectrum to become effective, the respective technical implementing measures need to be defined. The Radio Spectrum Committee supports the European Commission in their adoption with harmonised conditions for spectrum availability ensuring the efficient use of this scarce resource and the provision of the necessary information to all stakeholders.

CEPT (European Conference of Postal and Telecommunications Administrations) in Europe is an organisation with 48 members and is responsible for developing regulations on frequency usage. These regulations, however, are not legally binding for any member. They can take either of two forms: a *Decision* or a *Recommendation*. Decisions and Recommendations from CEPT and the European Commission embody different consequences. In general, frequency designation in Europe is subject to national legislation. Thus, the translation of Electronic Communications Committee (ECC) Decisions into national law is voluntary, but requires the national administration's commitment to their implementation. ECC Recommendations, on the other hand, only provide guidance to administrations as to how a certain part of the radio spectrum shall be used, without any commitment. As soon as the European Commission adopts a CEPT decision or builds upon it, these decisions, issued by the Radio Spectrum Committee of the EC, become legally binding for the 27 EU Member States.

The EC issued a mandate to CEPT on 5 July 2006, derived from Article 4(2) of the Radio Spectrum Decision to provide technical information about the spectrum requirements of safety-critical AutoNet applications, to clarify compatibility issues for these applications and to develop optimal channel plans for the targetted bands. The intention also was that the spectrum should be made available throughout the EU after both scope and conditions of usage had been clearly defined.

In response to this mandate, CEPT reported on 21 December 2007 (CEPT Report 20) that for the targetted ITS, or AutoNet applications, respectively:

- the foreseen 5GHz band in the range from 5.875GHz to 5.905GHz is appropriate;
- the systems will be compatible with the other services studied within the band and with all other existing services studied below 5.850GHz and above 5.925GHz when applying certain power emission limits;
- the corresponding usage of this band will not contradict the use of this spectrum in other regions of the world;
- protection from the existing fixed-satellite service earth stations is not possible, but also not necessarily required.

With the mandate from the EC, the ECC of the CEPT developed a decision on the harmonised use of the 5.875GHz to 5.925GHz frequency band (ECC/DEC/(08)01). Within this decision, the sub-band from 5.875GHz to 5.905GHz was assigned to ITS road safety applications immediately, and the sub-band from 5.905GHz to 5.925GHz for future extension.

The scope of the so-called ITS communications addressed includes vehicle-to-vehicle, vehicle-to-infrastructure and infrastructure-to-vehicle communications for the real-time transfer of information. Therefore, it comprises an important part of AutoNet applications.

In addition to this ECC Decision, an ECC Recommendation was published that suggests that the 20MHz band from 5.855GHz to 5.875GHz should be made available by CEPT administrations for non-safety ITS applications (ECC/REC (08)01), but without protection.

Based on the ECC decision, the European Commission regulated the use of the radio spectrum in the frequency band from 5.875GHz to 5.905GHz. The respective Commission Decision was approved and published in 2008. It was developed by the Radio Spectrum Committee of the EC and contains rules on the harmonised use of this spectrum for road-safety related applications. This means that the EU member states were then forced to designate the respective 30MHz of spectrum not later than six months after entry into force of this decision. Designation, however, does not yet mean that the spectrum may be used by the ITS applications. The Decision only stated that the spectrum needs to be made available *as soon as reasonably practicable*.

8.2.1.2   Protection Issues

Protection in this system can take different forms:

- Restrictions on types of systems that may use the frequency band.
- Restrictions on the applications that may use the allowed systems.
- Restrictions on the emitted signals inside the frequency band.
- Restrictions on the emitted signals outside the frequency band.

Only certain systems with certain services may be allowed to use a particular part of the radio spectrum. These systems have to follow certain rules on how to use this spectrum, for example maximum transmission power. Also, they have to ensure that their emitted signals do not breach the defined spectrum mask. This spectrum mask sets a maximum power level over the frequency range, including neighbour bands. The parts of the signal that reach into the neighbour bands, known as out-of-band emissions, must stay underneath this mask.

During a frequency allocation process, compatibility studies have to be conducted to ensure compatibility with existing systems. For the European safety part of the ITS spectrum, these studies proved that ITS safety applications would not suffer from interference resulting from the existing systems and other allowed services. Some fixed satellite service earth stations were already installed and operating in this frequency which meant that ITS safety applications could not claim protection from these existing services. But since the usage of these stations is very limited, ITS applications can develop ways to adapt to this. In the 20MHz of spectrum for possible future extensions, fixed services operating above 5.92GHz may interfere with future ITS services in this spectrum through out-of-band emissions.

Conversely, ITS systems may cause noise for other systems. In order to limit the interference, the allowed out-of-band emission levels in Europe have been set to the following, which will ensure compatibility with all other services according to the compatibility studies:

- Protection of the radiolocation services: less than -55 dBm/MHz are allowed below 5.850GHz.

- Protection of the RTTT applications: less than −65 dBm/MHz are allowed below 5.815GHz.
- Protection of the fixed service: less than −65 dBm/MHz are allowed above 5925GHz.

Also, both CEPT and EC regulation limit the allowed maximum transmission power of ITS systems in the whole band from 5.855GHz to 5.925GHz to 23 dBm/MHz e.i.r.p and absolutely no more than 33 dBm e.i.r.p. In order to protect broadband wireless access systems below 5.875GHz, additional mitigation is required, for example Transmitter Power Control (TPC).

A standard specification which fully complies with the restrictions set by the CEPT decision and recommendation is developed by the European Telecommunications Standard Institute (ETSI) as harmonised standard EN 302 571, respecting Article 3(2) of Directive 1999/5/EC of the European Parliament and the Council on radio equipment.

8.2.1.3 Licensing

Within the frequency band, AutoNet transceivers may be used both inside vehicles and in roadside equipment. Individual licensing will not be required for vehicle units. For roadside units, authorisation may be required by administrations to ensure a smooth system operation by more than one operator. It is interesting to note that it is very hard to give a clear and unambiguous definition of a roadside unit, especially with respect to licensing issues. Consider, for example, a unit in a movable road construction sign which might change its regulatory defined status depending on the situation it is used in (being carried along a road vs. static placement at a construction site).

Compulsive regulation on European level only applies to mobile equipment up to now. It is expected that the same rules will be set for roadside equipment, but at present this is a decision of each Member State. Also, it is not yet decided how device and service certification will be organised.

8.2.2   IEEE 802.11p

In the United States, after the frequency band at 5.9GHz was allocated, the American Society for Testing and Materials (ASTM) was asked to support ITS America with the development of the band usage rules, that is the physical and MAC layer specifications. ASTM evaluated three different technologies for their performance in AutoNets:

- IEEE 802.11a;
- ARIB T-55 (a Japanese standard);
- Free Space (a proprietary technology of Motorola).

After a series of tests, ASTM decided that DSRC systems at 5.9GHz shall use IEEE 802.11a based technology, and ASTM became the lead organisation to develop the respective DSRC standards.

Subcommittee E17.51 was formally assigned with the task and defined the Standard E2213-03: Standard Specification for Telecommunications and Information Exchange Between Roadside and Vehicle Systems in the 5.850–5.925GHz band – 5GHz Band Dedicated Short Range Communications (DSRC) Medium Access Control (MAC) and Physical Layer (PHY) Specifications. This standard was then referenced in the US Federal Communications

Commission's (FCC) DSRC spectrum usage rules. It was stated that the purpose of the specification was short range communications of less than 1000 m at line-of-sight (LOS), mainly from so-called roadside units to mobile units, where mobile units would generally be vehicles travelling at high speed along motorways. However, ASTM would also allow communications between mobile units and also include so-called portable units, originally mainly foreseen in the absence of infrastructure. The E2213 standard, which was published by the ASTM in 2003, was creating a backbone network for AutoNets based on infrastructure and could be used for commercial applications like highway tolling. In 2004, the ASTM standard, being a modification of the IEEE 802.11a standard, became an amendment to the IEEE 802.11 set of standards and was taken up by the new task group. Its main characteristics were the 10MHz channels (instead of 20MHz) and the modified MAC without the setup overheads. In May 2008 the IEEE 802.11p standard was accepted by the full IEEE 802.11 working group as a draft standard in a letter ballot [11]. To date, however, it is still awaiting its final approval as an amendment to the IEEE 802.11. Specifications for higher layers were assigned to the IEEE Working Group P1609, as described in Appendix A. As apparent from the IEEE system architecture, the IEEE 1609.4 specifications are relevant for the wireless technology [8]. They define multi-channel operations and IEEE 802.11 MAC adaptations as described in Appendix A.

In the following, we provide a short summary of the underlying IEEE 802.11 technology to better understand the adaptations defined for AutoNets and describe the AutoNet adaptations.

8.2.2.1 Modes of Operation

IEEE 802.11 defines two modes of operation, as further detailed in [13]:

- In the *infrastructure mode*, mobile nodes are connected to access points (APs). Each mobile node is associated with one AP at a time. Mobile nodes are organised in clusters known as basic service sets (BSS). In a BSS, the AP provides forwarding functionalities between the nodes and serves as an access device to other nodes in wired networks that the AP is connected to. In the infrastructure mode, all communication between the nodes inside the WLAN is controlled by the AP. The AP regularly broadcasts beacons to synchronise the devices with respect to slot start times and durations.
- In the *ad-hoc mode*, devices are able to communicate directly in a self-organising fashion without any infrastructure support. In this case, each device frequently broadcasts synchronisation beacons. If a beacon is received by another station, this station synchronises itself with that beacon.

In an IEEE 802.11p network, only the DSRC ad-hoc mode is allowed on the Control Channel (CCH). This DSRC ad-hoc mode only allows the broadcast of packets on the CCH without request to send and clear to send mechanisms, and without acknowledgement frames (these concepts are explained in Section 8.2.2).

8.2.2.2 WAVE Service Management

IEEE 1609.3 defines the WAVE Basic Service Set (WBSS) as a subtype of the 802.11 BSS [7]. A WBSS consists of a provider offering services and users of this service. In general, roadside units are foreseen as providers, however, vehicles are not excluded to take on the role of a service provider. A WBSS can be operated in two different modes. In the persistent mode, services are announced on the control channel during every cycle. In the non-persistent mode, services are only announced once. Only one service is allowed to be active at a time and a node can only join one WBSS at a time.

A WAVE node can be operated either in ad-hoc mode or in infrastructure mode. If in ad-hoc mode, communication is restricted to WAVE short messages to be sent via the control channel CCH. In a typical scenario, a WAVE application directly addresses a WSM (WAVE service management) to a broadcast MAC address. It selects appropriate radio channel settings (power level, data rate) to control the transmission and passes the request to the WME (WAVE management entity) for delivery to the lower layers and subsequent transmission on the control channel. The receiving nodes then pass the packet via the WME to the locally registered applications, identified through their application class identifier and application context mark (ACID/ACM). Since the WSM includes the address of the transmitting application, a subsequent communication between a receiving application and the sending application may continue on the CCH (via the same procedure).

In infrastructure mode, both IP-based and WSM communication is allowed. The station then belongs to a WBSS and can use the service channels for IP-based communication. Services running within a WBSS have to announce their availability on the control channel using the the WAVE short message protocol (WSMP). Applications will monitor these service advertisements on the CCH. The service itself can then be used in a point-to-point fashion via either IP or WSMP on the service channel SCH. In a typical scenario, an application that wants to offer a service sends a so called *WME-application request* to the WME with the following parameters: persistence of the service, unicast or broadcast address of the intended recipients of the announcement, number of advertisement repetitions and the desired SCH to be used. The service parameters are then transmitted on the CCH in a so called *WAVE-advertisement frame* or *WAVE service advertisement*.

The recipients of the service advertisement check if the service is of interest to any local application (registered within the WME). Decisions to join or reject a WBSS or a service offered, respectively, lie with the WME based on fixed priority values. It is under discussion at IEEE to add more flexibility there by allowing context-sensitive decisions. The decision of the WME to either join or reject the WBSS may be confirmed to the advertiser if this is required, which is specified within so-called service advertisements. The service, on the other side, can also reject users. Use of the service then starts with data exchange (IP or WSMP) on the respective SCH. A service usually terminates when the interaction is complete, signalled by an inactivity timeout. It may prematurely terminate if a higher priority service is offered or if security credentials fail.

8.2.2.3 Modulation

IEEE 802.11p is using direct sequence spread spectrum technology (DSSS) as defined for other IEEE 802.11 variants [13]. IEEE 802.11p is based on IEEE 802.11a, but the bandwidth of the channels was halved to 10MHz each.

8.2.2.4 Medium Access

As the basic medium access mechanism, all IEEE 802.11 WLAN devices as well as the IEEE 802.11p variant use carrier sense multiple access with collision avoidance (CSMA/CA) [13].

In wired networks, collisions on a wireline, that is two data packets sent simultaneously by two different nodes, mean none of the receivers can retrieve the original signal anymore due to the interference of the two signals. However, each node in the network, including the senders, can detect this situation. This, however, is different in an IEEE 802.11 WLAN network. Still, two simultaneously issued packets cause interference in the air. This usually means that receivers are also not able to retrieve the original signals anymore. However, since the receivers are distributed, the signal strengths of the two signals will differ. In some cases, if one signal is strong and the other is weak, the second signal may be interpreted as noise and the first signal may still be recovered. This will largely depend on the physical positions of the nodes, that is it may happen if two signals interfere at one receiver, but one of the two senders is close by while the other is far away.

Usually, two different cases are considered in wireless networks:

- Simultaneous transmission by two nodes which are within range of each other.
- Simultaneous transmission of two nodes which are not within range of each other, but a third node exists which is in range of both transmitters (this situation is generally referred to as the hidden node problem).

In both cases, there are a number of the potential receivers of the transmitters, which are not able to retrieve the signal. This number will generally be higher in the first case. Therefore, this has a higher relevance, and medium access technologies have to deal with this. The medium access mechanism for IEEE 802.11 for this case has to take into account another characteristic of this type of WLAN: the transmitters are not always able to detect collisions. This is handled by CSMA/CA. In order to deal with the second case, IEEE 802.11 defines two mechanisms: the so-called point coordination function (PCF) and the distributed coordination function (DCF).

Medium access control in IEEE 802.11 features a single transmission queue, which works according to the first-in first-out (FIFO) principle. Hence, packets that the station wants to send are put into the queue and transmitted in the scheduled order. When a packet arrives at the head of the transmission queue, the station first senses whether the channel is busy (i.e. if another station is transmitting) or idle (i.e. no transmission). If the channel is busy, the algorithm waits until the medium becomes idle. When the medium is sensed idle, the MAC defers the packet for a specific time interval before sending, called the distributed interframe space (DIFS). If the medium stays idle during the DIFS period, the MAC starts a back-off process. It selects a random back-off counter (BC). It then waits for a specified slot time interval and decrements the BC if the medium remains idle during that time. When the BC value is zero, the frame is transmitted. If the packet had arrived when the queue was empty before, and the medium had been idle for longer than the DIFS time, the frame would have been transmitted immediately (see also Figure 8.5).

Each station also maintains a contention window (CW) out of which the value for the back-off counter is randomly selected, that is the BC is a random integer variable uniformly distributed over the interval [0, CW]. The CW is variable and initially assigned a value of CW_{min}. Its upper bound is set to CW_{max}. The BC decrementing process pauses as soon as the medium is sensed to be busy during the back-off process. When the channel becomes idle again, the back-off process resumes.

In case of a unicast transmission, the receiver immediately acknowledges the correct reception of a packet with an acknowledgement frame (ACK). The ACK frame is transmitted

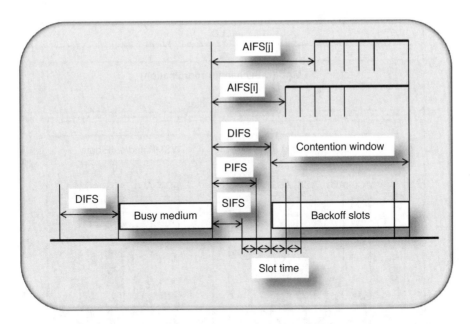

Figure 8.5   IEEE 802.11 Medium access control using CSMA/CA.

after a short IFS (SIFS). Since the SIFS time smaller than the DIFS time,[1] the ACK gets priority over other stations' contention for data transmission. If no ACK is received, the frame is retransmitted after another back-off phase. In this case, the CW is increased as $CW_{new} = 2 * (CW_{old} + 1) - 1$ if $CW_{new} \leq CW_{max}$ and $CW_{new} = CW_{max}$ else. If the transmission was successful, CW is reset to CW_{min}. The transmitter then performs another DIFS and back-off even if there is no other frame pending in the queue. This is often referred to as post back-off. It ensures at least one back-off interval between two consecutive frames.

This standard medium access is part of DCF and provides equal probabilities to all stations contending for the channel in a distributed manner. Figure 8.5 illustrates this algorithm, together with the IEEE 802.11e extensions. IEEE 802.11e [9] provides mechanisms for prioritised medium access with eight different priorities (from 0 to 7). A priority field is added to the MAC header which allows the setting of priorities on a per-packet base. The IEEE 802.11e MAC features four access categories (AC). Each packet is assigned to an AC depending on its priority. This mechanism is called enhanced distributed coordination function (EDCF). Instead of DIFS, CW_{min}, and CW_{max}, it uses AIFSD[AC], $CW_{min}[AC]$ and $CW_{max}[AC]$ values per AC. AIFSD[AC] is calculated as AIFSD[AC] = SIFS + AIFS[AC] * SlotTime. The back-off counter is selected from [1, 1 + CW[AC]] instead of [0, CW].

Smaller values for $CW_{min}[AC]$ and $CW_{max}[AC]$ lead to reduced channel access times for the corresponding priority, but the probability of collisions increases. Choosing appropriate values for these parameters is thus crucial for a particular network setting and can be used to discriminate between the ACs. The IEEE 802.11e MAC implements four transmission queues (see Figure 8.6). Each queue behaves as a single EDCF entity, that is an AC, where each queue

---

[1] DIFS = SIFS + 2 * SlotTime.

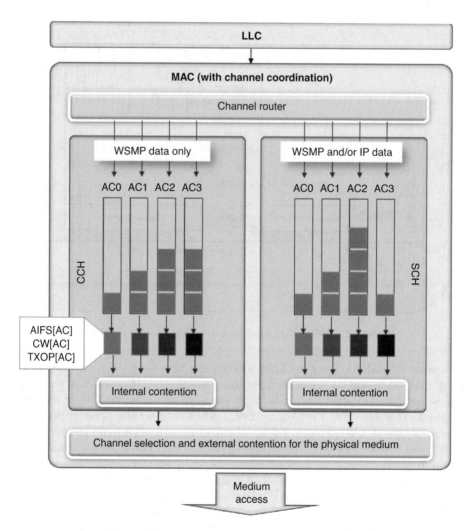

Figure 8.6 Medium access control in IEEE 802.11p.

maintains its own AIFS and BC. When there is more than one AC finishing the back-off at
the same time, collision is handled in a virtual manner. That is, the highest priority frame
among the colliding frames is chosen and transmitted, and the others perform a back-off with
increased CW values.

IEEE 802.11p uses the EDCF MAC of IEEE 802.11e with further adaptations. It specifies
separate queues for the control channel and the service channels and different sets of EDCF
parameters for each channel type. Figure 8.6 illustrates the design of the IEEE 802.11p MAC.

The values for the control channel are specified as shown in Table 8.1 and the default values
for the service channels are shown in Table 8.2. The values for the CCH are set to $CW_{min} =$
15 and $CW_{max} = 1023$. These values are fixed for all CCH transmissions. The values for the
SCHs can be adapted for each service and are announced during the service advertisements
for the respective channel.

Table 8.1 Default values for CCH queues

Access Category	CW_{min}	CW_{max}	AIFSN
0	aCW_{min}	aCW_{max}	9
1	$(aCW_{min} + 1)/2-1$	aCW_{min}	6
2	$(aCW_{min} + 1)/4-1$	$(aCW_{min} + 1)/2-1$	3
3	$(aCW_{min} + 1)/4-1$	$(aCW_{min} + 1)/2-1$	2

Table 8.2 Default values for SCH queues

Access Category	CW_{min}	CW_{max}	AIFSN
0	aCW_{min}	aCW_{max}	7
1	aCW_{min}	aCW_{max}	3
2	$(aCW_{min} + 1)/2-1$	aCW_{min}	2
3	$(aCW_{min} + 1)/4-1$	$(aCW_{min} + 1)/2-1$	2

The PCF mode requires an access point which is polling all stations within the network to check if they have scheduled packets for transmission. Each station replies with their request for bandwidth (and channel occupation time). The AP then grants access to each station separately.

DCF works both in infrastructure and in ad-hoc mode. It uses a request to send/clear to send (RTS/CTS) protocol to avoid collisions at receivers. Each sender, before transmitting a packet, requests access to the media with a short RTS message. The receiving station replies, if it has sensed the media being idle, with a short CTS message. We will not discuss the RTS/CTS mechanisms here further. They are most relevant for point-to-point communications [13]. They may thus be a useful mechanism in an AutoNet network with roadside stations that take the roles of access points. Most AutoNet applications, however, largely rely on broadcast transmissions.

8.2.2.5 Synchronisation

An important precondition for the CSMA/CA scheme – for synchronised time slots – are synchronised clocks in the nodes. In infrastructure mode, the AP broadcasts periodic beacon frames to all nodes with a fixed preamble and a time stamp. Each node can adapt its clock to this signal. In ad-hoc mode, each station has to maintain its own time. IEEE 802.11p therefore stipulates a GPS time signal for synchronisation.[2]

8.2.2.6 Multi-Channel Operation

As opposed to the other IEEE 802.11 technologies, the multi-channel operation defined by IEEE 1609.4 [8] specifies a characteristic synchronised time-triggered frequency hopping

---

[2] The GPS signal includes a Universal Time Coordinated (UTC) field.

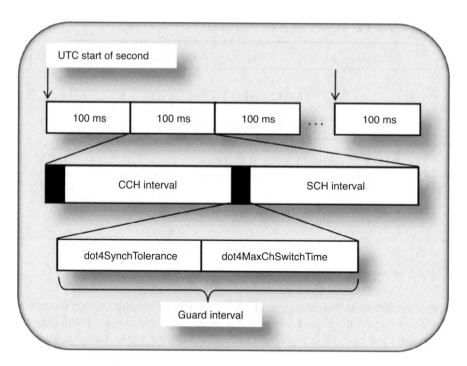

Figure 8.7   IEEE 802.11p channel synchronisation.

scheme. Every device is supposed to listen on the control channel periodically every 100 ms, synchronised by the start of a UTC second. A guard time precedes each CCH and SCH interval. During this time no transmissions are allowed and the MAC layer simulates a channel busy state. All devices that wish to send start with their contention for the channel after this guard interval to compensate for inaccuracies of the UTC signal. This scheme is illustrated in Figure 8.7.

Through this mechanism, it can be guaranteed that all devices monitor the CCH simultaneously during a defined time interval. Within this CCH interval, WSMP safety messages and WAVE Service Advertisements are sent. These service advertisements include information on which channel the service is offered on. After the CCH interval, every device may switch to an appropriate service channel and communicate within a WBSS or remain on the CCH.

After UTC synchronisation, the maximum time-allowed shift between the stations, induced by UTC errors and internal clock shifts, is less than 200 ns. A device is only allowed to operate as a service provider if its timing quality lies within a respective tolerance. If its clock shift exceeds this value, a device has to stop providing services and needs to switch to the control channel. This may be caused, for example, due to missing GPS signals in a tunnel. On the CCH, the device can use the time stamp fields of other WAVE service advertisement frames for synchronisation in order to keep its timing within the allowed limits.

8.2.3   ETSI G5A

Based on the IEEE 802.11p draft, standardisation in Europe at ETSI is currently developing a European profile, taking into account the European-specific spectrum regulations. It is thus dividing the profile into three parts:

- ITS-G5A: the core safety part working within the 30MHz band.
- ITS-G5B: the part for automotive efficiency applications.
- ITS-G5C: the 'consumer' part, which will be available for the remaining AutoNet applications.

In Europe, it is currently under discussion whether to divide the allocated frequency of 30MHz into three channels with 10MHz bandwidth each, one control channel and two service channels (CCH, SCH1, SCH2). The control channel shall be optimised for low latency transmissions of high priority safety messages (usually CAMs and certain DENs), high priority multi-hop messages (certain safety DENs) and even network layer beacons. The service channels may be used by all other services and for forwarding of packets in multi-hop fashion.

We will not further detail the current status of ETSI G5A, because it is subject to change until the final agreement of the profile.

8.3   Utility-Centric Medium Access in IEEE 802.11p

As described in Chapter 7, network utility maximisation techniques help to make AutoNets scalable. In order to do so, contention of communicating vehicles for the shared medium must be adapted in a way such that the most relevant packet of all nodes within mutual communication range is granted access to the medium [14]. For the purpose of data traffic differentiation, the dynamic adaptation of the contention window CW (see Section 8.2.2) within the MAC layer is considered a major lever.

8.3.1   Data Differentiation

Before any message transmission, each vehicle independently evaluates the benefit of the messages and applies the following three levers, which are part of the application layer and the data link-layer:

1. A variable called *benefit threshold* is introduced within the application layer. When active, only messages providing a higher benefit than the pre-determined threshold are allowed transmission. This ensures, for example, that the initiation of actually redundant messages is avoided.
2. A specific packet scheduling ensures that messages are broadcast in a sequence established according to their respective benefit values. All packets a node intends to transmit are buffered in the network interface queue. The functionality of this queue implements a message transmission sequence according to the respective benefit values. Packet dequeuing, thus, does not work according to the conventional first-in first-out (FIFO) principle.

In addition, packet enqueuing works as follows: in case of a full queue, a newly arriving packet can still enter the queue if its benefit value exceeds the one with the least benefit inside the queue. This lowest-benefit packet is then discarded.

3. In order to prioritise the packet with the most utility, a dynamic adaptation of the contention window CW within the MAC layer is considered a major lever. By selecting short contention windows in the event of highly relevant packet requests access to the medium, its likelihood of winning the contention process can be increased. This is due to the fact that both defer timers and back-off timers within the IEEE 802.11 MAC are determined based on the current magnitude of the contention window. Each station uses its individual CW to select both back-off and defer timers, which represent a key feature of the CSMA mechanism. The CW size is initially set to CW_{min}, and is increased each time a transmission attempt fails with an upper limit of CW_{max}. In this case, another back-off is performed using the new CW value. After each successful transmission, the CW value is reset to CW_{min}, and the station that successfully completed the transmission defers its activities for another DIFS and conducts a so-called post back-off.

8.3.2   Inter-Vehicle Contention

By selecting short contention windows in the event of a highly relevant packet requesting access to the medium, its likelihood of winning the contention process is increased. The smaller the contention window CW, the shorter the timers defining the period of time after a node may try or retry to access the shared medium. The following equation shows the benefit-based computation of a defer timer, that's starting integer value is randomly selected from a uniform distribution over the interval $[0, CW]$:

$$DeferTimer = (R \bmod (CW_{bb} + 1)) \cdot SlotTime$$

The length of SlotTime is set to 16 μs. R is a random number and CW_{bb} represents the node's current benefit-based CW. Then

$$CW_{bb} = f_a\{ut[p_{n_i}^{max}(t_i)]\}$$

with t_i as the point in time the contention is initiated, and f_a as a dedicated mapping function. The most simple example would be linear mapping of the expected utility ut and the CW, as, for instance

$$CW_{bb} = [1 - ut(p_{n_i}^{max}] \cdot (CW_{max} - CW_{min})) + CW_{min}$$

where CW_{max} (1023 slots) and CW_{min} (31 slots) are the minimum and maximum contention windows. Thus, the CW a node applies can always be adapted to the transmission utility ($ut \in [0, 1]$) of the currently handled message.

Besides the mere adaptation of the nodes' actual contention window according to estimated packet utility values, one could think about taking this one step further. Instead of passing on the remaining timer values into the next contention periods, timers are newly computed for each contention period, reflecting the expected utility (ut) of the currently treated data packets. This helps avoid the following situation: a node trying to transmit a rather redundant packet that has initiated a relatively long timer some contention periods ago. After waiting for a sufficient number of periods, its remaining timer is short and even may be shorter than the one started by another node which tries to transmit a highly beneficial packet. This ensures a medium access strategy that is fully utility-oriented and lacks any utility-agnostic, fair resource allocation tendency. Not the time a packet has already been waiting, but its up-to-date utility.

8.3.3 Cross-Layer Issues

In order to implement the three levers described above, the cross-layer design depicted in Figure 8.8 is proposed for each of the nodes participating in the AutoNet ad-hoc network: an ad-hoc agent in the application support layer is devoted to generating, broadcasting, receiving and storing messages. This agent has access to a set of context information through the management layer in the AutoNet Generic Reference Protocol Stack (see Section 10), such as geographic position and system time. It maintains a local memory containing previously received messages and also implements a benefit function for evaluating the benefit of all data packets before passing them down to the MAC layer.

The resulting values of the calculations are attached to each message's header. The so called benefit-based extension *BBE* within the data link-layer accounts for changing the functionality of the interface queue (modified enqueuing and dequeuing behaviour) and the medium access

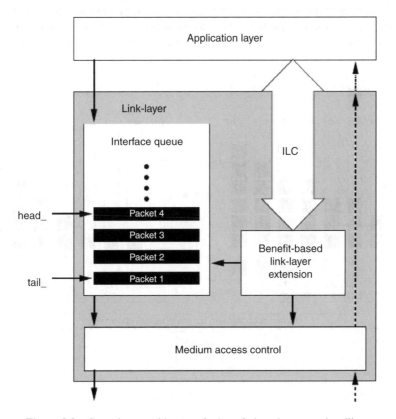

Figure 8.8 Cross-layer architecture for benefit-based message handling.

control mechanism as described above. The extension may thereby leverage the values attached to the packets' headers or continuously re-evaluate the benefit of all enqueued packets with the help of a benefit function. Again, the management layer as the means for inter-layer communication is used by the BBE as the means for acquiring all the necessary, mostly data offered by the application or application support layers, to calculate up-to-date benefit values for all packets waiting for medium access through.

An alternative to such a cross-layer design is the application of the already existing IEEE 802.11e standard to realise a benefit-oriented traffic differentiation. A great advantage of this available standard is that out-of-the-box network interfaces can be utilised instead of implementing a new cross-layer design. The IEEE 802.11e specification was originally developed to improve the quality of delay-critical applications such as Internet telephony (also called 'Voice-over-IP'). With the enhanced distributed channel access (EDCA) operating mode, it is fully functional in ad-hoc networks. To minimise delay and jitter of time-critical packets in situations of high overall network load, several traffic classes (TCs) are supported by IEEE 802.11e. Each category is assigned a certain priority determining the degree of favouritism with respect to access to the shared wireless medium. As visualised in Figure 8.9, each network node therefore implements up to eight packet queues to realise the prioritisation and sorts data into them according to the priority determined by the application layer before.

Nodes applying the EDCA scheme conduct a two-step contention process: in the first step, their different queues participate in an internal contention process which is conducted as if nodes contended for the real medium. As a consequence, virtual collisions occur frequently between the queues, upon which the highest priority packet is granted the chance to participate

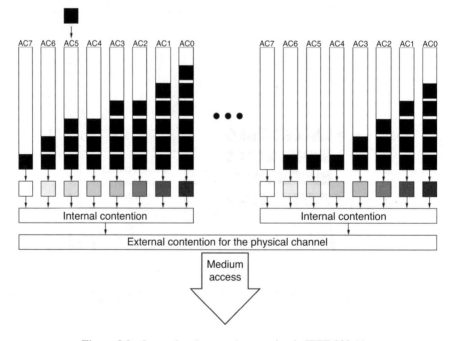

Figure 8.9   Internal and external contention in IEEE 802.11e.

in the external contention process while the colliding lower priority packets have to go into back-off. After doing so, the packet coming out ahead now has to contend with packets from other nodes for the real shared medium. This process is executed using the same parameters as before.

8.3.4 Evaluation of Utility-Centric Medium Access

The benefit-based scheme was evaluated by simulations. As the scheme is to maximise the overall benefit achieved by the communications, this overall benefit needs to be calculated based on a benefit model. We present an evaluation scheme and some evaluation results here. Within this scheme, each vehicle maintains a benefit variable, which is used to calculate the overall benefit within any simulated scenario. Upon message reception, the applications immediately evaluate the information contained in the message using the utility function ut. The resulting value can vary between 0 (no benefit) and 1 (highest benefit). This value is then added to the benefit variable, which is the sum of the overall benefit received from all messages at a single vehicle. The sum of all the individual benefit variables provides the overall, or global benefit. It represents the benefit provided to all AutoNet-enabled vehicles within a scenario and is the measure to be maximised.

For the evaluation scenario, we assume that vehicles initiate message transmissions in three different situations:

- Initiation of notifications about local dangers (distributed event notifications).
- Immediate retransmission or forwarding of messages in dense scenarios.
- Storage and forwarding of a set of messages (all messages within the evaluation simulation) in case of sparse scenarios when a new vehicle is met after a time of disconnection.

The queue assignment module ensures that packets are sorted into one of four queues according to their assumed benefit. Packets with benefit values between 1 and 0.75 are sorted into the first priority queue, packets between 0.75 and 0.5 into the second queue and so on. The queue resort module optionally ensures that the packet with the highest benefit is dequeued at the next transmission opportunity. Both internal contention between the packet queues and external contention for the real wireless shared medium support the transmission of the messages with the currently highest benefit values and thus are assumed to contribute to an improved global benefit.

The setup for the evaluation here is the ns2 network simulator with the extensions of the IEEE 802.11 standard (cf. Section 11) and the two ray ground model on the physical layer, where the radio channel is assumed to be error-free [3]. Within the ns2 simulation environment, we added a messaging functionality similar to the one described in Chapter 5 able to generate, broadcast, receive and store messages. Within this messaging agent, we implemented a store-and-forward communication scheme that broadcasts the whole message stack to every new neighbour vehicle entering the communication range. The messaging agent is also responsible for the computation of the message benefit ut which it adds to each message header together with the benefit threshold value. It thus needs access to the current geographic location, the current system time, data contained in the node's message stack and others. The AutoNet ns2 simulation environment allows two settings on the data link-layer within the ns2 environment:

the standard IEEE 802.11 specification and the IEEE 802.11e implementation. We added the possibility of re-sorting messages in the queues as an additional option. When activated, this mechanism changes the dequeuing sequence and the packet containing the message with the highest benefit value is chosen. For the standard IEEE 802.11 implementation chosen, the AutoNet implementation allows one to select mechanisms for a modified packet dequeuing, enqueuing and modification of the MAC functionality, each individually. In addition, the AutoNet simulation environment offers the option of a link-layer benefit re-evaluation of the packets. This allows one to re-assess benefit values that may be changing while packets wait within their respective queues. Vehicles' movements were generated with the CARISMA mobility simulator as described in Section 11.

Based on this simulation environment, we provide some simulation results to gain more insight into the way the benefit based system mechanisms work. The following results are calculated based on a simulation area of 8 km^2 in Munich downtown, 300 AutoNet-enabled vehicles with a communication range of 400 m running for 100 seconds in 50 independent iterations for each set of parameters. Interface queues by default could store up to 200 data packets. To simulate highly loaded network scenarios within manageable computing time, the nodes could only access a bandwidth between 0.1 Mbit/s and 0.5 Mbit/s. The message benefit was calculated using the distance from the source, message age, time since last reception, time since last broadcast and a general application relevance value. Simulations with other benefit functions and different parametrisation have proven that this particular improvement potential possesses a general characteristic.

Figure 8.10 shows the simulation results for a setup with the standard IEEE 802.11 MAC and a packet generation rate of 40 packets per second. The plots show how the overall achieved

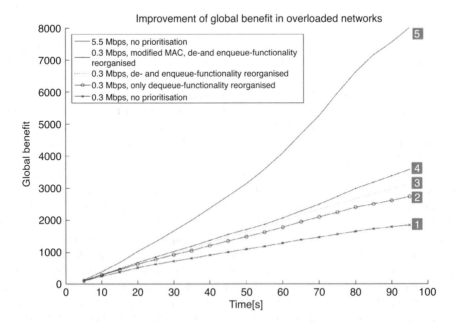

Figure 8.10   Effect of MAC and queueing modifications on overall information benefit.

information benefit develops over time when different mechanisms are applied on the MAC layer. Plot number five shows the benchmark of the benefit development when each node has a maximum data rate of 5.5 Mbit/s available. This ensures that almost all packets are sent immediately and queues do not build up. Thus, a reorganisation of enqueuing and dequeuing mechanisms would not show an effect. The overall benefit is calculated as the aggregated sum of the local benefit accounts of all vehicles over time. Plot number one shows the behaviour of a simple store-and-forward scheme in which nodes try to broadcast all packets contained in their send buffers with respect to the FIFO oder. Plot number two shows the results when the dequeuing mechanism is applied. In this case, the packet with the temporarily highest benefit is dequeued as the next one to be broadcast. Plot number three and four show that the global benefit can further be improved when in addition to the dequeueing mechanism the adapted enqueuing and adapted contention on the MAC are also applied. For the adapted enqueueing, the packet with the lowest current benefit existing in the queue is dropped instead of the newly arriving packet in case of a buffer overflow. For the adapted MAC, contention windows are adjusted continuously according to the benefit of the respective packets. This mechanism leads to a faster dissemination of packets with high benefit values. The magnitude of the improvement strongly correlates with the network load. In high load scenarios, long packet queues occur regularly within each node and benefit based sequence and contention modifications show a considerable impact on the global benefit.

Figure 8.11 shows the effect of a benefit re-calculation within the MAC layer. Since benefit calculation would generally need application knowledge, this mechanism needs a cross-layer information exchange implemented within the AutoNet management layer. This result is based on the same simulation setup with a simulated high network load and adapted message queueing. We compare the behaviour of the network with benefit calculation done by the

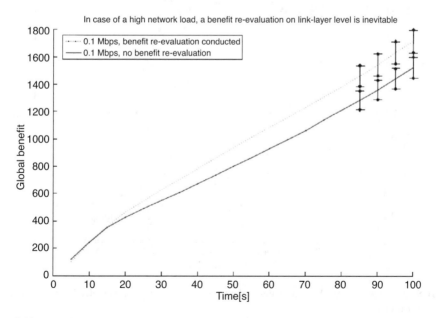

Figure 8.11   Development of global benefit over time with and without link-layer benefit re-evaluation.

applications when a message is generated with the behaviour in the case that the benefit can be recalculated over time within the different layers. In the first case, the BBE within the link-layer utilises the benefit values within the packets' headers to determine the broadcast sequence. In the second case, a re-calculation of all benefit values takes place each time the MAC layer has successfully finished the transmission of a packet and requests the next packet from the interface queue. Even though this process is quite complex and time-consuming, it significantly improves the global benefit over time (the 95% confidence intervals are shown in Figure 8.11). A simple example may explain this effect. While a congestion warning message may be stuck in a node's interface queue, another message might notify that the traffic congestion has dissolved. Thus, the congestion message in the queue is outdated and should be disregarded and not spread any more.

We finally illustrate the impact of contention window adjustments in highly loaded networks. Consider the following scenario: two different applications each generate a different type of message which is disseminated through the network with an unrestricted store-and-forward scheme. Each node is allowed a bandwidth of only 0.01 Mbit/s. The contention windows of the nodes are calculated dependent on the message benefit values. Messages of type-1 are always assigned a higher benefit value than messages of type-2 and thus they will receive shorter contention window times. Figures 8.12 and 8.13 show the behaviour of the benefit values received within this scenario using a standard store and forward scheme (Figure 8.12) and the benefit-based approach (Figure 8.13). When considering the Figures 8.12 and 8.13, the effect of taking into account benefit values for message dissemination is very obvious. For these simulations, each vehicle logged the benefit values of messages at the time of their

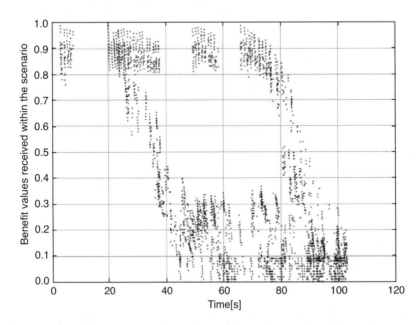

Figure 8.12 Total message benefit values at the time when messages are received with a simple store-and-forward scheme.

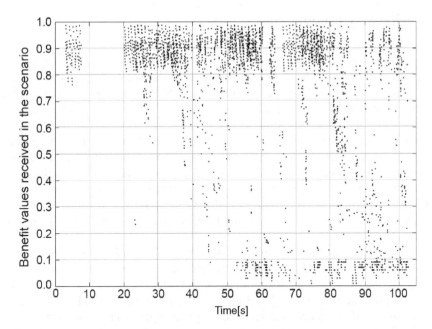

Figure 8.13   Total message benefit values at the time when messages are received with a benefit-adaptive scheme.

reception. The results shown in Figure 8.12 are based on a simulation with a simple store-and-forward mechanism. In this case, the benefit values of packets shortly after the first packet transmissions are relatively high. After some time when messages have been propagated by a high number of vehicles with varying interest in the information, many more packets with low benefit values are received. This leads to a waste of network resources by unnecessary information. A further effect can be taken from the graph: neighbouring vehicles estimate the benefit of the same message quite similarly. The numerous vertical chains of dots in the graph above represent the reception of one certain message by a number of cars at the same point of time. With increasing network load over time, the benefit of the information received by the different vehicles decreases further. The results in Figure 8.13 were based on the benefit-based scheme. While more messages with high benefit values are transmitted, some messages with relatively low benefit values are also still received by the vehicles in the scenario. We leave it up to the reader to think of explanations for this behaviour.

Besides efficiently disseminating data, the benefit-based mechanisms are also applicable to cooperative awareness scenarios. They can intrinsically deal with the challenges imposed by intersection collision avoidance applications at crowded intersections. The benefit evaluation approach leverages an optimum allocation of network resources. While the mechanisms on the physical, medium access control, and networking layers stay the same, it is crucial to compute the right benefit values. Those CAMs which are particularly important for avoiding accidents need to be favoured with respect to medium access, that is be attributed a higher benefit value, while others may be less time-critical.

8.4   Technology Comparison

In Chapter 2, we pointed out that several communication technologies co-exist together and, thus, can also be included in the physical communication technologies layer of the AutoNet Generic Reference Protocol Stack. We also pointed out the different communication characteristics: whereas, for example, cellular networks like UMTS or LTE provide a better coverage with a higher latency for broadcast transmissions, ad-hoc based networks like IEEE 802.11p provide for very low latency while requiring a sufficient penetration rate for proper functioning. Hence, there were several activities in the past focusing on the use of different communication systems for AutoNets. For example, in the project PRE-DRIVE C2X (see Appendix A) AutoNet safety messages between vehicles (V2X communication) were transmitted using IEEE 802.11p, whereas such safety messages were sent to neighbouring vehicles using UMTS or LTE in the project CoCarX (see also Appendix A). This raises the question of a technological comparison in order to create a future proof solution for AutoNets, especially with respect to system scalability.

In order to comprare cellular systems like UMTS and LTE with AutoNet ad-hoc networking technologies like IEEE 802.11p, intensive technological studies were performed [15, 16], especially for intersection scenarios. For example, Mangel et al. showed that by using the broadcast mechanisms of UMTS and LTE on the downlink within one cell, UMTS is barely able to handle a load of 1500 AutoNet messages per second, whereas LTE should provide sufficient bandwidth. In order to investigate the technologies in real AutoNet scenarios, Mangel et al. investigated in [17] typical intersection scenarios in the city of Munich, Germany, especially with respect to multi-channel and multi-carrier aspects, and with the number of communicating vehicles within a cell. Mangel et al. analysed that the average cell size in Munich is roughly 0.41 km^2, and the authors computed for each street segment per cell the number of vehicles by combining the vehicle frequency information and average speed with length. The sum over all street segments within a cell leads to the respective number of vehicles per cell.

Figure 8.14 shows the average number of vehicles per UMTS cell for all streets, for 42 m to the intersection, and 21 m to the intersection. Both values are derived from [18], representing the driver notification 42 m to the intersection, and the warning 21 m to the intersection. The results reveals that varying vehicle densities do not siginificantly reduce the deviation that can be found in cell sizes, with an average number of vehicles per cell on a workday between 10 and 100 vehicles. Using this result, Mangel et al. showed that a typical cross-traffic assistance scenario might introduce 1715 AutoNet messages per second in a cell. Therefore, the available bandwidth of UMTS is not sufficient to handle the data traffic generated by one single AutoNet application during rush hour in a typical city. On the other hand, Mangel et al. showed in [16] that IEEE 802.11p would be able to deal with such scenarios due to the nature of ad-hoc networking. However, it needs to be mentioned that in rural areas, it may be completely different since there are fewer communicating vehicles having sufficient coverage to perform the functionality of cross-traffic assistance, although cell sizes are potentially much larger in rural areas.

As a result, we can conclude that there is no single AutoNet communication technology that is able to handle all types of AutoNet applications in every type of scenario. Instead, the design of the AutoNet system has to choose suitable communication systems for different AutoNet application types carefully, and it needs to be flexible enough for future extensions and adaptations.

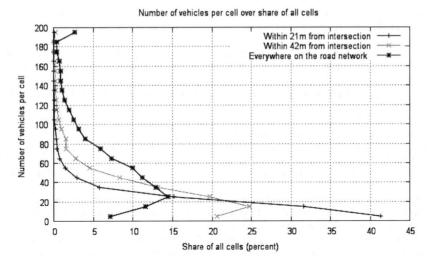

Figure 8.14 Relative histogram of average amount of vehicles per cell, based on [17]. Reproduced with permission of © 2011 IEEE.

8.5 Conclusion

The physical transmission of bits among the AutoNet entities are realised in the physical communications technology layer, thus forming the basis for communicating entities. We therefore take a closer look on IEEE 802.11p, which is the upcoming communication technology for both vehicle-to-vehicle and vehicle-to-infrastructure communication, especially for safety-related AutoNet applications. The most important conclusion from this chapter is the fact that there is not one single technology suitable for all AutoNet applications. Instead, there is a plethora of different communication technologies, and it greatly depends on the requirements of the application, the characteristics of the environment, and the current context in the vicinity of a vehicle, which communication system is currently best suited. In this way, the system design of AutoNet applications has to consider these issues in order to provide the best possible system for all AutoNet applications deployed. We can also conclude from this section that existing communication technologies can be further improved with respect to AutoNet scenarios. We showed that rather simple mechanisms are able to improve a utility-based message exchange significantly, and we also showed that new and upcoming technologies like LTE have better performance by far than the existing ones, making them interesting for other AutoNet applications too.

References

[1] Dietz, U. (2009) CoCar Feasibility Study – Technology, Business and Dissemination. Public Report by the CoCar Consortium.
[2] Li, X., Nguyen, T. and Martin, R. (2004) An analytic model predicting the optimal range for maximizing 1-Hop broadcast coverage in dense wireless networks. *Proc. 3rd International Conference on Ad Hoc Networks and Wireless Networks (ADHOC-NOW 2004)*.

[3] Torrent-Moreno, M., Jiang, D. and Hartenstein, H. (2004) Broadcast reception rates and effects of priority access in 802.11-based vehicular ad hoc networks. *Proc. ACM Workshop on Vehicular Ad Hoc Networks (VANET 2004)*.

[4] Takai, M., Martin, J. and Bagrodia, R. (2001) Effects of wireless physical layer modeling in mobile ad hoc networks. *Proc. ACM/IEEE International Symposium on Mobile Ad Hoc Networking and Computing (MobiHoc 2001)*.

[5] IEEE (2005) Draft Standard for for Wireless Access in Vehicular Environments – WAVE Resource Manager. IEEE P1609.1/D12.

[6] IEEE (2005) Draft Standard for Wireless Access in Vehicular Environments – Security Services for Applications and Management Messages. IEEE P1609.2/D3.

[7] IEEE (2005) Standard for Wireless Access in Vehicular Environments (WAVE) – Networking Services. IEEE P1609.3/D18.

[8] IEEE (2005) Standard for Wireless Access in Vehicular Environments (WAVE) – Multi-Channel Operation. IEEE P1609.4/D07.

[9] IEEE Standards Board (2005) Local and Metropolitan Area Networks – Specific Requirements Part 11: Wireless LAN Medium Access Control (MAC) and Physical Layer (PHY) specifications Amendment 8: Medium Access Control (MAC) Quality of Service Enhancements IEEE Standard 802.11e.

[10] IEEE Standards Board (2006) Draft Amendment: Local and Metropolitan Networks – Specific Requirements, Part 11: Wireless LAN Medium Access Control (MAC) and Physical PLayer (PHY) specifications: Amendment 3: Wireless Access in Vehicular Environments (WAVE). IEEE Standard 802.11p/D1.0 WAVE.

[11] IEEE Standards Board (2008) Standard Specification for Telecommunications and Information Exchange Between Roadside and Vehicle Systems – 5 GHz Band Dedicated Short Range Communications (DSRC) Medium Access Control (MAC) and Physical Layer (PHY) Specifications. IEEE Standard 802.11p.

[12] Menouar, H., Filali, F. and Lenardi, M. (2006) A survey and qualitative analysis of MAC protocols for vehicular ad hoc networks. *IEEE Wireless Communications*, October 2006.

[13] Schiller, J. (2003) *Mobile Communications*, 2nd Edition. Addison Wesley.

[14] Moustafa, H. and Zhang, Y. (2009) *Vehicular Networks – Techniques, Standards, and Applications*. Auerbach Publications 2009.

[15] Mangel, T., Kosch, T. and Hartenstein, H. (2010) A comparison of UMTS and LTE for vehicular safety communication at intersections. *Proc. 2nd IEEE Vehicular Networking Conference (VNC 2010)*.

[16] Mangel, T., Schweizer, F., Kosch, T. and Hartenstein, H. (2011) Vehicular safety communication at intersections: buildings, non-line-of-sight and representative scenarios. *Proc. 8th International Conference on Wireless On-Demand Network Systemsand Services (WONS 2011)*.

[17] Mangel, T. and Hartenstein, H. (2011) An analysis of data traffic in cellular networks caused by inter-vehicle communication at intersections. *Proc. IEEE Intelligent Vehicles Symposium (IV2011)*.

[18] Klanner, F. (2008) Entwicklung eines kommunikationsbasierten Querverkehrsassistenten im Fahrzeug. [Development of a communication-based crossing-traffic assistant in vehicles.] Dissertation. [German]

9

Security and Privacy

Alongside the ISO/OSI reference model, the AutoNet Generic Reference Protocol Stack includes a security subsystem, as illustrated in Figure 9.1. Like the management subsystem, security flanks the overall AutoNet Generic Reference Protocol Stack, because the security and privacy issues address all layers and thus need to be coordinated respectively. This way, security as well as privacy is considered as a cross-layer functionality, which needs to interact with all other communication layers.

This chapter provides insights on the mechanisms required to secure AutoNets. While robustness requires a system to stay operational even under harsh conditions, security addresses the protection of the system against deliberate attacks and fraud. Based on typical application scenarios for AutoNets, this chapter provides an overview of the assets in danger, the threats upon them, and ways to secure them – both by preventing attacks and enclosing attacks where prevention may fail (intrusion management). Providing security for a system is not only a technical challenge but also a management task. Based on the assets under threat, the likelihood of attacks and the techniques available for protection, the security manager needs to find a solution that decides upon the achievable level of security under certain conditions. For example, it has to decide on the use of resources such as bandwidth or processing time, which have to be weighed against cost and impact on system behaviour (e.g. latency) caused by attacks.

In order to access the services of the security layer, every layer of the AutoNet Generic Reference Protocol Stack uses the security service access point (S-SAP) as illustrated in Figure 9.2. Due to its cross-layered functionality, the security layer controls – as well as the management layer introduced in Chapter 10 – security aspects of the overall system. This cross-layered functionality allows for an overall system optimisation with regard to the security aspects of communication. Therefore, the security layer is able to control the security settings in every other layer using their respective SAPs.

In this chapter, we present AutoNet security along a defined framework. The framework is a basic security system model which can be implemented with different security technologies to achieve different levels of protection at different expenses. The implementation may vary and can be adapted to changes on the threat side as well as changes on the

Automotive Internetworking, First Edition. Timo Kosch, Christoph Schroth, Markus Strassberger and Marc Bechler.
© 2012 John Wiley & Sons, Ltd. Published 2012 by John Wiley & Sons, Ltd.

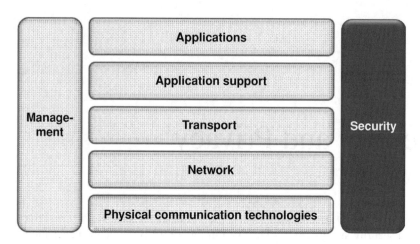

Figure 9.1 Security subsystem in the AutoNet Generic Reference Protocol Stack.

technology side, for example when new technology becomes available. This chapter comprises three parts:

- In the first part, we focus on AutoNet-based security from a general perspective. We discuss assets and threats that are of particular importance for AutoNets. Afterwards, stemming from these characteristics, challenges and requirements of a security system for AutoNets are emphasised, followed by the introduction of the AutoNet security framework.
- The second part introduces basic security elements and models and discusses their use in the AutoNet context. It starts with the basic cryptographic toolset required (Section 9.4), followed by a discussion of trust models (Section 9.5.1). With an introduction to certificates and certificate handling, it lays the ground for the ability to communicate securely in AutoNets.

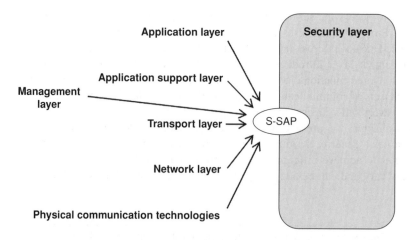

Figure 9.2 Service access points of the security layer.

- The third part is dedicated to more specific mechanisms to secure AutoNets, both the mobile entities and the communications. This part starts with a discussion on how to protect the vehicles against both physical attacks and remote attacks using wireless access (Section 9.6), that is in-vehicle security as a prerequisite for trust establishment in cooperative systems. It then looks at AutoNet-specific solutions for the typical AutoNet-based communication aspects. Furthermore, privacy will be emphasised in this context in Section 9.8.

Like the previous chapters, this chapter is also summarised by the conclusion in Section 9.9.

9.1   Stakes, Assets, Threats and Attacks

9.1.1   Stakeholders and Assets

In terms of the assets at stake, we distinguish between assets to be lost by normal AutoNet entities and assets to be gained by malicious subjects. The different owners of the assets and the attackers interested in (some of) them, may have differing orders of how important or valuable an asset is to them. For which asset would you spend the most resources for protection: the asset that is most valuable to you or the asset that is most endangered because it is most valuable to an attacker?

In many typical AutoNet scenarios, the systems have to protect the greatest asset, the life of the system user. No malign intervention shall threaten the health of either the drivers or other traffic participants. In addition, the value of the vehicles equipped with such a system and that of other goods endangered by the operation of these vehicles must be protected from theft or damage.

Within the AutoNet system, one of the main assets used as the basis for correct operation is the content generated and used by the applications. Thus, the authenticity of safety and traffic message data is among the most important objectives of automotive internetworking security systems. At the same time, solutions should protect the privacy of the users of the system in the best possible way.

Overall, the major interest of owners and users of AutoNet systems concerning security is that the system is guaranteed to work, untampered with and exactly as specified, in a reliable way. Additionally the privacy of the users (and owners) must be protected. Data access needs to be controlled accordingly. Apart from this fundamental protection role, AutoNet security systems provide added-value services for users such as vehicle tracking and immobilisation in case a stolen vehicle is driven by an unauthorised person. Besides the owner, other stakeholders of the system also have viable interests in security, which must be addressed and balanced by the security system solution.

Vehicle manufacturers are interested in a low negative societal impact of their vehicles, out of their own ethos, because of market requirements or because they have to fulfil a number of regulations. Therefore, vehicle manufacturers share with vehicle owners the goal of a securely and robustly working system. This is also important for liability reasons. If anything goes wrong due to reasons not accountable to the vehicle, it is important for the vehicle manufacturer to be able to prove that the incident has not been caused by vehicle system malfunction. Naturally, the vehicle manufacturers also have an interest in a high benefit/cost ratio. This is true not only for current systems, but also for future versions. Therefore, the system needs to be highly scalable and extendable. The system should also be flexible, that is easily adaptable to the specific requirements of a certain set of uses in a certain vehicle model sold in a certain market. This is necessary to meet regulatory requirements and to be able to adapt the security level according to the user requirements.

The interests of *suppliers* are similar to those of the vehicle manufacturers. However, there may be a difference with respect to the type of system a supplier offers. It is reasonable for a security solution provider to sell a more complex solution, while a system integrator prefers to avoid too much complexity. Moreover, suppliers have certain interests in common (and standardised) vehicle interfaces, but possibly not in standardised security solutions. This is due to the fact that suppliers typically provide solutions for several vehicle manufacturers. An individual (security) solution is likely to be sold for a higher price than a standardised one.

Service providers need a secure system environment for their offerings as well as means for billing and accouting. Moreover, they have a vital interest that their services are not used illegally – and naturally the services must be protected against attacks and manipulation of attackers outside the AutoNets. Moreover, they are interested in a high service availability and thus others must not be able to interrupt service availability.

Road authorities are responsible for overall safety on roads and therefore require high reliability paired with the possibility of detecting misbehaving entities and the means to stop faulty or malicious behaviour. To do so, road authorities will work together with enforcement authorities, which need the ability to identify, access and possibly control entities.

Apparently there are different expectations from the different stakeholders for security issues. As a result, system design must resolve some of the potential conflicts in the goals of the stakeholders. Transparent identity usage and individualisation will allow the users to configure the system such as to protect their desired level of privacy while at the same time being offered easy to use services.

9.1.2 Threats and Attacks

Attacks on AutoNets can be manifold [17]. We distinguish two separate classes of attack:

- Attacks with physical access to a vehicle (physical attack, PA).
- Attacks without physical access (over-the-air or wireless attack, WA).

The focus of our discussion will be on WA types of attacks, but we will also discuss some basic security measures to protect the vehicle physical infrastructure from illegal tampering. With automotive internetworking, the effect of such tampering may not be contained within the attacked vehicle, but may affect other entities as well, for example by initiating false safety messages. In the following, we provide and discuss a comprehensive but not necessarily exhaustive list of attacks.

- PA: *Tampering* with the system of a vehicle, roadside unit or central management entity, for example changing sensor values or forced malfunctioning of (safety critical) in-vehicle control units.
- WA: *Denial of service (DoS)* attacks aim at disturbing or bringing down an AutoNet entirely or in parts. Network jamming can be considered a DoS attack. In this case, an attacker jams the network by generating a number of transmissions that disturb or prevent communication

for other nodes, for example by sending large numbers of packets, thereby reducing the availability of the network for other nodes or even breaking it down completely. Local jamming is an easy, low cost task for any network node. It is, however, also easy to detect.

- WA: *Alteration attacks* are also likely in AutoNet scenarios. Examples include packet delaying or packet rejection by malicious nodes. Node selfishness falls under this type of attack, too.
- WA: *Fake message injection or forgery* includes advertising of fake events by malicious nodes, influencing application behaviour. Message replay belongs to this kind of attack. Messages sent by honest vehicles in this case are caught and sent out again at some future point in time without any further modification. Both message fabrication or alteration may also be misused for impersonation attacks where attackers take on certain roles, for example of public authority vehicles like the police or an ambulance. Message alteration can also be used to change the behaviour of the system. An example is injecting false positions in CAMs, leading to false warnings or even automated breaking in other vehicles.
- WA: *Privacy infringement* is possible in AutoNets by associating any kind of identifier, contained in a data packet or message, to the vehicle and in turn to a user profile. If this is possible when simply collecting information by overhearing communications, it is an easy task.
- WA: Via *Masquerading*, an attacker makes use of the rights of another entity, for example by using false identities or stealing identification credentials, either to gain some of its assets or to damage its reputation.

For all types of attacks, the following two methods can be applied:

- *Extraction and modification of secret material.* Here, the attacker extracts or modifies secret information (e.g. private key, digital signature, certificate, confident information) of a vehicle, roadside unit or infrastructure entity. The attacker is then able to perform identity spoofing by sending messages, which will appear as sent by the unaware victim vehicle. The attacker can generate and validly sign fake messages.
- *Sybil attack.* The sybil attack breaks the one-to-one relationship between physical and logical entities. A malicious node tries to obtain many different identities in order to control a consistent part of the network (e.g. by owning different pseudonyms at the same time and then claiming to be many vehicles) [16].

Along the taxonomy of [1, 5], attackers can be classified in several ways: (i) according to their network membership as outsiders or insiders of the system, (ii) according to their motivation as rational or malicious, (iii) according to their methods as active or passive, and (iv) according to their scope as local or extensive. An insider can be understood as an authenticated node within the network, and may possess a certified public key. An outsider can be viewed as an intruder without the usual standard credentials an insider would have. Thus, an outsider will usually use different means to attack. While a rational attacker is most interested in their own benefit, a malicious attacker's greatest interest is in causing damage to others or the network as a whole. An active attacker will perform malicious actions while a passive attacker is content with eavesdropping. A local attacker is limited in the consequences of their attacks or reach into the network while an extensive attacker could either cause considerable damage on a wider scale or be able to interfere with a large part of the network. As examples of types

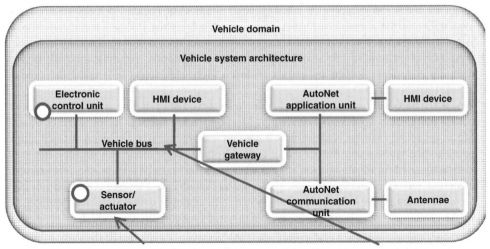

I. Furgel and K. Lemke, A review of the digital tachograph   M. Wolf, A. Weimerskirch and C. Paar, Security in
system, in: Proceedings of the Workshop on Embedded   automotive bus systems, in: Proceedings of the
Security in Cars (escar)'04, 2004.   Workshop on Embedded Security in Cars (escar)'04,
                                                                                2004.

Figure 9.3   Possible attacks and countermeasures in AutoNet settings.

of attackers and their motivations, consider drivers who want to gain a comparative advantage in traffic over other drivers, criminals interested in harming another person or organisations who would like to get access to driver profiles.

For an illustration of the threat and attack model, consider the following two examples. They show which assets are in danger or what damages are caused by a specific means of attack. In some cases, countermeasures are noted as illustrated in Figure 9.3.

1. *Introducing false traffic information.* The asset to be gained by the attacker in this case is a faster driving time by clearing their route [2]. The attacker would typically be a system insider (generating system traffic messages), acting in a rational way (for their own benefit) and behaving actively (by message generation), the attack being local.
2. Changing sensor values. The probability of accidents rises with false driver information caused by an attacker changing sensor values [3]. In this case, the attacker can be classified as a malicious, active, local insider. Possible countermeasures are plausibility checks.

The reader might be interested to know that there are ISO standards dealing with attacks on computer systems: ISO 26262 classifies severity of attacks while ISO/IEC 15408 & 18045 define the probability of success of an attack.

9.2   Challenges and Requirements

AutoNet characteristics form a specific environment for security solutions. We presume the availability of position-knowledge for all entities. This information can be used for security solutions (e.g. for plausibility checks) but needs additional protection as well, as we will see later on in this chapter. Vehicles are already equipped with means to protect them from physical access and theft. They undergo periodic maintenance and checks of their systems. Vehicles usually possess fixed IDs (e.g. vehicle identification numbers (VINs)[1] and number plates), which are physically attached to the vehicle. Also, owners of vehicles are usually registered. Raya and Hubaux [5] suggest vehicles should store electronic counterparts of these physical IDs, that is an electronic license plate and an electronic chassis number. These IDs can be used for law enforcement purposes for which the security system would need to provide special mechanisms to allow authorised retrieval of the vehicle's or owner's identity.

AutoNets pose some specific security system challenges. They require system design decisions with respect to their conflicting goals. Among these are the following:

- *Mobility versus trust.* Mobile AutoNet nodes exchange data with thousands of other nodes, but with almost all of them only very few and for a very short time. Thus, it is difficult to establish trust relationships.
- *Liability versus privacy.* Vehicle drivers need to be liable for their actions and this means linking data to vehicles or natural persons. This creates the challenge of protecting privacy.
- *Real-time communications versus secure cryptography.* Applying hard cryptographic mechanisms requires computing time and produces communication overheads, which need to be balanced not only against the cost for additional hardware resources, but also against the real-time requirements of many safety and traffic efficiency applications.
- *Global mobility versus trusted authorities.* This challenge can be described by the following question: how can a system be achieved in which a vehicle potentially driving around the globe can always trust others' credentials?

Depending on the system requirements, the architecture decision and the implementation solution need to be balanced within this area of conflict.

The threat analysis shows that it is imperative to provide protection against software attacks as well as a certain level of protection against physical attacks by hardware tamper-protection mechanisms. Based on the identification of the assets to secure and the possible threats and attacks, we derive the following requirements for the AutoNet security system:

- *Data and message integrity.* Integrity ensures that data or messages have not been altered on their way from source to destination.
- *Data consistency or plausibility of data.* This requirement checks whether data received from different sources is consistent, This can be achieved by verifying logical plausibility to identify fake messages or invalid data, for instance.
- *Data confidentiality.* Confidentiality includes as the most important feature the provision of secure communication channels, for example between vehicle and operation support system.
- *Privacy protection.* Privacy protection of drivers and/or vehicles prevents location tracking or illegal access to any type of private information.
- *Prevention of unauthorised access.* It must not be possible for unauthorised parties to access services within the infrastructure or the vehicle.
- *Detection of malicious behaviour and containment of a security attack.* The security architecture has to provide features to detect malicious nodes within the AutoNet, which also

---

[1] VINs are globally standardised in ISO 3779.

includes measurements to identify security attacks. Therefore, it is of particular importance that the system has to support the exclusion of (malicious) nodes from communications in the AutoNet.

- *Resistance against software attack and tamper.* Apparently, the system must be designed in a way that software attacks will not cause a total breakdown of service availability or, even worse, a breakdown of the overall network.
- *Authentication of entities.* Authentication of entities is required for any personal service use or payment systems. It increases trust in the origins of data since it provides additional trusted information about the different entities.
- *Authorisation.* Apparently, robust authorisation mechanisms must be supported to ensure the different entities in the AutoNet perform respectively defined actions (and will be prevented from performing other not-allowed actions).
- *Availability.* System availability should be ensured by robustness measures, redundancy and alternatives, for example other communication channels if one is jammed.
- *Non-repudiation.* The system has to provide mechanisms such that malicious or unlawful behaviour can be identified and the subjects acting that way are not able to deny their behaviour. This certainly needs to be ensured for proper participation in the AutoNet system.
- *Hardware/software configuration attestation.* A reliable proof of the hardware and software setup is an important precondition that the system will behave correctly. This addresses security aspects as well as functional aspects of the different components. This measure avoids the fact that a misconfigured component will disrupt or disturb AutoNet communication, resulting, for example, in a degradation of performance or scalability of the overall system.

On an abstract level, the security requirements can be classified as *(i) secure communications*, including integrity, consistency, authentication and non-repudiation, *(ii) secure system*, including resistance, intrusion detection, authorisation, availability, hardware/software configuration attestation, secure key storage and detections of manipulations, and *(iii) privacy provisions*, including confidentiality, access authorisation and other information access prevention mechanisms. Proposals on how to address the respective requirements and achieve the desired system behaviour are presented in the third part of this chapter, with respect to a secure (vehicle) system in Section 9.6, with respect to secure communications in Section 9.7 and with respect to privacy in Section 9.8.

Besides the functional requirements, the security system additionally has to provide several non-functional requirements. Among the most important non-functional requirements are the following:

- *Scalability.* An important characteristic of AutoNets is the unpredictable number of vehicles participating in the AutoNet. Numbers may range from none or a few vehicles in rural areas at night, up to hundreds or thousands of vehicles on a large congested motorway. It is crucial that all system requirements continue to be met with an increasing number of system participants.
- *Efficiency.* Due to the unpredictable number of communicating vehicles, timing constraints of applications should be met accordingly. Hence, efficiency has to provide dynamically

adjustable mechanisms to ensure that these constraints are met in sparse as well as in dense AutoNets.
- *Costs.* Security mechanisms require additional hardware resources, which must be available in the different entities. Resources may range from additional memory required, faster processors to compute the security algorithms, up to completely new hardware such as dedicated crypto processing units, which may be needed to compute security algorithms in a very short time. In AutoNets, it is also important to consider that hardware cost in vehicles is typically high compared with hardware costs in infrastructure components.

It is important to mention that some requirements may be difficult to meet at once, or they even may be conflicting. For instance, conflicts may appear between the required high level of security for integrity, while efficiency needs must also be met. In this example, a stronger encryption requires more computation time for the encryption, resulting in higher communication latency. Many of the proposed security solutions are thus either not meeting all security requirements, fail to meet performance constraints or pose unduly high implementation cost or operational challenges. This is where important design choices have to be made and different mechanisms have to be combined in a meaningful way. While meeting these functional requirements, it is important to design an efficient security system. The overheads on communication must be low and the processing time short.

9.3   AutoNet Security Architecture and Management

In this section, we describe the security system components and their interaction in more detail. The security solutions are based on certain system assumptions like the use of a certain wireless technology, the availability of connectivity, secure positioning, secure global time, tamper resistant hardware, trusted authorities, fixed number of message types or contents and so on. Where it is not clear from the context or it deviates considerably, we explicitly denote the assumptions made by a system.

Recall the generic AutoNet architecture from Chapter 3. We will focus on security within the AutoNet mobile domain and the corresponding domain interfaces in the following because this domain is the part of the AutoNet system architecture that poses special requirements not easily met with known security solutions, as mentioned before. In particular, we will concentrate on the AutoNet messaging required for realising the driving-related applications. This is especially challenging with the large number of vehicles participating in the system all generating messages. Vehicles receiving these messages must be able to trust their content and to assess the quality of the information contained in them. To achieve this, the following parts are combined: AutoNet systems in vehicles are secured and receive a certification when able to prove that they implement a certain minimum of data sensing, information processing and communication capabilities together with a minimum set of security mechanisms. When participating in message exchange, vehicles need to be able to sign these messages and prove that they possess certain certified properties. They need not be required to disclose their identity though, unless under clearly defined circumstances for non-repudiation. Also, there is typically no need to encrypt the message content.

To realise the desired functionality and address the requirements, the AutoNet security architecture is based on public key infrastructure (PKI) [18]. In addition to the in-vehicle

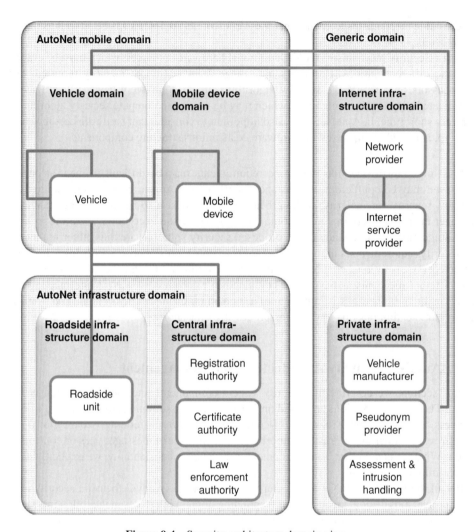

Figure 9.4   Security architecture domain view.

security system, certain infrastructure entities are dedicated to the security of the overall system, especially certificate authorities (CAs) as so-called trusted third platforms (TTPs). Figure 9.4 provides a view on a possible organisational architecture. Registration authority, certificate authority and law enforcement authority are all assigned to the central infrastructure domain. For registration and law enforcement, this is straightforward. The certificate authority may as well be part of the private infrastructure domain. However, since the certification process requires checking AutoNet-specific behaviour, we place it into the AutoNet infrastructure domain. The provision of pseudonyms, on the other hand, is depicted as a service in the private infrastructure domain as is assessment and intrusion handling as an additional, optional service that we include here according to the organisational proposal of [15]. The registration authority is responsible for providing (electronic) identifiers for vehicles and binding these identifiers with other information, especially about the owner of the vehicle. The certification authority checks the compliance of an AutoNet system implementation with defined rules, for example compliance with a standard, and manages the overall certification process (issuing, checking, updating, revoking certificates). The pseudonym service provider in this view is a separate entity. Vehicles can make use of this service to acquire pseudonyms with which they are able to generate validly signed AutoNet messages and hand out certificates without disclosing their identity. The assessment and intrusion handling service provider is a private entity that supports vehicle systems in the assessment if data received is correct, for example by additionally using information available in the infrastructure, and by identification and management of intrusions, such as providing information on known intrusion risks. The law enforcement authority is responsible for detection and punishment of misbehaviour. To do so, it must be able to resolve pseudonyms and link systems and messages to individuals. We believe, however, that this linking should only be possible when two or more authorities, possibly together with the vehicle manufacturer, combine their knowledge.

Security in an AutoNet entity is a cross-layer approach rather than a specific AutoNet security layer. Nevertheless, each layer provides their own security mechanisms interacting with the cross-layer security system. AutoNet communication makes use of both standard security mechanisms of the different layers as well as additional security mechanisms specifically addressing further AutoNet requirements, as depicted in Figure 9.5.

The standard security mechanisms and protocols of the different layers are quite common and will not be detailed here further. They are only of minor relevance for the AutoNet messaging system for driving applications. They are, however, required for other communications in AutoNets, for example for providing entertainment services to passengers or for secure remote access to vehicles. Interested readers are referred to the respective literature and Request for Comments (RFCs) of the Internet Engineering Task Force (IETF). On the application layer, a typical application-specific protocol is HTTPS for securing HTTP traffic. On the transport layer, security protocols like SSL (Secure Socket Layer) or its successor TLS (transport layer security) are used to secure data exchange. On an IP-based network layer, a typical security protocol is IPSec (IP Security), and the lower layers for IEEE 802.11 based wireless communication are secured by WPA or WPA2 (WiFi protected access). The IEEE 802.11i standard defines additional security solutions for the IEEE 802.11 wireless LAN systems, namely WPA PSK (WiFi protected access pre-shared key) authentication, WRAP (wireless robust authenticated protocol) and CCMP (counter mode with cipher block chaining message authentication code protocol). A RADIUS (remote authentication dial in user service) server is required for the use of these techniques and needs to reside somewhere in the AutoNet infrastructure. AutoNet nodes need access to it either through an RSU or a cellular communication link.

AutoNet messaging includes additional specific security fields and uses AutoNet security services through the AutoNet security component interface. The AutoNet system can decide at each layer which type of security it will use. This way, the system chooses the corresponding available cryptographic primitives, that is it decides for each message or packet whether its payload will be encrypted, whether it will be signed or whether it will not be secured. In the case of encryption or signature, the system may select between different cryptographic methods.

Figure 9.6 provides an overview of services and related security data stored in the AutoNet security subsystem. While not exhaustive, it contains the most important features. Policies,

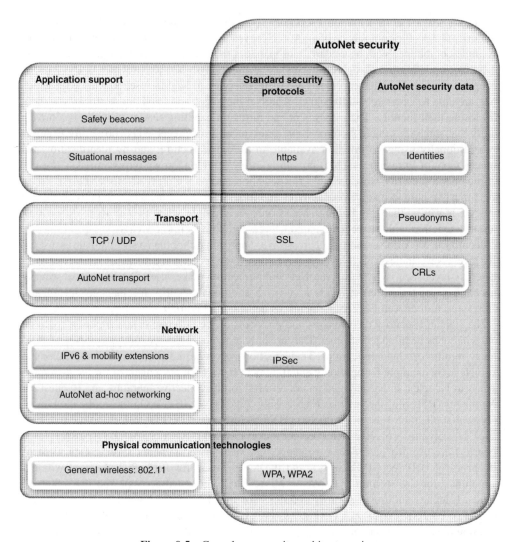

Figure 9.5 Cross-layer security architecture view.

profiles and rules are required to control access to the vehicle system. Based on successful authentication, an object may be authorised to gain a certain degree of system access depending on the policies defined. Rules are specified to be used by the vehicle firewall, for example to filter out certain data or allow or block data to pass according to these rules.

9.4   Security Services

9.4.1   Cryptographic Mechanisms

The requirements introduced in the previous section need to be translated into corresponding security services. All security services use some underlying security primitives, that is

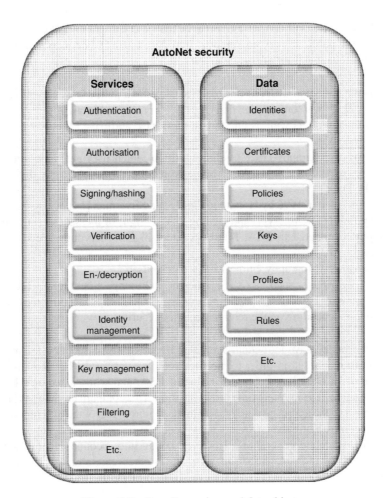

Figure 9.6   Security services and data objects.

primitives for cryptographic key management [19], encryption, signatures, certificates or filters. In general, cryptography distinguishes symmetric and asymmetric encryption mechanisms. For automotive internetworking, cryptographic primitives for asymmetric cryptography are of particular importance. Symmetric cryptography based on pairwise keys is not suitable for vehicle communications due to the very large number of nodes and the quickly changing communication partners. Although symmetric cryptography is typically faster to compute, and it generates less transmission overhead, it has one serious disadvantage: every two communication partners need to compute their own pair of session keys which can then be used to calculate (hashed) message authentication codes (HMACs) for subsequent messages. In an AutoNet, each vehicle cannot store the keys of all other vehicles. Dynamically computing session keys between any two neighbouring vehicles is also not an efficient solution since messages are frequently addressed to all neighbours and neighbours are changing all the time.

In the following, we will concentrate on asymmetric cryptography. There are several approaches to realise public key cryptography, as introduced in [11]. For encryption in AutoNets, the following mechanisms are of particular interest:

- Rivest, Shamir, Adleman (RSA) is one of the most popular public key encryption algorithms nowadays [12]. It was one of the first available cryptographic algorithms that could be used for both signing as well as encrypting messages, and paved the way for public key cryptography.
- Elliptic curve cryptography (ECC) is an algorithm which uses algebraic structures over elliptic curves using finite fields. An example protocol is the elliptic curve digital signature algorithm ECDSA [13].
- NTRU was invented in 1996 by Hoffstein, Pipher and Silverman [14]. It also can be used for both encryption (NTRUEncrypt) and signing (NTRUSign) of messages using lattice-based cryptography. Compared to RSA, NTRU provides higher encryption speeds at the same security level.
- For group-based cryptography, bilinear maps can be used.[2]
- Many other algorithms, such as (XTR (efficient and compact subgroup trace representation), cryptography based on hyperelliptic curves, Braid groups, Merkle trees, etc.) can be considered too, depending on the requirements for security.

The mechanisms all have individual strengths. Hence, for each scenario the most suitable algorithm has to be chosen in order to meet the requirements in the best possible way. The signature size of ECDSA is a smaller than that of RSA, and while it requires less time for signature generation, it needs more time for verification. This comparison shows the importance of the assessment of the approach with respect to the requirements of the different scenarios. In typical AutoNet messaging scenarios, a vehicle needs to process more signature verifications per second than signature generations. However, taking into account signature size as well, apart from the application scenario it may depend on the vehicle density which algorithm performs better.

9.4.2 Digital Signatures

The possibility of authenticating groups, entities or properties of these entities is a fundamental capability for security. Authentication is required as a basic mechanism for authorisation. It is also required to verify that received data has been generated by a trustworthy entity with a system that fulfils certain requirements. Authentication is guaranteed by the employment of a digital certificate, which 'binds' a certain public key to an identity.

Authentication for messaging is realised by sending entities signing all generated messages with their private keys and attaching these signatures to the messages. Receivers can then verify that a message has indeed been sent by an authenticated entity with the corresponding key, where such a key can only be owned by entities that fulfil certain basic requirements and, thus, the message has been generated by such a trustworthy and capable sender.

---

[2] For using bilinear maps in cryptography, interested readers may be refered to http://www.umiacs.umd.edu/partnerships/lts/LTS_Report_Jan04.pdf.

This way, a signature for a message validates that the sender of the message is known by the receiver, and that the message was not forged on its way from the sender to the receiver. It is worth mentioning that with all private key-based techniques, signatures are protected by the strength of the signature algorithm as long as their private key is not disclosed. In Europe, Directive 1999/93/EC of the European Parliament and of the Council defines legal measures for electronic signatures.

Signatures can be used in AutoNets to increase trust in messages or in data received by some nodes [6]. This is achieved by signing the messages, thus ensuring that the message cannot be modified by intermediate nodes and increasing the likelihood that the sender is a trustworthy entity. Vehicles require private/public key pairs to be able to compute message signatures with one of their private keys and for other vehicles to verify the signatures with the corresponding public keys. These key pairs need to be generated by a trusted third party, thus a public key infrastructure (PKI) is required.

Usually, the signed message consists of the original message Msg, which is enhanced by the digital signature and the certificate of the vehicle:

$$\text{Msg}|\text{Signature}_{\text{PrivateKey}(v_i)}([\text{Msg}, \text{Time}])|\text{Certificate}(v_i)$$

Here, v_i denotes some vehicle i sending the message. Thus, the vehicle sends the message without encryption, followed by the signature computed over the message itself (or over the hash value of the message) concatenated with the current time stamp. For the calculation of the signature, the private key of the vehicle v_i is used.

Nodes receiving this message use the public key of the node contained in and verified by its certificate to decrypt the signature. How certificates are managed and how keys of the certificate authorities are distributed will be discussed in the following section.

Group signatures allow nodes to generate signatures on behalf of a group they belong to, through which they can prove membership with respect to a public key used by all nodes within the group. Such public keys are contained in respective group certificates, which must be available to all communicating nodes within the group. The verifying node cannot infer any information about which member of the group has generated the signature, thus providing some degree of privacy to the individual node.

The drawbacks of group signatures are that node exclusion is only possible by re-keying other nodes, which does not scale well. Additionally, group signatures are big and can easily reach more than thousand bytes in AutoNets. Mechanisms aiming at smaller signature sizes are, for example, the use of bilinear maps, reducing signature length to 192 Bytes with a security level comparable to a 1024 Bit RSA signature. In addition to the size of the signature, group signatures are complex to compute, allowing only a few signatures and verifications per second. One approach for re-keying is described in [4].

9.5   Certification

9.5.1   Trust

A trust model is a set of rules, which can be used by applications to decide the legitimacy of an entity. Existing trust models usually use different co-domains to represent different levels of trust, ranging from probabilities over pre-defined values for different classes to open intervals including negative values.

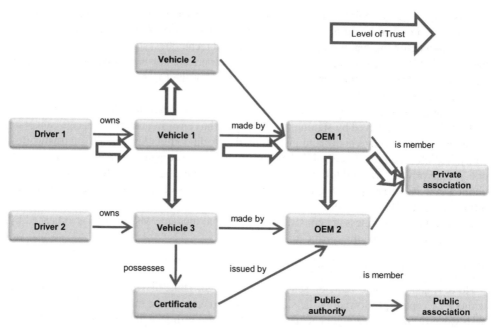

Figure 9.7 Initial trust model for AutoNets.

For AutoNets, we define a *level of initial trust* that one party has in another as depicted in Figure 9.7. Relations in this figure are denoted by thin arrows while the level of trust – as an example – is denoted by the thickness of the bounded arrows. One vehicle may have some trust in another vehicle, based on the behaviour of that vehicle or messages received by that vehicle. Similarly, a vehicle – or more precisely, its systems when making decisions – may have some trust in its manufacturer, and possibly less trust in another vehicle manufacturer.

While trust in authorities may be higher than in private organisations,[3] it will not usually be without some doubts. In some cases there may even be less trust in public authorities, for example if they have been proven or 'are known' not to be trustworthy. AutoNets therefore should be designed such that no one party alone (apart from the owner) can gain access to stored information of a participant or the true identity of a participant. Only when several (two or more, e.g. judge and police) authorities agree on the access or disclosure should this be possible. Secret sharing is a way to technically solve this problem [11].

The level of trust in another party may change if this party can somehow prove that it is trustworthy, for example by possessing a certificate claiming that this party shows a certain behaviour. For example this party provides messages of a certain quality. Moreover, the level of trust in another party also changes with the behaviour of this party. Many systems provide information about the trustworthiness of a party, for example reputation and recommendation

---

[3] This is because the authority has been proven to behave in a certain way in the past, and any entity representing this authority has been charged with that behaviour and can prove its authenticity in some form (e.g. a certain uniform ID).

systems. For the remainder of this chapter, we assume a sufficient level of trust in certificates that AutoNet vehicles provide. Note that the amount of trust will probably depend on the certificate authority that issued the certificate.

In order to increase trust in certificates of vehicles, they should be initialised and maintained by making use of situations in which the vehicle is in a physically trusted and controlled environment, such as during production, at certified garages or at registration facilities. In such an environment, the level of trust in both the infrastructure and the vehicle systems are higher compared to a distant wireless connection.

9.5.2 Trusted Third Platforms: Certificate Authorities

Each AutoNet mobile node needs a set of public/private key pairs, which must be authenticated by a trusted third party (TTP). The role of certificate authorities (CAs) is to issue such certified public/private key pairs to vehicles required for message signatures. CAs may be organised publically or privately, but they must be absolutely trustworthy. In the case of public organisation, governments may control the certification process. In the case of private organisations, large and trusted companies like vehicle manufacturers may take this role. Due to the potentially large number of AutoNet-enabled vehicles, it will hardly be possible to run just one global CA or some kind of central global infrastructure for the large number of vehicles all over the world. Hence, CAs, the relationship between them and the relation of the vehicles to the CAs needs some form of decentral organisation on a global level while regionally, centralised subsystems may exist.

One possibility for organising vehicle-CA assignments is with respect to their geographical homezone (just like they are usually assigned to a local vehicle registration agency). Another way to structure these systems is to have vehicles retrieve their certificates from a CA run by their manufacturer. When vehicle manufacturers take the role of CAs, this has the advantage that there is a limited number of them but the disadvantage that they cannot employ the power of law enforcement agencies. Vehicle owners may be forced to register with a certain CA (in a public organisation) or may be able to choose one (in a private, market-based organisation.)[4] This will be a matter of (a certain amount of) regulation. Operation of a CA certainly costs, this means that a CA will offer its services at a price. Depending on the organisation and the level of service, these costs will vary. At the same time, the main products of a CA, their issued certificates, have a value. While it appears that this would allow a market-based system, the problem remains that usually, in an AutoNet setting, the receivers of messages benefit from their content. As long as safety beacons and situational awareness messages are provided for free, there is hardly an incentive for an entity to provide higher trust in their messages, especially if this means higher cost and their is no guarantee of getting anything in return.

With several distributed CAs in an AutoNet system and vehicles communicating with each other carrying certificates from different CAs, it is necessary that CAs mutually trust each other, that is they must be cross-certified. Hence, respective processes need to be established

---

[4] Can CAs be organised in a purely market-based approach, that is CA services being offered in a competitive manner?

for such cross-certifications. The vehicles will need to carry the public keys from all CAs or retrieve them immediately on-the-fly for as soon as they need to verify issued certificates.

There are different proposals for how vehicles that need to authenticate messages from other vehicles whose certificates are issued by a different CA can do so and trust the corresponding public keys. One possibility is to organise the system in a hierarchy. Here, nodes add all certificates along the certification chain up to the root CA to their messages. This approach, however, is not very applicable to AutoNets due to the high message overhead and expensive verification processing in terms of computing time.

A more practical solution for AutoNets is to organise the CAs geographically. These CAs embody regional responsibility and issue certificates to all vehicles with their home base inside that region. In this case, a regionally responsible CA needs to verify the original certificates of visiting vehicles' homezone CAs. They can check demand or hand out locally valid certificates after successful verification. This procedure requires mechanisms to ensure that the vehicles receive the necessary certificates in time before entering a new region. If they are only equipped with AutoNet-based ad-hoc communication capabilities, then either Roadside Stations need to be installed along the borders of each region at every street, or some preloading algorithm is required to ensure that vehicles cannot enter another region without possessing the required credentials. If a vehicle features a cellular connection, it can request the necessary certificates from some provider before crossing a regional boundary.

9.5.3   Certificate Generation and Distribution

With privacy as an important requirement for AutoNets, the key pairs being certified need to be anonymous, that is they should not allow the retrieval of the actual identity of their owner, neither the mobile node nor its user, that is neither the vehicle nor the driver. Therefore, each vehicle is assigned a set of anonymous key pairs before participating in an AutoNet system. This first assignment may be processed by a public authority like an appropriate CA or by a private provider like the vehicle manufacturer or a pseudonym service provider, possibly in turn certified by the CA. To prevent tracking and because keys expire after the end of their lifetime, the vehicle needs to regularly receive new sets of keys. Therefore, it is necessary that the vehicle has access to the respective authority or service provider located in the infrastructure. Cellular communication systems seem appropriate for this link, but access via V2R communication is possible as well, if the RSU provides the required protocol implementations.

More formally, each vehicle v uses a set of anonymous private/public key pairs (PrivateKey$_{v_i}$ and PublicKey$_{v_i}$) at any point in time to sign their messages. These keys are certified by a CA with the corresponding certificate Cert$_v$, which contains the following information:

$$\text{Cert}_v(\text{PublicKey}_{v_i}) = \text{PublicKey}_{v_i} \mid \text{Sign}_{\text{PrivKey}_{CA}}(\text{PublicKey}_{v_i}, \text{ID}_{CA})$$

A public key infrastructure (PKI) digital certificate conveys the signature of a TTP on the certificate fields. The TTP can also keep a certificate revocation list (CRL), so that one can verify the certificate validity before trusting it. However, CRLs are difficult to apply to AutoNets due to the very large number of vehicles and their mobility.

X.509 certificates are widely used in the Internet [7], but with their size of about 1kB, they would generate a considerable overhead when used in AutoNets. In addition, some data fields of the X.509 certificate contradict the anonymity requirement. While they are readable for humans, this is not necessary in AutoNets. A suitable pseudonymous certificate for AutoNets needs to contain the following information:

- *Random vehicle ID (6 Bytes)*. This ID is used to uniquely identify a certificate. The size of this random ID needs to be large enough to ensure a negligible probability of ID collisions for physically close vehicles.
- *Algorithm ID (1 Byte)*. The Algorithm ID identifies the cryptographic algorithm used by the TTP to sign the certificate's fields. A size of 1 Byte allows 256 different algorithms.
- *Issuer (1 Byte)*. The issuer indicates the identifier of the TTP, which has signed the certificate. A size of 1 Byte allows for 256 different issuers.
- *Validity period (12 Bytes)*. This field indicates the certificate's validity period.
- *Entity's public key algorithm (1 Byte)*. This field identifies the algorithm used by the vehicle to sign messages.
- *Entity's public key*. A vehicle's public key. Its size varies and depends on the adopted algorithm.
- *TTP's signature*. This field contains the TTP's signature on the certificate's fields. Its size depends on the adopted algorithm.

Fixed – that is endlessly valid – certificates bound to vehicles fail to meet the privacy requirement, because vehicles using the same certificate all the time are easy to track. Therefore, time-bound certificates are required. Usually, long-life certificates (LLCert) and short-life certificates (SLCert) are distinguished. In addition, we distinguish between vehicles carrying a single certificate and vehicles carrying a certificate pool. In the latter case, each node manages a set of pseudonymous certificates so that it can sign outgoing messages by choosing between several private keys related to pseudonyms. A drawback of this approach is that – at a certain moment – a vehicle is in possession of several identities, represented by the simultaneously valid certificates. This may lead to sybil attacks. In order to avoid sybil attacks, vehicles may only own one certificate at a time. The necessarily short-lived nature of such certificates requires the vehicle to refresh its certificate quite frequently. This approach satisfies the node exclusion requirement: instead of managing certificate revocation lists (CRLs), the TTP can simply decide not to provide a new SLCert to an excluded vehicle.

In the following, we discuss certificate management schemes, especially certificate distribution algorithms and certificate revocation. In [8], we suggest the following certificate issuance protocol description for AutoNets:

1. A vehicle v generates a new public/private key pair (NKpub and NKpri), which will be related to the new issued certificate.
2. v connects to the TTP and sends following message:
   (a) TTPKpub(original vehicle ID, random number): the original vehicle identifier is kept confidential and is only known to the vehicle and the TTP. For this reason, it is encrypted with the TTP's public key so that only the TTP can disclose its value. The random number is added in order to avoid vehicle tracking. It aims to make the requests of the same vehicle always completely different and, hence, preserves privacy.
   (b) Sequence number: this number is used to avoid replaying the request.
   (c) NKpub: the public key to be certified by the TTP.

(d) Signature: the signature (with the original private key) on ETTPKpub (original vehicle ID, random number), sequence number and NKpub in order to prove the possession of the private key related to the original vehicle identifier (authenticity) and integrity of the data.

3. The TTP performs the following actions:
   (a) It decrypts the vehicle identifier with the TTPKpri (the random number is skipped).
   (b) It verifies the signature through the public key associated with the sent original vehicle identifier.
   (c) It performs a plausibility check on the sequence number in order to avoid the replay attack.
   (d) It checks whether the vehicle associated to that original vehicle identifier is authorised to obtain a new certificate, that is that it has not been dismissed, stolen and so on.
   (e) It verifies that the vehicle – identified by the original identifier – does not already own a valid certificate in order to prevent the same vehicle from having multiple identities.
   (f) It generates a random pseudonym for the new certificate.
   (g) It stores the certificate issued for that vehicle.
   (h) It issues the certificate and signs it with TTPKpri.

4. The TTP sends the new short-lived pseudonymous certificate to v.
5. v verifies the TTP's signature on the certificate fields and stores the certificate.

Based on the issued pseudonym certificate, the vehicle signs messages with NKpri and sends them along the certificate to other vehicles. We presume that the TTP root certificates are known to all vehicles. The root certificates can be updated during the issuance of the pseudonymous certificate.

Since messages issued by single vehicles in many cases are only of interest to other close-by vehicles and since short-life certificates expire quickly, it is sensible that the vehicles distribute their own certificates themselves. Adding certificates to every message is a simple solution, but at least for heartbeat safety messages consumes unnecessary network capacity. Separate distribution techniques at suitable, adaptable rates, possibly using piggybacking, have shown better performance. With this, certificates may be a one-hop or two-hop broadcast with appropriate forward suppression techniques [9] when WLAN communication is used, or a cell broadcast when cellular communication is used.

The drawback of sending certificates at a certain frequency is that it causes the problem of missing certificates. In this case, vehicles receive messages from other vehicles whose certificates have not yet arrived. Since these messages cannot be trusted, policies are necessary to decide whether to throw them away or to delay their processing until the corresponding certificate has been received. The behaviour of different relations of fixed message and certificate issuing frequencies with which each vehicle periodically sends beacon messages and certificates has been analysed in [8]. It is worth mentioning that even in sparse networking scenarios, the mean value for certificate misses was always below 10% for all cases when certificates were distributed with a frequency of 10Hz of the messages or higher and that the vehicles missing the certificates were always further away than 140 m from the issuer. It appears that one-hop techniques perform best in sparse networking scenarios, while intelligent two-hop techniques perform better in denser scenarios. It proved better to send the certificate along with some messages instead of sending it separately. A frequency ratio of 0.3 of certificate sending versus message sending frequencies leads to a good trade-off between bandwidth saving and missing certificate probability.

9.5.4   Certificate Revocation

Certificates need to be revoked not only in case of compromision, but also in case of system malfunction inside a vehicle due to technical defects or in case of wrongly issued certificates. Ideally, revoked certificates could not be used anymore. If they are used, though, other nodes need to be aware that the certificate used to sign a message is not valid anymore. Several procedures have been proposed to deal with certificate revocation:

- *Certificate revocation lists (CRLs).* With an increasing number of vehicles, CRLs do not scale well for AutoNets. CRLs can be considered when locally and temporally bound, that is only containing very recently revoked certificates, possibly through the lifetime of the certificates.
- *Short-lived certificates (SLCert).* If certificates expire after a short time automatically, their misuse is limited.
- *RTPD: revocation of tamper-proof devices.* Using RTPD, the CA revokes all keys of a given vehicle. This is only possible with a tamper-proof device that is guaranteed to still work in the original way and will thus react on the CA's call. The tamper-proof device will cease to sign messages and erase all keys upon receiving the signed revocation message of the CA [5].
- *RCCRL: revocation by compressed CRLs.* With RCCRL only a subset of the keys of a vehicle are revoked. In contrast to other approaches, this protocol warns other vehicles rather than asking the originator to stop sending messages [5]. This way, such an approach contradicts the scalability requirement, because the overhead generated for the warning of the vehicles depends on the number of communicating vehicles in the AutoNet.
- *DRP: distributed revocation protocol.* DRP defines a protocol to detect misbehaviour of communicating nodes. DRP is used to report the reception of false or fake messages to the CA [5].

9.6   Securing Vehicles

Availability and integrity of the in-vehicle security system are a fundamental prerequisite to fulfil generic security requirements. To ensure this, the vehicle system must be protected against denial-of-service attacks, against the execution of malicious code, against the tampering of vehicle sensor data and against the processing of manipulated incoming data and unauthorised access to vehicle data and resources.

AutoNets need to provide some basic security mechanisms and functionalities to implement this protection. From a design perspective, the separation of different functional domains supports the possibility of running less-trusted code or to process less-trusted messages in better protected environments or in environments in which possible damage is limited.

Both tamper-proof devices (TPDs) and trusted platform modules (TPMs) feature hardware security modules (HSMs) and provide hardware security means to securely and secretly store identities and keys, as well as to securely perform cryptographic operations in a trusted environment. In order to be able to operate autonomously and, thus, reduce the risk of compromisation, tamper-proof devices should feature their own clock and their own battery. In an AutoNet setting, the battery can be recharged using the electrical power supply of the vehicle and the clock can be regularly synchronised with a trusted reference time. TPDs are more expensive than TPMs, but they also provide higher resistance against physical attacks.

In general, we distinguish three types of hardware security in an electronic control unit:

- *Separate CPU and HSM.* This architecture is costly, because two chips are used. The connection between CPU and HSM is not secure. Therefore, this solution is only reasonably applicable to low-security applications with only a few units needed (and therefore an extra design is not sensible).
- *HSM realised in the same chip as the CPU with its own state machine.* While highly secure, this solution still requires hardware redesign of a CPU and has a fixed behaviour, thus it is not flexible and is therefore difficult in a long-lifetime automotive setting.
- *HSM realised in the same chip as the CPU with a programmable secure core.* This is a cost-effective, secure and flexible solution. Hence, respective control units will most likely feature this type of hardware security if needed.

It is important to mention that current TPDs – as they are used for financial applications, for instance – are not yet automotive grade. They do not meet the required level of vibration-resistance or operating temperature range.

9.7   Secure Communication

9.7.1   Secure Messaging

For messaging, any data payload is precluded by a security header and concluded by a security trailer. When transmitting, messages or packets are passed to the AutoNet security component to add the requested security information, and possibly to encrypt the data. On reception, messages are handed to the security component to parse the security fields, assess how much its content or its sender can be trusted, possibly decrypt the data and respond with the respective results. A generic representation of the corresponding service invocations may look as follows in an object-based programming language:

```
SecureMsgResult = AutoNetSecurity.SecureMsg
(MessageSend, CryptoMethod, [boolean Time, boolean Position, boolean Speed])
SecurityAssessmentResult = AutoNetSecurity.AssessMsg(MessageRcv)
```

Both *SecurityMsgResult* and *SecurityAssessmentResult* are objects which define the result of the respective security operations by providing information on the success of security operations. Their operations are typically provided in an object named 'AutoNetSecurity', which provides two important methods: SecureMsg() and AssessMsg(). Besides the message being sent (MessageSend) and the cryptographic method being used, SecureMsg() optionally takes boolean arguments on whether to include time, position or speed information. It reports the result of its processing and the secured message. AssessMsg() also responds with the result of its processing, including not only the decrypted messages stripped of its security fields but also the assessment result, possibly including detection of tampering or a likelihood value of tampering and a corresponding value of how much the message content and or its sender can be trusted based on a trust model. The result of AssessMsg() may contain additional information such as position if provided by the sender.

9.7.2   Secure Routing and Forwarding

For protocols with some form of routing or forwarding, they need to be secured so that these protocols work correctly and no attacker can change their behaviour, such as by redirecting packets. The objectives of secure end-to-end communication in an AutoNet are thus the integrity and non-repudiation of packets, authenticity and authorisation of source and router or forwarder, respectively, and plausibility and freshness of data. Authorisation of forwarders may be needed to grant them certain access to network resources. For geocasting, the plausibility of position data may be needed to avoid unfair forwarding behaviour. Network packets contain fields that shall not be changed by any intermediate node and fields that require updates like time-to-live values or positions of forwarders in geocasting scenarios.

One-hop network layer control packets like network layer neighbourhood beacons are secured by the signatures of their originators. Multi-hop control and payload packets require a differentiation of the fields that need to stay unchanged from source to destination, and those that will be updated or added by intermediate nodes. The latter will be signed by the intermediate nodes for each hop, while the former will be signed by the source.

9.7.3   Secure Group Communication

Whereas symmetric cryptography based on pairwise keys does not fit the broadcast nature of communication in AutoNets and does not scale well, it appears more natural to establish secure communication within a (stable) group of vehicles. Vehicles travelling along one road in the same direction are good candidates for form such stable groups. Also, group membership can be determined by geographical position, if the total system area is divided in non-overlapping regions and all vehicles in one region belong to the same group. In the literature, different mechanisms for node organisation including group establishment and management have been proposed, usually referred to as clustering. To establish a common key for a group of nodes, the clusterhead could simply generate the key and distribute it among the group members. If there is no clusterhead, distributed key establishment schemes are needed [4]. Groups in AutoNets are naturally dynamic. Hence, nodes regularly join and leave the groups (or clusters). A new (and trusted) node can simply be presented with and use an established group key. Depending on the scheme being used, no operation is necessary after a node leaves a group. Vehicles belonging to neighbouring groups also need to be able to check the validity of messages generated by some group. Group keys do not allow one to securely identify the source of malicious data and do thus not meet the requirement of non-repudiation.

9.7.4   Plausibility Checks

Among the most important data for AutoNet applications is the position and movement data of the vehicles. Since an AutoNet node does not need to base its decisions on a single foreign provider of this data, but rather receives data from different sources and additionally has access to its own sensor data, it is possible to validate this data for consistency (see for example [20]).

Data providers in an AutoNet are required to add some information on the quality of the data they provide. For security, this means adding a confidence value for the respective data. This confidence value thus becomes part of the signed message. The certificate in this case contains information on the confidence model and its generation process. Typically, the confidence value would contain information on the accuracy of the data and the type of sensor used (e.g. including its failure rate) to obtain the data.

9.8   Privacy

According to Wikipedia, privacy is 'the ability of an individual or group to seclude themselves or information about themselves and thereby reveal themselves selectively. The boundaries and content of what is considered private differ among cultures and individuals, but share basic common themes. Privacy is sometimes related to anonymity, the wish to remain unnoticed or unidentified in the public realm. When something is private to a person, it usually means there is something within them that is considered inherently special or personally sensitive. The degree to which private information is exposed therefore depends on how the public will receive this information, which differs between places and over time. Privacy is broader than security and includes the concepts of appropriate use and protection of information.'[5]

Directives 95/46/EC and 2002/58/EC of the European Parliament and of the Council concern the processing of personal data and the protection of privacy in the electronic communications sector (Directive on privacy and electronic communications). Communication service providers are obliged to take appropriate technical and organisational security measures, and so automotive internetworking systems in Europe must comply with these rules. The European eSafety Forum has one Working Group dealing with security in *Cooperative Systems*, called eSecurity WG. It works together with experts from the Data Protection Offices. Art. 29 defines a code of practice with recommendations on how to deal with privacy and data protection issues in the design of in-vehicle telematics and cooperative systems.

9.8.1   Secret Information

For people, privacy also means that they are empowered to control their personal data, especially its disclosure. A major prerequisite is that people are able to correctly judge which personal data they hand over to whom when using a system.

A Eurobarometer survey was carried out in 2006 addressing privacy concerns with respect to the use of intelligent systems in vehicles. The survey shows that people weigh privacy against other assets. For eCall systems, privacy was not a major concern. On the other hand, real-time traffic and travel information systems were considered a privacy intrusion threat, especially by French (26%) and German (25%) respondents.

When communicating, people may be concerned that the content of their transmitted information is overheard by somebody else other than the correspondent. Apart from information directly issued by talking or writing, sensors may also detect and possibly collect personal information. Hence, this information must be in control of the person it belongs to. People may also not want information about their location, their looks, health state, clothes and so on to be accessed by others, possibly including a remote communication partner. For AutoNets, this means that both identity information and personal data must be kept secret. This does not only refer to direct identifiers like names, number plates, vehicle IDs or network addresses,

---

[5] http://en.wikipedia.org/wiki/Privacy.

but also to information like vehicle brand, colour, location, travel routes or destinations that can be used to figure out an identity and link other information to that identity.

In order to preserve privacy, two basic concepts are fundamental:

- Concealment of identities, which is typically realised using pseudonymisation of digital identities.
- Anonymisation of data.

The AutoNet security system provides the services that implement these concepts as well as their configuration and management. Different AutoNet usage scenarios require these services. We distinguish solutions with respect to their application context, which is defined by four parameters: (i) whose personal data is dealt with, (ii) who is the communication partner, (iii) what is the communication channel, and (iv) which information is sent. In the following, we consider solutions for the following cases: data of a vehicle relating either to driver, to the owner or to a passenger when communicating inside the AutoNet mobile domain or with the AutoNet infrastructure domain, that is the V-V, V-M, V-R and V-C cases. We do not consider the V-I and V-P case and other communications that do not involve the vehicle. With respect to the information that is sent, we focus on identities and identity-related information issued either by an AutoNet application or AutoNet system component. Solutions are required for all cases where this data needs to be communicated, either for the functionality of the system (like communication addresses or positions for geocasting), or for the functionality of the applications (like positions for safety or user identities for payment). We can assign certain techniques to the type of information that needs protection: encryption for application payload data and access control data (e.g. for payment), pseudonymisation for all kinds of addressing data and obfuscation for all kinds of data that may allow inferences about its issuer (e.g. location, navigation destinations).

Obfuscation is the concealment of the meaning of privacy-related information in order to make the information difficult to interpret. Obfuscation needs to address all information contained in an outgoing data packet, included in payload and headers. Obfuscation should be realised during design of the respective security modules. In addition, runtime functionality may be provided as a system service and applied to outgoing data packets. In this latter case, consequences for the behaviour of the affected modules need to be taken into account.

Using *pseudonymisation*, any identity-related information is replaced by some artificial identity. This artificial identity changes over time in order to avoid a vehicle that can be traced by its pseudonym for example. In order to avoid any two pseudonyms used by an identity being linked to each other, it is important that when pseudonyms are changed, all other identifying information used in combination with that pseudonym at the same time is changed as well. For key management, pseudonyms and anonymous (pseudonymous) keys which are not linked to one's identity have to be changed over time. In this context, the intervals for changing pseudonyms need to be chosen carefully. This is due to the fact that changing pseudonyms from time to time makes it harder for some subject to track an AutoNet entity. Assuming that a vehicle's messages can be overheard by some privacy infringing party only at spots where this party can receive the vehicle's signals, it will be necessary to change pseudonyms in between these spots. If any party can receive all messages in the network, more sophisticated pseudonym change algorithms are required.

Privacy solutions must meet the requirement of non-repudiation, that is for the pseudonyms the possibility of linking the keys with the identities that use them, at least in certain situations.

Besides the technical mechanism allowing one or a group of subjects access to pseudonyms to identities, it is a political issue to decide who should be allowed to be able to access the identity and under which conditions. Proposals usually suggest executive forces be allowed to do so in law enforcement cases with prior clearance by some judicial institution, for example a judge.

For many of the different algorithms used in networking scenarios, the use of identifiers is essential. Since at the same time, many of the applications require open and regular communication, that is data cannot be exchanged in private and must be transmitted within short time intervals, it would be easy for an observer to figure out the positions of vehicles (and potentially their owners) and track identities. As a possible countermeasure, the use of pseudonyms has been heavily discussed. Pseudonyms conceal the real identity of the vehicle or any human inside the vehicle and, thus, are means to protect their privacy, both against possibly unintended casual tracking by data collections and against malicious trials.

In order for this mechanism to be effective, all layers from the physical layer over network and transport and facilities to applications have to use pseudonyms.

While pseudonyms conceal real identities, their use may still allow tracking of vehicle paths. To avoid this, it has been suggested to use pseudonyms which are only temporarily valid. Researchers still discuss how frequently pseudonyms will need to be updated. Moreover, in AutoNet scenarios it is important to avoid a change of pseudonyms in some situations. For example, in the event that a vehicle sends important driving related information, for example at an intersection, a change of the pseudonym must be prevented since the vehicle then disappears for the other vehicles and a new vehicle is potentially available at this position.

9.9   Conclusion

This chapter offers only an introduction and a brief glance into the field of security for AutoNets. While there has been a lot of work on securing AutoNets, sometimes with unrealistic assumptions, such as complete absence of infrastructure, experts still seem sceptical that security has reached the level required for the deployment of safety applications. We believe, however, that the necessary mechanisms are there, but some sacrifice in terms of cost, risk of successful attack and/or level of privacy will have to be made and weighed against the benefits of the applications. Different security mechanisms for AutoNets will have to be assessed with respect to the level of security and the level of privacy they provide versus their level of system efficiency, that is computation cost, storage size and communication overhead, implementation resource requirements and operational cost. Scalability is another important assessment factor.

It is a challenge for security systems that have been designed to operate in a certain legal framework to offer their services in different regions with different legal requirements (e.g. on privacy of data). Thus, it appears currently unlikely that there will be one global security solution for automotive internetworking. Local specialities are likely to remain and require some form of adaptation.

References

[1] Hu, Y.-C., Perrig, A. and Johnson, D. (2002) Ariadne: A secure on-demand routing protocol for ad hoc networks. *Proc. ACM International Conference on Mobile Computing and Networking (MobiCom) 2002.*

[2] Wolf, M., Weimerskirch, A. and Paar, C. (2004) Security in automotive bus systems. *Proc. Workshop on Embedded Security in Cars (escar) 2004.*

[3] Furgel, I. and Lemke, K. (2004) A review of the digital tachograph system. *Proc. Workshop on Embedded Security in Cars (escar) 2004.*

[4] Bechler, M., Hof, H.-J., Kraft, D., Pählke, F. and Wolf, L. (2004) A cluster-based security architecture for ad hoc networks. *Proc. IEEE International Conference on Computer Communications (Infocom) 2004.*

[5] Raya, M. and Hubaux, J.-P. (2005) The security of vehicular ad hoc networks. *Proc. ACM Workshop on Security of Ad Hoc and Sensor Networks (SASN) 2005.*

[6] Gollan, L. and Meinel, C. (2002) Digital signatures for automobiles. *Proc. Systemics, Cybernetics and Informatics (SCI) 2002.*

[7] Housley, R., Ford, W., Polk, W. and Solo, D. (2009) Internet X.509 public key infrastructure certificate and CRL profile *Request for Comment 2459, Internet Engineering Task Force.*

[8] Weyl, B., Tornesello, S., Vögel, H.-J., Kosch, T. and Eckert, C. (2009) A security protocol for scalable and privacy-preserving car-to-X eSafety applications. *VDI Berichte*, Vol. 2016.

[9] Briesemeister, L., Schafers, L. and Hommel, G. (2000) Disseminating messages among highly mobile hosts based on inter-vehicle communication. *Proc. IEEE International Symposium on Intelligent Vehicles, October 2000.*

[10] Parno, B. and Perrig, A. (2005) Challenges in securing vehicular networks. *Proc. of HotNets: Fourth Workshop on Hot Topics in Networks.*

[11] Shamir, A. (1979) How to share a secret. *Communications of the ACM*, Vol. 22(11).

[12] Rivest, R. L., Shamir, A. and Adleman, L. (1978) A method for obtaining digital signatures and public-key cryptosystems. *Communications of the ACM*, Vol. 21(2).

[13] Robshaw, M. J. B. and Yin, Y. L. (1997) Overview of elliptic curve cryptosystems. An RSA Laboratories Technical Note. Revised June 27, 1997.

[14] Hoffstein, J., Pipher, J. and Silverman, J. H. (1998) NTRU: A ring based public key cryptosystem. *Algorithmic Number Theory (ANTS III), in Lecture Notes in Computer Science* 1423.

[15] Gerlach, M., Festag, A., Leinmueller, T., Goldacker, G., and Harsch, C. (2007) Security architecture for vehicular communication. *Proc. 4th International Workshop on Intelligent Transportation (WIT).*

[16] Douceur, J. R. (2002) The sybil attack. *Proc. 1st International Workshop on Peer-to-Peer Systems (IPTPS02).*

[17] Aijaz, A., Bochow, B., Doetzer, F., Festag, A., Gerlach, M., Kroh, R., and Leinmueller, T. (2006) Attacks on inter vehicle communication systems - an Analysis. *Proc. 3rd International Workshop on Intelligent Transportation (WIT).*

[18] Adams, R., Lloyd, S. (1999) *Understanding Public-Key Infrastructure: Concepts, Standards, and Deployment Considerations.* Macmillan Technical Publishing

[19] Schwingenschloegl, C., Eichler, S. (2004) Certificate-based key management for secure communications in ad-hoc networks. *Proc. 5th European Wireless Conference: Mobile and Wireless Systems beyond 3G.*

[20] Friederici, F., Gerlach, M. (2008) Towards RSSI-based position plausibility checks for vehicular communications. *Proc. 4th Intl. Workshop on Vehicle-to-Vehicle Communications (V2VCOM 2008).*

10

System Management

The system management is responsible for allying applications, networks and interfaces in a specific implementation of an AutoNet station. This implementation may range from a simple stand-alone unit in a vehicle to a complex router/host interaction for large roadside networks. The main challenge for the system management is to combine the contradictory requirements from safety applications with requirements from other applications, while considering possibly contradictory cross-layer congestion control requirements and potential fall-back communication solutions. In the AutoNet Generic Reference Protocol Stack, system management is realised by the management layer, which is addressed in this section.

In this chapter, we first start with on overview of the management layer functionality and its integration into the AutoNet Generic Reference Protocol Stack. Therefore, we will separate the management functionality into building blocks, which are needed to describe the correlations and interactions within the management layer and between the management layer and the other layers in the AutoNet Generic Reference Protocol Stack. Based on this classification, we will introduce important management issues and describe their realisation in the classification concept. Finally, we outline potential realisation scenarios from an implementation perspective, which will be completed by a summary of this chapter.

10.1 System Management in the AutoNet Generic Reference Protocol Stack

The management layer resides outside the communication stack and provides management functions for all other layers. Its role is to amalgamate AutoNet applications, facilities, networks and access technologies. It also covers the profiling concept, which basically determines the overall functionality of an AutoNet station. Hence, the core tasks of the management layer are the following:

- Managing policy and profile setting, including maintenance for each layer in an AutoNet station.
- Managing (dynamic) access technology selection per application. The selection is based on policies, application requirements and access technology performance and availability, and may include user preferences.

Automotive Internetworking, First Edition. Timo Kosch, Christoph Schroth, Markus Strassberger and Marc Bechler.
© 2012 John Wiley & Sons, Ltd. Published 2012 by John Wiley & Sons, Ltd.

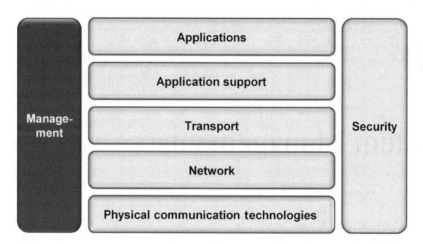

Figure 10.1   Management layer in the AutoNet Generic Reference Protocol Stack.

- Managing transmit permissions and synchronisation based on physical cross-interference between air access technologies combined with information priorities.
- Managing security and privacy functions under consideration of application types and available access technologies.
- Providing information to each layer to optimise performance and efficiency of of the overall AutoNet station as well as of the overall AutoNet.

Figure 10.2 illustrates the integration of the management layer, which flanks the communication-based layers in the communication reference system. Due to this superordinated management functionality, the management layer can be accessed by every other layer

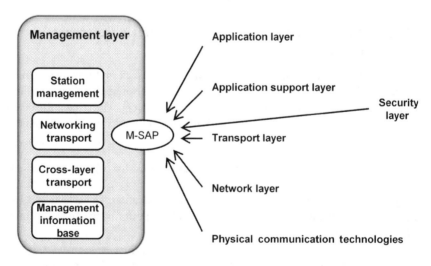

Figure 10.2   Service access points and functionality of the management layer.

in the AutoNet Generic Reference Protocol Stack via the M-SAP. Likewise, it can also access functionality of the remaining layers via their SAPs. The access points are used to transfer status and configuration information from the respective layer to the Management Layer. Vice versa, the management layer sends commands via the respective SAPs to the different layers in order to control the operation of the layer accordingly. It is important to mention that most of the management functions will require a high level of security. As long as critical functions are performed outside of a trusted environment, there is a need to protect against threats. This includes configuration and initialisation protocols on a global level (e.g. new channel usage rules from national spectrum authorities), and down to local station-internal protocols between routers and hosts. It is up to the responsible system integrator to decide on the level of trust and protection needs, but a defined minimum level of security protection must be present, in particular for real-time authentication of new and unknown AutoNet stations. These functions are currently under (joint) development in ISO TC204/WG16 and in ETSI TC ITS WG5.

10.2   Functional Management Building Blocks

According to Figure 10.2, the management functionality provided by the management layer is split into four functional building blocks: station management, networking management, cross-layer management, and management information base (MIB). These function blocks are described below, followed by a selection of typical and important management functions for an AutoNet station and some implementation issues for the management decisions.

Station management. Station management is the high-level management of an AutoNet station. It therefore handles the following issues:

- Internal control over multiple routers and hosts that belong to one AutoNet station.
- External control and communication between AutoNet stations as far as is needed for management purposes.
- Initialisation and dynamic configuration of (parts of) an AutoNet station, including status reporting from each available physical communication technology.
- Interference mitigation between multiple physical communication technologies within the same AutoNet station.
- Interference mitigation and load reduction between nearby AutoNet stations.
- Decision making on which application shall use which AutoNet access technology, including dynamic configuration of all involved layers.

Networking management. Networking management will handle aspects of various network functions. These functions include (i) routing table updates for Mobile IPv6 as defined in ISO 21210 [2], (ii) general geonetworking management and optimisation, (iii) medium-specific Georouting as being defined in ETSI TC ITS, and (iv) optimised non-routing networking for low-latency single-hop scenarios. Most medium-specific functionality is defined in ISO 29281 [3] and is under study in ETSI TC ITS.

Cross-layer management. Cross-layer management handles initialisation as well as dynamic configuration and status reporting from each available layer of the

AutoNet Generic Reference Protocol Stack. It has the superior functionality to optimise the performance of the overall AutoNet station (and even the AutoNet itself). This is achieved by adjusting all layers according to the information of all layers rather than an optimisation of each layer based on its own information.

Management information base (MIB). The MIB is a (virtual) data store inside the management layer. The purpose of this entity is to define important variables, profiles, policies and data sets, which typically need to be present for managing purposes. This way, the MIB can be considered as a repository of all information that is relevant for the management of an AutoNet station, in particular for the cross-layer optimisations. A set of definitions can be found in ISO 29281 as well as in ISO TC204/WG16 documents. Besides, the MIB may also contain vendor-specific information, which can be used to customise management functionality to the differing requirements of different vendors or vehicle manufacturers.

These functional building blocks provide the basic mechanisms to perform the necessary management tasks for an AutoNet station. Apparently, the feature set for management depends on many factors; especially on type of the AutoNet station, profiling and the functionality being supported. The following section describes selected (and important) management issues of an AutoNet station. This list of issues is not exhaustive; in a final deployment scenario, there may be additional management tasks being implemented in an AutoNet station.

10.3   Selected Management Issues of an AutoNet Station

Besides the controlling functionality, the managment of an AutoNet Station also comprises optimisation and efficiency issues. In the following, we will glance at four important management tasks for an AutoNet Station: cost/benefit management, congestion control, mobility management and TCP management. Besides these 'general purpose' management tasks, there may be additional vendor-specific management issues, which will not be detailed here.

10.3.1   Cost/Benefit Management

A typical management mean of an AutoNet Station is to ensure the requested quality of service levels for AutoNet applications. However, quality of service is only one aspect. In the context of an AutoNet application, cost /benefit management is even more important; in fact, it is often crucial for the success of an AutoNet application. This is due to the fact that costs as well as benefit are immediate aspects dealing with user perception and expectation. Apparently, there is a valid and general correlation between costs and the benefit generated: the higher the costs, the higher the benefit. However, benefit is rated differently among AutoNet applications as well as among different users. The following example illustrates this heterogeneity in cost/benefit management decisions. Currently, there are only few drivers willing to pay a noticeable amount of money for pure safety applications: drivers typically expect to get the highest standard of safety for free. In contrast, drivers are typically willing to spend more money for information and entertainment services. Combined with the quality of service requirements of an AutoNet application and the variety of different user preferences, this results in an area of conflict.

From an architectural management viewpoint, this area of conflict must be addressed appropriately. Besides a 'common basis', which is most acceptable for all drivers, there must be the possibility of considering user requirements and user types accordingly for the management decisions. One solution could be to customise AutoNet stations to the bias of the users by considering user interaction at runtime. This information would be stored in the MIB immediately from respective AutoNet applications via the M-SAP. Additionally, the MIB also contains information about the costs of respective services. Such costs can be calculated locally based on available information about costs. Examples include the costs for data transmission, service costs and so on. Based on such information, an intelligent cost/benefit management has to deduce an appropriate decision that fits best to the expectations of the current driver.

10.3.2 Congestion Control

Congestion control ensures that radio resources are used in the best possible fashion while preventing network overload. Congestion control mechanisms include priority assignments or changes, transmission delay and redirection of messages to a different channel. A number of indicators can be used as input parameters for congestion control decisions, such as the number of neighbours in the radio transmission range, packet transmission delays or the current utilisation of message queues. Since both input values as well as controlled system variables can belong to different communication layers, congestion control is a cross-layer task. Therefore, careful consideration is necessary for the specifications and for the implementations in order to avoid overloading the scarce radio spectrum that has been allocated, for example, in the 5.9 GHz communication spectrum.

For a proper congestion control, AutoNets need to assign priorities to all messages being transmitted. These priorities may be composed of a static and a dynamic priority. The static priority depends on the content of the message, for example the quality of the service requirements of the application which has sent the message. The dynamic priority is calculated based on the vehicle and AutoNet context parameters, such as current vehicle speed or number of retransmissions. A simple calculation scheme would be the following: the dynamic priority increases with an increased vehicle speed. If the packet was not received before because it has been transmitted by another network node, the dynamic priority would decrease. The overall priority is then calculated using both priority values. Other variables that define the priority of a message may include the size of the message and the age of the message.

In addition to priority assignment, messages in the queue will be re-ordered according to their priorities. This means that high-priority messages can preempt the transmission of low-priority messages. Therefore, message queues need to be channel-specific (e.g. for control and service channels) or they may be based on cross-channels.

10.3.3 Mobility Management

Mobility management is a typical management issue since it addresses almost every layer in the AutoNet Generic Reference Protocol Stack. Consider Figure 10.3 as an example. In a real-world deployment scenario, an AutoNet Station will likely provide several communication technologies, especially for IP-based networking. Besides a combined GSM, GPRS, and UMTS interface, the AutoNet station may additionally support LTE, IEEE 802.11 a/b/g,

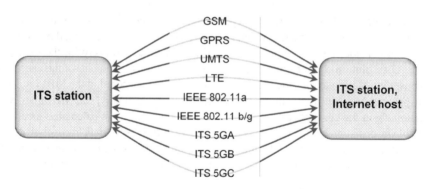

Figure 10.3 Communication scenarios in AutoNets.

Bluetooth and even wired network interfaces like Ethernet to connect a vehicle to an in-house communication infrastructure. Each of these communication systems has its own specific communication capabilities and characteristics. Moreover, some communication systems may only provide temporary connectivity. The following two examples illustrate the wide spectrum of the different characteristics:

- GPRS or UMTS may provide instant and continuous connectivity with a high coverage, which can be used at almost every location in Europe. However, the data rate is rather limited: 9.6 Kbit/s for GSM connection-oriented services, and theoretically up to 384 Kbit/s for connectionless GPRS services.
- IEEE 802.11a provide data rates of up to 54 Mbit/s with pretty low latencies at low costs. But, access is limited to respective hotspot-areas only.

Besides this heterogeneous communication environment, AutoNet applications have different quality of service requirements – including the cost/benefit issues detailed in Section 10.3.1. These requirements are also important for IP-based communication technologies since it is possible in AutoNet scenarios to transfer AutoNet messages via an IP-based (back-end) communication infrastructure to other vehicles. Mobility management has to handle these aspects in order to bridge the gap between these different issues by addressing several aspects in the different layers. Hence, mobility management is a typical example of cross-layer management: it provides the required information and mechanisms for physical communication technologies, networking layer,[1] and security layer.

In order to ensure mobility management, a number of mechanisms need to be defined for the related layers in the AutoNet Generic Reference Protocol Stack. Figure 10.3 illustrates these mechanisms together with their interactions. This approach was originally proposed by Bechler in order to integrate vehicular networks into the Internet [1]; Figure 10.4 extends this approach for the mobility management of AutoNet applications. For the physical communication technologies, a small monitoring component manages the available communication technologies. Therefore, the monitoring entity collects relevant information from

---

[1] As we will see in the next example, the transport layer may also benefit from this basic mechanism via cross-layered optimisation.

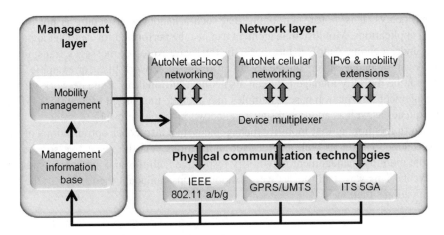

Figure 10.4 Mobility management within an AutoNet station according to [1].

each available communication technology and feeds it via the M-SAP into the management information base of the management layer. Moreover, pre-configured static information such as the transmission costs of a communication system or the gross data transfer rate, is also available in the management information base. This way, the information about the supported access technologies is available for other layers in the AutoNet Generic Reference Protocol Stack and, thus, can be used for cross-layer optimisations. The other way round, the mobility management in the management layer can be implemented to optimise the behaviour of the different communication technologies. In the event of an impending connection loss of, for example, IEEE 802.11, the mobility management can prepare an alternative communication technology to be used, for example it can establish a GSM connection, which can be used some moments later.

At the network layer, Mobile IPv6 realises the basic mobility support for IP-based networking. The mobility management supports the network layer by deciding which communication system will be used for communications. It therefore queries the management information base for respective information considered for the decision process. This way, the mobility management defines the point in time to trigger the handover procedure. This means connections can be handed over transparently and with low delays using Mobile IPv6. In order to provide seamless mobility with transparent connection handover, the applications always have to use the same global IPv6 home address (using bi-directional tunnelling). As a result, Mobile IPv6 can only be bound to one specific network interface at a time, and all applications using Mobile IPv6 have to use this network interface. In order to overcome this inflexibility and to ensure a flexible simultaneous use of networking technologies, Bechler proposed a management entity called a *device multiplexer* within the network layer. The device multiplexer decouples the IPv6 home address of an AutoNet station from the network interfaces being used. This means IP packets from different applications to (and from) the home agent can be transmitted using different communication systems. Therefore the choice of which communication to use depends on the information in the MIB; based on this information, the mobility management controls the device multiplexer via the N-SAP. Measurements in [4] showed that such a transparent handover procedure can be performed without introducing significant additional delays. It is worth mentioning that such a handover procedure is not limited to IP-based applications. Mobility management can also be performed for AutoNet applications sending and receiving AutoNet messages. In this way, AutoNet messages transmitted via ITS 5GA can be handed over 'on the fly' to an IP-based network like, for example, UMTS. Vice versa, the mobility management can also collect information of the network service for the management information base. Examples may include the current utilisations of different priority queues as well as additional statistical networking measures [1].

It is worth mentioning that mobility management may also cover optimisations of the applications support layer as well as AutoNet applications. Via the respective service access points, the mobility management may notify respective facilities and applications about different events. This way, facilities as well as applications can adapt themselves to the current communication situation.

10.3.4   TCP Management

A third interesting option of cross-layer management is the optimisation of the transport protocol functionality called 'TCP management'. IP-based applications are typically based on TCP (transport layer protocol), which is optimised for wired communication scenarios. However, the performance of TCP is degraded significantly in AutoNet environments, as we have already described in Chapter 6:

- The conservative flow control used in TCP approximates itself to the available bandwidth using slow start and congestion avoidance. However, the varying available data rates in AutoNets requires continuous adaptation, and TCP therefore has to perform slow starts frequently. Jitter has a similar impact on the congestion control since the round trip time is an important input parameter for the calculation of the congestion control algorithm.
- TCP does not distinguish between congestion in the network and transmission errors. TCP interprets transmission errors as congestions in the network. Hence, TCP will barely leave the slow start phase, which significantly reduces data throughput although enough bandwidth would be available. This effect is amplified by the higher jitter in AutoNets.
- TCP also reacts in a conservative way to temporary disconnections, which also may occur frequently in AutoNets. Due to 'Karn's algorithm', the recovery time of a TCP connection is increased with the duration of the disconnection. This means it can take up to minutes to detect a reconnection after a longer period of disconnection. In this time, TCP does not transmit any data although the sender and receiver are able to communicate with each other.

In order to optimise transport layer functionality while maintaining the compatibility with TCP, TCP management controls the TCP implementation in the AutoNet station using the T-SAP. The general idea is to use the available information stored in the management information base – and provided by other management entities – to optimise the performance of TCP at the transport layer. This interaction is illustrated in Figure 10.5. For example, if transmission errors or a disconnection occurs for a physical communication technology, this information is propagated to the MIB. In this case, TCP management allows TCP to react appropriately according to the current situation. Following this approach, the TCP interface for the application does not change; hence, applications will implicitly benefit from the improvement without any

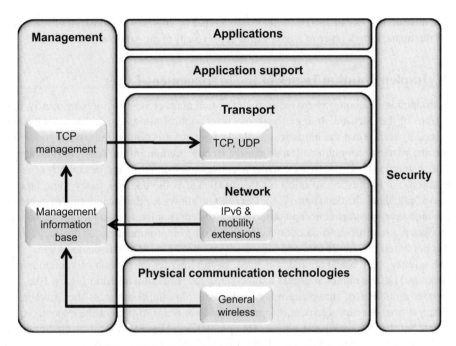

Figure 10.5   TCP management in an AutoNet station.

modifications. However, such an optimisation requires the interoperability with common TCP used in the Internet. From an architectural point of view, this can be achieved in the following two ways:

> **TCP-friendly extensions.** This architectural approach follows the idea of implementing standard TCP compatible extensions within the AutoNet station only. As a result, the AutoNet station is able to communicate with standard TCP hosts, but it benefits from the respective optimisations of improved communications. A prominent example of a TCP-friendly extension is a 'freezing' of TCP as described in [11], which freezes the current TCP state in case of a disconnection. After the re-connection, TCP is recovered and restored immediately in order to continue without limitations.

> **Performance-enhancing proxies.** A more complex alternative is the introduction of a transport-layer proxy, which separates the continuous TCP connection into two parts: a standard TCP part for communication with Internet hosts, and a 'mobile' part for communication with the AutoNet station. On the mobile part, a highly optimised TCP can be used to adopt TCP performance to the characteristics of the respective communication systems. Such an approach is proposed in [1]. An optimised transport protocol for such a system architecture is outlined in Section 6.2.4.

Measurements of a proxy-based architecture showed that there was a significant performance gain in communication efficiency for vehicular communication environments [6] – a snapshot of these measurements can also be found in Section 6.2.5. This also demonstrates the important role of the management layer in a real-world deployment of an AutoNet.

10.4   Implementation Issues of the Management Layer

The introduction of a comprehensive AutoNet station management requires respective implementations of the decision strategy for (cross-layer) optimisations. Such decisions cannot be performed by traditional mathematical methods, because it is almost impossible to formalise mobile and wireless communication scenarios. Hence, predictions are only possible at a high level of abstraction and need to cope with the current unavailability and inaccuracy of relevant information. An interesting solution for this challenge is the use of a fuzzy logic based approach to 'calculate' the decisions [7, 8]. Fuzzy logic allows for the formulation of coherences at very high levels of abstraction by describing a system intuitively using linguistic variables. This makes it a suitable candidate for such contexts and environments since the decision logic itself is not based on simple true and false decisions but on more sophisticated logic.

Such a fuzzy logic based approach is used for the Internet integration of vehicular ad-hoc networks by [1]. The author suggests a three-step process as illustrated in Figure 10.6. In the first phase, a number of fuzzy engines called *prediction engines* process the available state information from the management information base in order to predict the expected quality of service for the individual alternative. This prediction is based on a set of prediction rules stored in the MIB, which need to be defined accordingly. The second phase of the decision process implements *evaluation engines* based on fuzzy logic to evaluate the predicted quality of service by means of the application requirements. Therefore, the management information base provides relevant application profiles to be considered. The evaluation engines also perform

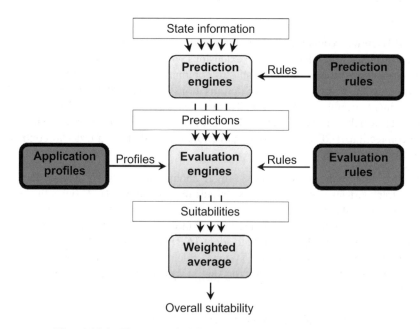

Figure 10.6   Three-step decision process based on fuzzy logic [1].

a defuzzification to map the suitability for each application profile onto a numerical result between 0 and 1. Depending on the decision, an optional third step calculates the weighted average representing the overall suitability for each alternative being considered.

The result of this process is a list of available alternatives ordered by the overall suitability. This result is used for the following mechanisms of the respective management activities. For deployment, it is important to mention that prediction rules, application profile specification and evaluation rules may differ from AutoNet station to AutoNet station. This allows a product differentiation of different suppliers and vehicle manufacturers, because the rules itself determine the customisation to the preferences of vehicle manufacturers and users.

10.5 Summary

In this chapter, ambitious readers should have gained an impression of the need (and the potential) to manage both an AutoNet station as well as an AutoNet itself. The message is that the management layer provides important mechanisms and functionality to improve the efficiency of many aspect in AutoNets. Therefore, the management layer has to interact with every layer in the AutoNet Generic Reference Protocol Stack. The examples given above also showed that there is a variety of aspects, which need to be managed – ranging from costs and benefits to mobility and communication efficiency. However, the most important message is that these examples only give a small insight into this wide field. For a long-term (and scalable) deployment of AutoNets, there will be by far more management issues that need to be addressed by the management layer. However, both the basic functionality and potential realisation approaches described in this chapter should be a solid basis to implement this value added management functionality.

References

[1] Bechler, M. (2004) *Internet Integration of Vehicular Ad Hoc Networks*. Logos Verlag.
[2] ISO/DIS 21210 (2009) Intelligent transport systems – Communications access for land mobiles (CALM) – IPv6 Networking. ISO Standard 21210.
[3] ISO/DIS 29281 (2010) Intelligent transport systems – Communications access for land mobiles (CALM) – Non-IP networking. ISO Standard 29281.
[4] Bechler, M., Hurler, B., Kahmann, V. and Wolf, L. (2001) A management entity for improving service quality in mobile ad-hoc networks. *Proceedings of the International Conference on Wireless LANs and Home Networks*.
[5] Schiller, J. (2003) *Mobile Communications*, 2nd Edition. Addison Wesley.
[6] Bechler, M., Jaap, S. and Wolf, L. (2005) An optimized TCP for internet access of vehicular ad hoc networks. *Proceedings of the IFIP Networking Conference*.
[7] Mukaidono, M. and Kikuchi, H. (2001) *Fuzzy Logic for Beginners*. World Scientific Publishing Company.
[8] Leondes, C.T. (1998) *Fuzzy Logic and Expert Systems Applications (Neural Network Systems Techniques and Applications)*. Academic Press.
[9] Paxson, V. and Allma, M. (2000) Computing TCP's Retransmission Timer. RFC 2988, Internet Engineering Task Force (IETF).
[10] Bechler, M. and Wolf, L. (2005) Mobility management for vehicular ad hoc networks. *Proceedings of the 61st IEEE Semiannual Vehicular Technology Conference (VTC 2005 Spring)*.
[11] Goff, T., Moronski, J., Phatak, D. S. and Gupta, V. (2000) Freeze-TCP: A true end-to-end TCP enhancement mechanism for mobile environments. *Proceedings of the 19th IEEE Conference on Computer Communications (Infocom)*.

11

Research Methodologies

In order to evaluate different applications, architectural approaches and technical implementations, a consistent research methodology is required. In this context, real-life experiments with properly equipped vehicles and road-side infrastructure represent an interesting research methodology. As outlined in the Appendix A, different research and industry consortia have been conducting so-called field operational tests for a few years now. For example, the German research initiative simTD is currently developing an integrated system with 21 different AutoNet applications, which will be tested with up to 400 vehicles in the area of Frankfurt/Main (Germany). While such tests have the advantage of exactly reflecting the real behaviour of vehicles, drivers and communication networks, they imply huge costs. Particularly when it comes to an evaluation of applications involving a significant number of vehicles, computer simulations are a valuable alternative. These are usually less cost-intensive and also offer deterministic behaviour, that is simulation results can be exactly reproduced when using the same input parameters and boundary conditions. As a major drawback of simulations, they lack a perfect reflection of reality and thus need to be built and adjusted in a way to minimise any deviation from real-life behaviour. In general, both research methodologies are required for the introduction of AutoNet-based cooperative systems. Therefore, the simulation methodology generates fundamental results in a first phase, whereas field operational tests validate such systems in a second phase and show the typical limitations of real applications as well as their impact.

This chapter follows exactly the methodologies used in the two phases described. It first of all outlines a selection of early existing simulation setups used to investigate AutoNets and their implications on road traffic, followed by a short introduction on the two research methodologies mentioned above in Section 11.2. Based on this initial preparation, the research methodology for evaluation and analysis by simulation is introduced, especially with respect to the important characteristic results that are generated by respective simulations. This research methodology is completed by proposing an integrated simulation model that aims to mitigate limitations of existing approaches and serves as the basis for statistic analysis. Section 11.4 introduces the field operational trial methodology from a project's perspective. It therefore emphasises the typical activities and their interactions in such a project and provides some important issues that need to be considered to run the field operational trial methodology successfully. Finally, a summary on the research methodologies concludes this chapter.

Automotive Internetworking, First Edition. Timo Kosch, Christoph Schroth, Markus Strassberger and Marc Bechler.
© 2012 John Wiley & Sons, Ltd. Published 2012 by John Wiley & Sons, Ltd.

11.1   Early Activities to Investigate AutoNets

There are lots of different simulation frameworks and analysis tools for investigating AutoNets. Today, even different simulation environments with different AutoNet extensions are used in research in order to analyse different aspects of AutoNets. In the following, we will present two approaches as an example, which illustrate the complexity of the different research challenges in the field of AutoNets.

11.1.1   Activities at the University of Duisburg

In 2000, a team at the University of Duisburg investigated the effects of disseminating dynamic traffic information among vehicles on the stability of traffic patterns [4]. In order to investigate the effects, a simple route choice scenario was proposed to investigate these impacts. Figure 11.1 shows vehicles entering the scenario on the left-hand side. They have to choose whether to drive on route A or route B in order to get to their destination on the right-hand side in Figure 11.1. Some of them do not have access to information about the current traffic situation on the roads and randomly choose their route; other vehicles do have access to this information and thus know about the actual travel time on each road and decide with regard to this information. These vehicles, which are also called *floating cars*, also measure their own travel time and store it within a central database when arriving at their destinations. Assuming a perfectly rational driver model, all floating cars will choose the faster road. One could believe this process helps single cars choose the faster option and thereby saves time.

The simulation environment implemented for analysing the effects of this application on traffic patterns is rather simple. One single simulator accounts for both vehicle movement and information dissemination among vehicles, where network constraints like message delay or medium access collisions are not considered at all.

Simulation runs led to the following results: since the informed drivers choose their route according to the actual data stored in the database and also store back their own travel time, a feedback loop is closed. In the case that road B was reported to be currently very crowded, all vehicles deciding on which road to take enter road A. As a consequence, after a while travel times on road A rise due to the high number of vehicles on it. This is reported to the database and causes most of the vehicles at the starting point to choose road B again. Hence, an

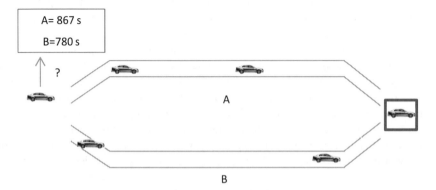

Figure 11.1   Simple traffic scenario consisting of two roads, based on [4].

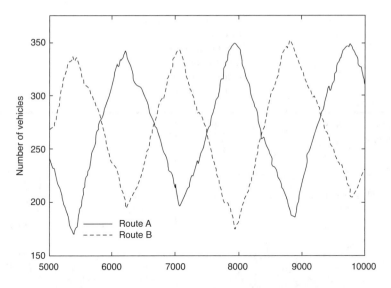

Figure 11.2 Oscillating process on a two-road scenario [4].

oscillating process with a characteristic cycle length starts, which is shown in Figure 11.2. The number of cars driving on road A is represented by the curve starting at about 270 vehicles. Oscillation characteristics heavily depend on parameters like the percentage of floating cars, the vehicles' physical properties, the lengths of the optional roads, and – most important – the driver model applied. Summing up, introducing dynamic traffic information does not necessarily benefit the traffic throughput.

Although these activities assume very simplified pre-conditions, they show the variety of different aspects one has to face with in research when analysing different aspects of AutoNets.

11.1.2 Activities at the Ohio State University

Another very interesting project was started in 2002 at the Department of Electrical Engineering and Computer Science at The Ohio State University [5]. It highlights the enormous potential of using AutoNet ad-hoc networks for increasing traffic safety, especially in the context of intersection collision warning systems. It is assumed that all vehicles approaching an intersection automatically connect to an inter-vehicular ad-hoc network, that is an AutoNet ad-hoc network. They have access to local geographic data and their GPS-based position. By exchanging messages containing position and driving direction, the vehicles are notified about the existence of each other. Every vehicle continuously runs a specific algorithm evaluating whether it is likely to collide with another vehicle or not. If so, it warns the driver via display or voice output. The driver can then take appropriate actions like conducting an emergency break or simply evading the other vehicle. The main goal of this project was to quantify the benefits of the intersection collision warning system implemented.

For simulating different scenarios, a unique environment has been designed and implemented. It features graphical user interfaces for both adjusting parameters before a simulation run and displaying traffic scenarios afterwards. Its major parts are the integrated *vehicle traffic*

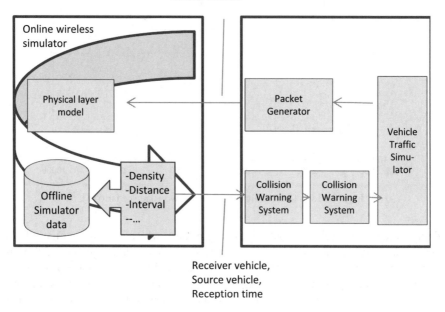

Communication protocol
Protocol parameters
Scenario parameters
Vehicle density
Vehicle positions
Vehicle sources

Online wireless
simulator

Physical layer
model

Offline
Simulator
data

-Density
-Distance
-Interval
--...

Packet
Generator

Collision
Warning
System

Collision
Warning
System

Vehicle
Traffic
Simu-
lator

Receiver vehicle,
Source vehicle,
Reception time

Figure 11.3 Simulation environment consisting of traffic and network simulator, based on [6].

simulator (VTS), which generates the vehicle location patterns by imitating natural driver behaviour, and the real-time *wireless simulator* (WS), which represents the network layer and contains proprietary physical communication layer models. VTS provides traffic flow information at an extremely detailed level. It is responsible for representing driver behaviour, vehicle velocity and acceleration, different kinds of vehicles and for choosing vehicle destinations and the routes to these destinations. VTS operates at a millisecond-level in order to enable a slow-motion analysis of car positions, velocities and directions on a crossing. WS, which has been developed at the university as well as VTS, works on a microsecond-level, and accounts for the functionality of the MAC layer as well as the radio wave propagation models. It also comprises the representation of shadowing through buildings along the road and also some other effects, which are typical for wireless communication in urban areas.

The most important characteristic of this simulation environment is the feedback loop between the wireless network simulator and the traffic simulator. Thereby, the effects of message exchange in the AutoNet ad-hoc network can be taken into account for vehicle movement during simulation time. This 'on-the-fly feedback' enables – together with a certain driver behaviour – a realistic evaluation of whether the collision warning system will help avoiding incidents or not. This model has already been implemented in some prototypes, but there are still some problems with matching simulation data with results from experiments on the road, especially regarding the functionality of the physical communication layer [6].

Summing up, both presented simulation environments propose models for analysing the effects of inter-vehicular communications on certain traffic patterns. The team at the University of Duisburg rely on a single simulator, which disregards wireless network constraints, whereas a rather complex environment has been set up within the second approach conducted at The Ohio State University: two basically independent simulators have been coupled in order to establish a feedback loop.

11.2   Methodologies

The introduction of a cooperative transportation system based on AutoNets requires a thorough investigation of several relevant and crucial aspects. These aspects have to be evaluated and – to some extent – tested before the cooperative system will be ready for the market. This is of special importance for AutoNets since the introduction of a new technology like AutoNets into new vehicles requires long development and integration processes, which in turn is very expensive, and the technology itself has to be available and supported for a rather long time due to the long production and usage cycles of vehicles. Due to this long period of time for the market introduction, combined with respective market penetration, potential refunding will start very late. This also increases the risk for vehicle manufacturers since the success and acceptance of a new technology is hard to predict in advance for such a long period of time. Moreover, these basic investigations are important for the market introduction itself. Cooperative systems in general, and AutoNets in particular, will only be successful if all vehicle manufacturers collaborate and cooperate together. This collaboration requires profound and intensive standardisation activities, because there are a lot of stakeholders involved and all stakeholders are seeking to meet their requirements of the system in the best way.

In general, there are two important methodologies to investigate and to prepare both for the development and integration processes as well as the market introduction scenarios of AutoNets.

- *Analysis and evaluation by simulations.* In the early phase, a system evaluation is required using simulations. This helps to investigate the difficult and complex aspects that come along with AutoNets, such as scalability issues and the very basic understanding of AutoNet applications in different road traffic and penetration rate scenarios.
- *Analysis and verification by field operational testing.* An important methodology for the market introduction preparation is the verification by testing such a system. For AutoNets, so-called field operational tests (FOT) allow the adjustment and evaluation of AutoNet applications in real-world scenarios.

The results generated by both methodologies correlate with each other. For example, large-scale testing requires a profound analysis by simulations for several reasons: (i) the development and deployment of such a system is expensive, (ii) it requires an agreed system architecture that (iii) will be flexible enough to be supported for the next decades combined with different market introduction scenarios and highly varying penetration rates. Vice versa, testing provides important input for refining the simulation activities since the assumptions and requirements used in the different simulation environments have to be adjusted and verified.

In the following subsections, we will first introduce the different model domains for AutoNets. This structure is the essential basis for setting up respective methodologies. The remainder of this chapter will then discuss simulations and field operational tests from the methodology's point of view.

11.2.1   Model Domains for AutoNets

The investigation and evaluation of AutoNets has to address different aspects of AutoNets. An AutoNet has to be considered as a large and complex overall system model, which consists of different loosely coupled sub-models for the different aspects. Figure 11.4 shows the different aspects (called *model domains* further on) of an AutoNet model:

> **Driver and vehicle model.** From an abstract point of view, this model reflects the behaviour of a single vehicle. This behaviour is typically determined by the driver and by the capabilities of the driver's vehicle. Drivers are usually differentiated according to their driving style. For example, an aggressive driver usually travels at higher speeds while using the vehicles' brakes rather often. On the other hand, a passive driver typically travels cautiously, which also results in a decreased fuel consumption of his or her vehicle. Moreover, a driver's behaviour is influenced by different additional aspects, such as the current time, the current mood or time schedule of the driver, the environment and potentially even the presence of vehicle

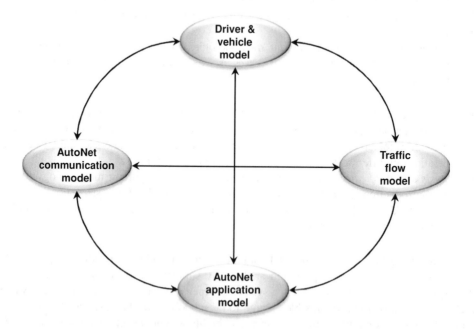

Figure 11.4   Model domains for research methodology.

assistance systems. The vehicle characteristics are also important for the driver and vehicle model. On the one hand, they have an impact on the driving style itself. For example, powerful sports cars are typically driven at higher speeds. On the other hand, a model of the vehicle is an important input factor to determine the impact of a crash on the vehicle and its passengers.

Examples for vehicle and driver models were introduced by Treiber et al. [19] or Bayliss [18]. The importance of the vehicle characteristics for this model is emphasised by the fact that insurance companies differentiate their insurance rates based on vehicle models and types. The rates are determined based on highly complex mathematical and statistical calculations, which also consider the frequency a vehicle type is involved into accidents. For example, the Insurance Institute for Highway Safety in the United States of America annually publishes the 'Top Safety Picks'.[1]

In general, the driver and vehicle model allows us to investigate vehicle safety aspects with respect to the damage of the passengers. It is also an important input parameter to determine the efficiency (e.g. in terms of fuel consumption) in different movement scenarios. Furthermore, the driver and vehicle model is also required to measure driving pleasure, which is an important factor especially for vehicle manufacturers.

Traffic flow model. According to Wikipedia, the term *traffic flow* is defined as 'the study of interactions between vehicles, drivers, and infrastructure (including highways, signage, and traffic control devices), with the aim of understanding and developing an optimal road network with efficient movement of traffic and minimal traffic congestion problems.'[2] The traffic flow model therefore combines several instances of the driver and vehicle model in order to reflect typical traffic situations. Based on respective algorithms and theories that combine the different micro-mobility patterns of the vehicles (based on the respective driver and vehicle models), it allows us to model the characteristic macro-mobility patterns of typical traffic situations on the roads [20–22]. Therefore, the traffic flow model has to consider additional factors. Examples include road conditions, legal and regulatory factors such as, for example, speed limits, or weather conditions. However, it is also important to consider experiences or heuristic factors since there are lots of different phenomena on the road that are non-linear and cannot be explained in a mathematical way. Such phenomena include different formations of vehicle clusters or shock wave propagation, especially on roads with a high density of vehicles.

In general, the traffic flow model provides the basis to evaluate traffic efficiency aspects as well as safety considerations with respect to the potential to avoid accidents in different traffic situations.

AutoNet application model. The AutoNet application model addresses the behaviour and quality of cooperative AutoNet applications. This model is necessary

---

[1] The Top Safety Picks 2010 can be found at http://www.iihs.org/ratings/default.aspx.
[2] Wikipedia: http://en.wikipedia.org/wiki/Traffic_flow.

since there are two dynamic and unpredictable factors that need to be considered for the design of cooperative systems:

- Different vehicle manufacturers provide different functionality and visualisations for cooperative applications. Hence, the same application in different vehicles may have a different impact on the driver since it interacts with the driver in a different way.
- The simultaneous existence of several cooperative applications requires a prioritisation of the information and warnings generated by them. For example, important warnings of critical safety application will preempt information of other cooperative applications. Moreover, different vehicle manufacturers will have different concepts for the prioritisation, resulting in a completely different behaviour of the cooperative system that needs to be modelled appropriately.

As a result, the AutoNet application model has to consider several aspects that have an impact on the behaviour of the system. Examples include, among many other factors, the AutoNet application integration into the vehicle – especially with respect to human machine interface (HMI), the strategies and requirements of the vehicle manufacturers with respect to the prioritisation concept, the input data from the vehicle itself, the message fusion strategy of the respective vehicle or even the current time of day.

The AutoNet application model provides the fundamentals for the situation awareness, the prioritisation strategies and, finally, the acceptance of cooperative systems by road users.

AutoNet communication model. An important model for cooperative and intelligent transportation systems is the AutoNet communication model, which basically addresses the data exchange among the road users. This data exchange depends on many factors, such as the characteristics and performance of the different communication layers (namely physical layer, medium access control and upper-layer data exchange protocols), the environment and its impact on mobile communication, and the routing strategies used to forward messages in a multi-hop fashion. Communication models played an important role in research in the past. Nowadays, there are simulation models for almost every communication system and routing strategy available, which were – together with the results generated – accepted in the respective research community. For example, Hartenstein et al. give a sound overview in this field of research [23, 24].

The AutoNet communication model provides the prerequisites to adjust the communication system parameters onto respective traffic scenarios, to perform optimisations with respect to spectrum and communication efficiency, and also to fulfil necessary quality of service requirements for different types of AutoNet applications.

11.2.2 Dependency Examples

Besides the different factors that need to be considered by the different model domains, a model domain also mutually interacts with the other model domains, as illustrated in Figure 11.4.

This way, a minimal modification of one model may have an immediate effect on the behaviour of the other models, resulting in completely different behaviour of the overall cooperative system. In this section, we will emphasise this interaction by describing the interactions using typical example flows. Such example flows give a notion of the variety and complexity of such a system. Ambitious readers may additionally figure out lots of other examples for the interactions.

Example interactions among the four model domains are summarised in Figure 11.5. The *driver and vehicle model* has an immediate impact on all other model domains. Since it models the movement of vehicles, the driver and vehicle model provides the micro-mobility patterns of the vehicles as an important input factor for the traffic flow model. The driver also configures and controls respective driver assistance systems and applications in the vehicle. Application control also affects the behaviour of the AutoNet application model; for example, the driver may deactivate cooperative applications, or may provide information such as the final destination for the cooperative systems. Finally, the antennas mounted in the vehicle and their directional characteristics result in antenna characteristics, which are an important input factor for the AutoNet communication model.

The *traffic flow model* also has an immediate effect on the other three model domains. For example, it provides macro-mobility patterns such as the traffic density in the respective traffic scenario to the driver and vehicle model. This factor thus influences driving behaviour: for example, an increased traffic density usually results in a lower average speed and lower distances between the vehicles. The traffic flow model also provides the expected traffic flow to the AutoNet application model, which we call 'macro context'. This way, the AutoNet applications may consider this information in their calculations in order to adapt the cooperative

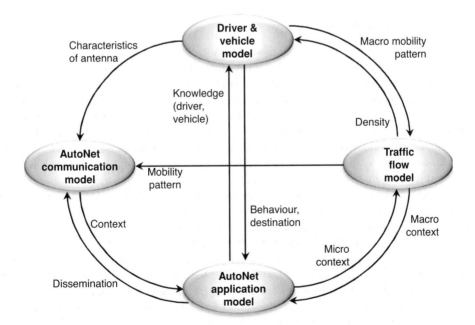

Figure 11.5   Example dependencies between the model domains.

system to the traffic situation in the optimal way. Finally, the macro-mobility pattern is also an important impact for the AutoNet communication model. It is basically used to determine the potential information dissemination strategies in different traffic scenarios.

Like the driver and vehicle model and the traffic flow model, the *AutoNet application model* also has an immediate impact on the other three model domains. This is obvious since AutoNet applications provide the necessary information and functionality for cooperative systems. Therefore, the AutoNet application model provides important input about the context of a vehicle and its vicinity, called micro-context. For the driver and vehicle model, it provides its knowledge by 'representing' the cooperative system to the driver and the vehicle. This way, drivers and vehicle control systems can react accordingly to adapt their driving style to this knowledge. Finally, the AutoNet application model also affects the AutoNet communication model: the functionality and processing of the AutoNet applications are an important input for the message dissemination and, thus, the expected network load, the network density of communicating nodes and potentially the transmission powers of the communicating nodes.

In contrast to the other model domains, the AutoNet only has an immediate impact on the AutoNet application model, since it determines the dissemination of context among the vehicles with an expected service quality. However, this immediate effect indirectly has an impact on the driver and vehicle model as well as the traffic flow model. For example, the dissemination of messages relevant for one particular cooperative AutoNet application may not affect temporarily disconnected vehicles, resulting in different behaviour of the cooperative AutoNet application. This means that both driver and vehicle behaviour as well as the traffic flow will be different, resulting in a completely different behaviour of the overall cooperative system.

11.3 Simulation Methodology

In order to work with the methodology model for AutoNets described in the previous section, simulations are an important methodology to evaluate the cooperative system and to validate respective assumptions. This is of particular importance in the early development phase, since the interactions between the model domains have to be evaluated in order to estimate the potential of cooperative systems. Therefore, the four model domains have to be simulated appropriately, whereas the four simulators have to be coupled in order to share the different information and results among the model domains. An important integral approach to simulate such an overall cooperative system was introduced by Schroth et al. [25–27], which we will detail in the following since it already covers the four different domain models.

The AutoNet simulation environment proposed by Schroth et al. basically consists of three major components. The first is represented by a road traffic simulator, which periodically computes new positions for a certain number of vehicles within a specific geographic scenario, as visualised in Figure 11.6. This way, the traffic simulator unites both the driver and vehicle model as well as the traffic flow model. A wireless network simulator constitutes the second major component, which is dedicated to imitating the full functionality of a real wireless AutoNet ad-hoc network with all its complex effects of mobile communications. Only a proportion of the vehicles defined by the traffic simulator also participate in the AutoNet ad-hoc network, whereas others play a certain role for traffic considerations, but are not able to send or receive data. The wireless network simulator must be continuously notified about the

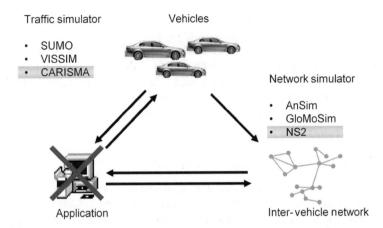

Figure 11.6   Two different simulators account for message dissemination and vehicle mobility.

positions of the vehicles that participate in the network in order to have the current connectivity pattern available. The third and last component is the AutoNet application itself, which actually accounts for modelling driver behaviour and controlling the whole simulation environment. It evaluates received messages and decides upon the action to be taken by the traffic simulator. For example, it can make single vehicles move around a danger zone in the event that they have already been warned about it. Apart from this, the application is able to generate new messages and broadcast them via the network. The application relies on up-to-date information about current vehicle positions, which are provided by the traffic simulator.

For the AutoNet simulation environment, the application component has been integrated into the network simulator as an additional module. This simplifies implementation work, since communication links are only necessary between two simulators. Figure 11.7 visualises the coupled simulation setup. Two communication modules handle the data exchange between the simulators. This provides the possibility of running the simulators on two different machines to increase simulation performance. The environment could also be split into three parts (as shown in Figure 11.7), adding an application simulator to the setup. This creates the possibility

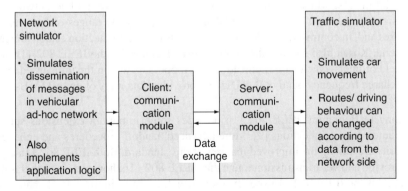

Figure 11.7   System overview of the two main components of the simulation environment.

of using different programming languages and allows for the reuse of real-life prototype applications in the simulator. The simulation environment consists of both a network and a traffic simulator. A feedback loop ensures periodic data exchange between them. The network simulator will act as a client requesting information from the traffic simulator. For instance, it needs to know both the exact number of vehicles that are currently part of the AutoNet ad-hoc network and their geographic positions within the scenario. The traffic simulator, on the other hand, acts as a server and sends all the requested information to the network simulator.

11.3.1   Communication Network Simulation

As mentioned above, one of the two simulators used will only account for imitating the functionality of a wireless ad-hoc network which is connecting the vehicles. There are certain important criteria when choosing an adequate tool. First of all, it should feature a comprehensive model for mobile wireless networks. A representation of specific effects like shadowing caused by buildings in urban areas or radio wave reflections due to roads and building walls must be readily available and adjustable as well. Another crucial element would be proper antenna models with adjustable transmit powers and reception thresholds in order to reproduce the properties of the real wireless modules built into prototyping vehicles. Apart from that, efficient routing protocols like AODV (Ad-hoc on demand distance vector) must be available [17]. Furthermore, the simulator should be widely used and accepted in research and known in the community.

A network simulator also has to implement widely accepted communication standards such as the different IEEE 802.11 specifications. IEEE 802.11, in the following simply called WLAN, specifies both the physical and the MAC layer of the OSI layer network architecture. WLAN has become the prevailing wireless transmission technology in the field of AutoNet ad-hoc networks for several reasons:

- Its sensitivity to high velocities is relatively low. Kosch already proved that there is only a very little correlation between vehicle velocity and data transfer rate [8]. Even in the case of two cars driving at a speed of 120 km/h in opposite directions – that is, the cars have a relative speed of 240 km/h – almost the maximum data rate is achieved.
- Transmission ranges are in the order of several hundreds of metres, for IEEE 802.11b even up to 1000 metres, which satisfies the requirements of AutoNet ad-hoc networks. Other wireless communication standards like Bluetooth only provide ranges of between 10 and 100 metres and, in contrast to WLAN, require long times for connection establishment of up to a second. Kosch also showed the applicability of especially the IEEE 802.11b standard in the field of inter-vehicle communication [8]. It operates within the unlicensed and, thus, free-of-charge frequency band of 2.4GHz and provides data rates of up to 11MBs.

According to Benjamin, IEEE 802.11p provides some important extensions specifically aiming at automotive applications [9]. More advanced handoff procedures and enhanced network security are among the improved features of this standard. Both IEEE 802.11a and IEEE 802.11p provide a lower transmission range than IEEE 802.11b, however. A further, relatively new standard is the IEEE 802.11g specification, which operates in the 2.4GHz band and provides rather long communication ranges of up to 1000 metres. Due to the implementation of orthogonal frequency division multiplexing (OFDM) [13], the same modulation scheme as the one applied in IEEE 802.11a, data rates of up to 54MBs are realised.

The latest research focuses on the development of the IEEE 802.11s specification, which defines efficient ways for wireless nodes to discover each other, set up connections and determine the most efficient route for a specific task in meshed networks. Such a self-configuring and topology-aware standard for wireless communication could also become interesting for AutoNet applications as soon as it is released and widely accepted. Another wireless communication standard whose applicability for wireless ad-hoc networks has been considered is WiMAX (worldwide interoperability for microwave access), which comprises the diverse sub-specifications of IEEE 802.16. WiMAX supports both licensed and license exempted frequency bands ranging from 2GHz up to 66GHz. It was mainly designed for bridging the last mile from wired network infrastructure to end-users and providing a broadband connection, especially in rural areas, where communication infrastructure is sparse. Typical communication ranges are at several kilometres and, thus, remarkably higher than those of WLAN. Since it was originally not designed for mobile applications, resource allocation algorithms are not suited to scenarios with fast network topology changes. Hence, the WiMAX specification is not adequate for use in AutoNet ad-hoc networking. Summing up, IEEE 802.11b, IEEE 802.11g, and IEEE 802.11p can be considered as particularly suitable candidates for AutoNet applications due to their relatively high communication ranges and their ability to handle fast network changes.

Besides a comprehensive model for mobile, wireless networks and the implementation of the WLAN wireless communication standard, interoperability is crucial for an appropriate network simulator. Its software architecture must be open and accessible in order to realise data exchange with the traffic simulator during simulations. Extensions like the implementation of a completely new network node behaviour must be possible as well, and a simulator written in a standard programming language with a structured and well-documented source code is preferable, too.

There are numerous network simulators available. For instance, the ad-hoc network simulator (ANSim)[3] is a possible approach to representing AutoNet ad-hoc networks. Developed at the International University (IU) in Germany, ANSim has been realised in Java and is basically a tool for visualising ad-hoc networks. It is well suited for simple connectivity considerations, but is not adequate for simulating message dissemination in AutoNet ad-hoc networks. The simulator passes on the implementation of the OSI layer model, on routing and wave propagation models. It features simple mobility models for simulating network node mobility. Node movement is tracked and stored in files that can be used with several relevant road traffic simulators.

The second considered option is the wireless network simulator GloMoSim (global mobile information systems simulation) [14], which features the free space model and the two ray ground model for representing radio wave propagation. Unfortunately, no model for shadowing caused by buildings is provided, which is a remarkable drawback especially in urban environments. The IEEE 802.11 communication standard determining both the physical layer and the MAC protocol is available within this simulator. Also, AODV is provided as a routing protocol. GloMoSim is for academic use only, however. A similar version, which is called QualNet, can be bought for pretty high license fees.

---

[3] ANSim project homepage: http://www.ansim.info.

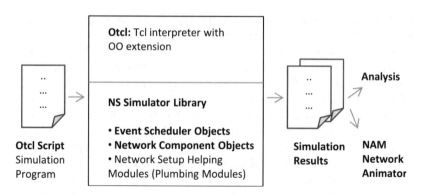

Figure 11.8   Simplified ns2 system architecture, based on [15].

One of the most renowned network simulators for IP-based wired and wireless networks worldwide is the so-called network simulator 2, in the following simply called ns2. The development of this rather complex tool started in 1989 at the University of California at Berkeley. Figure 11.8 provides a simple overview of the ns2 system architecture. Basically, ns2 is written in C++ and OTCL (MIT object tool commandlLanguage), a TCL script language with object-oriented extensions developed at MIT. ns2 is started by loading an OTCL script containing the network topology, the data traffic pattern within this network and other parameters like the simulation duration. Simulation results can be easily viewed with the built-in tool network animator, which must be characterised as rather unimpressive, however. ns2 can be downloaded for free and is also open for commercial use. Its software architecture is well structured and enables the attachment of software modules for data exchange with other programs and is thus ideal for setting up a coupled simulation environment.

In the late 1990s, the Monarch research group at the Carnegie Mellon University developed and published a path breaking extension for ns2, which enabled the simulation of multi-hop wireless networks. The most important achievements of this extension were comprehensive models for the physical layer, the data link-layer and for MAC protocols such as the IEEE 802.11 specification. These modules are of crucial relevance for simulating mobile wireless ad-hoc networks. In fact, such extensions are also used as a basis for current simulations of AutoNet ad-hoc networking aspects. Another very important point is the implementation of the most relevant wireless routing protocols such as dynamic source routing (DSR), destination sequenced distance vector (DSDV), temporally ordered routing algorithm (TORA), and ad-hoc on demand distance vector (AODV). Later versions of ns2 also feature modules for simulating satellite communication networks.

In terms of modelling radio wave propagation, ns2 provides several possible approaches. The first option is to assume the free space model, which completely neglects effects like reflections on the street surface, radio waves that possibly interfere with each other or reflection and scattering of waves incident on different materials. Another possible approach is the so-called two ray ground model (cf. Figure 11.9), which not only considers the direct path between two vehicles (which is not likely to be the only means of wave propagation), but also a ground reflection path.

But still reflections and shadowing effects caused by buildings are not considered at all. In order to achieve realistic simulation results, these effects must be taken into account. One possibility is to use a readily available model provided by ns2: it is based on a simple

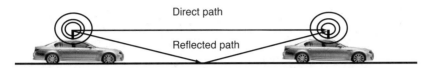

Figure 11.9   Direct communication path and a ground reflection path between two vehicles.

line-of-sight approach. Data concerning the geographic position of buildings and streets are read out from a bitmap file, which should be black (for streets and areas without buildings) and white (representing buildings). The algorithm periodically evaluates the nodes' positions and updates connectivity information. Only nodes with a line-of-sight connection between them are able to communicate, which is still a simplification, since effects like wave reflections on building walls are neglected, for example.

Another way of determining signal attenuation due to obstacles has been proposed by a team at the University of California at Santa Barbara [16]. Their so-called obstacle mobility model is a comprehensive set of different components that enables users to model a terrain defining the obstacles, to setup node movement patterns within this terrain in a realistic manner (not random movement without destination) and to model the radio wave propagation including all relevant effects like reflection, diffraction and scattering. Empirically acquired data is used to represent these effects, changing signal properties during a simulation. In contrast to the ns2 built-in approach mentioned above, within the obstacle mobility model radio waves can reach a receiver, even if there is no direct line-of-sight connection, due to reflection off walls of obstacles. This model represents all the different effects radio waves undergo in a mobile ad-hoc network in an extremely detailed way. C++ files containing the full functionality can be downloaded from a website [16] and compiled into ns2. Regarding the simulation of a relatively big scenario (several square kilometres) with a significant number of network nodes, this model is not adequate, however. It requires the user to design all the obstacles (buildings) manually, including all their corners and wall penetration characteristics for radio waves. The obstacle mobility model is an outstanding tool for a detailed simulation of radio wave propagation in urban areas where obstructing buildings have to be considered, but it is hardly applicable for a larger scenario with lots of network nodes.

Summing up, both the ns2 built-in line-of-sight approach and the externally provided tool enables users to model radio wave shadowing caused by buildings in urban areas. The two ray ground model satisfies the need for simulating both a direct and a reflected communication path between two vehicles. Apart from a proper model for radio wave propagation, ns2 also features the widely accepted IEEE 802.11 standard as a MAC protocol. Different antenna models can be chosen, and all routing protocols, which are relevant in the field of ad-hoc networks, are readily available too. ns2 provides a comprehensive platform that can easily be modified and extended and is therefore the ideal solution for the requirements mentioned above. This is especially approved by the fact that with the continuous development using ns2, the simulation environment was extended even for automotive simulation aspects.

11.3.2   Traffic Simulation

Depending on the type of analysis, different types of mobility simulation are used in research. We categorise the existing mobility and traffic simulation models with respect to their level of

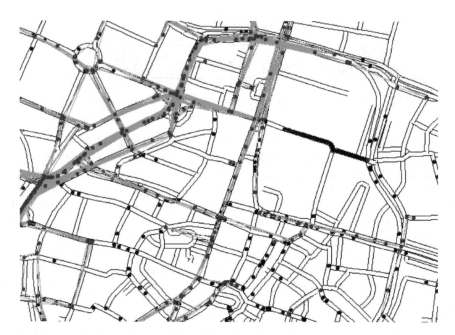

Figure 11.10   Graphical user interface of the CARISMA road traffic simulator.

detail into microscopic and macroscopic models. Microscopic models describe the behaviour of each individual vehicle and the interactions between single vehicles. Macroscopic models describe road traffic with respect to certain aggregated system properties.

For the analysis of AutoNet properties and system behaviour, that is network characteristics and protocol behaviour, usually a microscopic simulator is needed to generate position and movement information for single vehicles. This information is needed on a per-vehicle basis since the vehicles act as nodes in the AutoNet communication networks. Different simulators have been used for this purpose, some specially developed, others originally developed for road traffic analyses. BONNMOTION [11], METROPOLIS[4], CARISMA [8] and SUMO [12] belong to the set of specifically developed tools. VISSIM[5] is a popular commercial microscopic traffic simulator that has been used for AutoNet simulations but was originally developed for road traffic simulation only.

The CARISMA road traffic simulator features a simple visualisation tool. A snapshot of a scene is shown in Figure 11.10. Vehicles with and without AutoNet capabilities can be participating in a simulation and are visually represented with different colours. Lines between vehicles visualise potential AutoNet ad-hoc network connections between vehicles, that is they are within mutual communication range. In CARISMA, a symmetric communication range model is used. The simulator works on a time-discrete basis and updates vehicle positions and directions every second. CARISMA reads in data regarding the geography of the scenario from ESRI shape files containing information about the street network for a scenario. Before

[4] http://www.adpc.be/software/metropolis/description.html.
[5] http://www.ptv.de/.

a simulation run, the scenario size and the number of vehicles moving can be adjusted. Parameters concerning velocity and acceleration, the simulation duration and other vehicle properties have to be predetermined as well. The mobility model implemented is relatively simple, but provides some useful options. Its three main characteristics are the way the driving directions of vehicles are determined, the process of calculating and updating vehicle velocities, and the model for conflict management at crossings:

- Regarding the vehicle driving directions, CARISMA offers two basic possibilities. After being uniformly distributed across the streets during the simulation initialisation phase, vehicles either select a new direction each time step randomly, which is called the random drunken model, or they select one of the crossings within the scenario and drive there sticking to the shortest path policy. This is called random waypoint model, which was an important model for the simulation of common ad-hoc networking issues in research. Destination crossings are either randomly chosen among all those existing within the scenario, or crossings located next to a place, which is in random distance, and in a random direction when seen from the current vehicle position are selected.
- The second major characteristic, the way vehicle speeds are computed, basically follows the so called Krauss model [10]. It takes into account physical properties of the vehicles such as length, maximum acceleration and deceleration. Apart from these, physiological and psychological models are considered: for example, human beings lack accuracy when estimating the distance to preceding vehicles. As a consequence, spacing between vehicles must be larger than it would be when only taking into account mere physical limits.
- Crossing conflict management, the third main characteristic of the CARISMA traffic sim-ulator, works as follows: in the event that several vehicles approach an intersection, the one arriving and stopping first is allowed to drive on first as well. This process follows the American first-come first-serve (FCFS) principle.

It is crucial to understand the basics of the routing algorithm implemented in the CARISMA simulator. Basically, a street network is represented by a number of crossings, street points and connections as illustrated in Figure 11.11. A connection, within this context, is a piece of a road located between two crossings. Street points define the bending of a connection. The same priority is assigned to all connections, which is a simplification of real-world street networks, where main roads, side roads and others exist. Apart from this, all connections provide the same traffic capacity, regardless of their breadth or geographical location. Besides bending, position and direction, a certain length, which represents the real geographic length, is assigned to all connections. Vehicles move along these connections and choose their routes dynamically by running through a search algorithm (based on the well-known Dijkstra algo-rithm) and comparing the overall lengths of possible routes. Eventually they wind up with the shortest possible path. By weighting the length of a certain connection (e.g. the connection *incident_conn*, see Figure 11.10) with a number greater than one (e.g. 5) during this route search algorithm, the path to the destination can be influenced, because this piece of road is suddenly regarded as longer than it actually is. Vehicles will seek routes not containing this specific connection. When using a very high number (e.g. 1000) as a weighting coefficient, the possibility that this connection will be part of a vehicle's route can almost be excluded. Being able to influence the routing of certain vehicles is crucial for showing the impacts of inter-vehicular message dissemination on traffic scenarios.

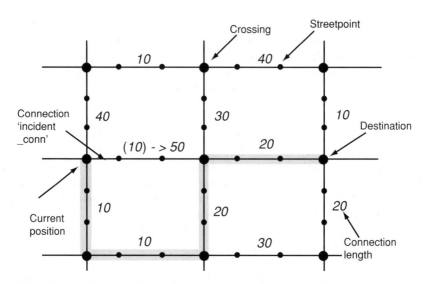

Figure 11.11 Simplified CARISMA street network consisting of connections, crossings and street-points.

CARISMA users have to be aware of several simplifications: the simulator only considers two lanes per street and prevents vehicles from overtaking each other. Apart from that, vehicles do not leave or enter a simulation scenario, but simply choose a new destination when arriving at the border of the simulation scenario.

11.3.3   Implementation Issues

After defining adequate components of the simulation environment, new functionality has to be implemented into each of them. The network nodes will have to act according to specific rules while sending, receiving or forwarding messages. Apart from this, the traffic simulator needs new functionality for influencing vehicle behaviour during the overall simulation time depending on the data received from the network layer. Finally, a certain data exchange protocol must be designed in order to ensure an efficient coupling between the simulators.

11.3.3.1   The AutoNet Agent for ns2

Most ad-hoc networking aspects are simulated using the network simulator ns2, especially for the analysis of the AutoNet message dissemination schemes. The implementation of the legacy IEEE 802.11 standard has been used as a basis for the representation of the link-layer functionality. In terms of modelling radio wave propagation, we chose the two ray ground model, where the radio channel is assumed to be error-free.

Figure 11.12 provides an overview of the configuration possibilities of this simulation environment. Benefit thresholds explained above can be activated or not. Concerning the standard underlying the data link-layer functionality, the user may choose between the standard

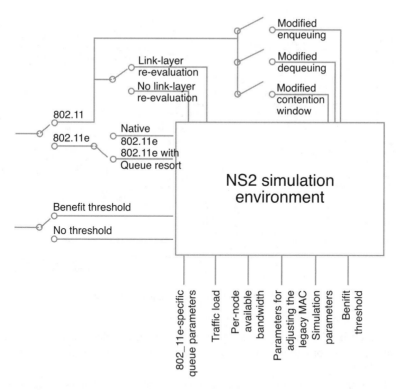

Figure 11.12   Configuration of an AutoNet ns2 simulation environment.

IEEE 802.11 and the IEEE 802.11e specification. When performing simulations based on the IEEE 802.11 standard, one may optionally activate mechanisms such as link-layer benefit re-evaluation, modified enqueuing and dequeuing, and the benefit-based contention-window adjustment. For IEEE 802.11e, users may additionally add the queue resort functionality (remember the usage of these handles from Chapter 7). The environment also comprises numerous configuration parameters for adjusting the behaviour of the proposed communication scheme, such as the IEEE 802.11e specific control parameters for inter-frame spacing (e.g. the arbitrary inter-frame spacing AIFS), the min/max ration of the contention window $CW\,min/CW\,max$, the performance factor PF of IEEE 802.11e, the data traffic load, the per-node available bandwidth, parameters for the MAC layer and the optional benefit threshold. Various other key parameters such as road traffic density and simulation duration are adjustable, too.

Within the ns2 environment, a message agent takes care of generating, broadcasting, receiving and storing AutoNet messages. This message agent is part of the application support layer of the AutoNet Generic Reference Protocol Stack. Figure 11.13 illustrates its basic functionality. The agent maintains local data storage to retain and gather information received via the wireless networks. It also accounts for realising the store-and-forward communication schemes specified for AutoNets. The message agent has access to all data, which is crucial for computing the estimated benefit of a message: such data includes the current geographic location, the current system time, data contained in the node's message storage and much

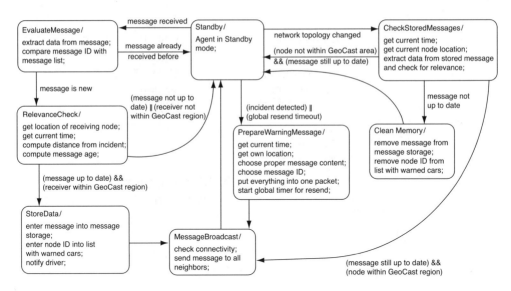

Figure 11.13   AutoNet ns2 message agent functionality.

more application level data. As a consequence, the agent is able to compute the estimated benefit messages and provide the resulting value to potential recipients and attaches in the packet's header.

As illustrated in Figure 11.13, there are three possibilities for the message agent to leave its standby state and become active:

1. A DEN message is received, whereupon the benefit provided to the recipient (ur) is calculated. Before initiating any forwarding, the benefit that may be provided to neighbours by re-sending of the message is computed (ut) and attached to the message header. If the benefit threshold is activated, messages are only relayed if the resulting value is higher than the threshold.
2. In a disconnected situation, upon the detection of a new neighbour (e.g. via a network beacon or a CAM), the previously received and stored DEN messages are read from the local memory, evaluated with respect to their current benefit for recipients and passed down to the lower layers for transmission.
3. An application generates a message and also evaluates its current benefit before passing them down for transmission.

Regarding the implementation of the data link-layer in the extended ns2 environment, two possibilities can be chosen. First, the legacy representation of the IEEE 802.11 specification may be activated and adjusted in various ways. Second, a representation of the IEEE 802.11e standard is alternatively available in the simulation environment, which has been set up for this work. The IEEE 802.11e implementation ensures that each node operates four packet queues of which each applies different parameters regarding medium access. For the purpose of comparison, a link-layer level resorting of messages can be enabled additionally. When active, this mechanism changes the dequeuing sequence such that not necessarily the

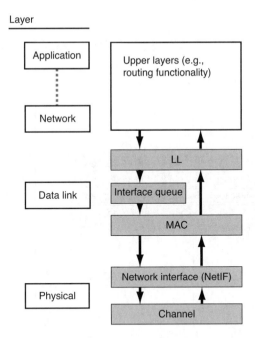

Layer

Figure 11.14 Network node layer model in ns2.

head-of-line packet is the next one to be dequeued, but the packet with the highest benefit value. In the event that the legacy IEEE 802.11 implementation is selected, mechanisms for the modified packet dequeuing/enqueuing and the modified MAC functionality can be activated or deactivated independently. Additionally, the simulation environment provides the possibility for adding a link-layer benefit value recalculation of the packets. This allows for investigating the improvement potential due to a continuous benefit re-evaluation while packets wait within their respective queues.

Minor deviations from the IEEE specification within the ns2 implementation have been corrected. Figure 11.14 shows the composition of the two lowest layers of the OSI model as realised in the ns2 environment. The data link-layer conventionally consists of the logical link control (LLC) and the medium access control. The MAC layer and the functionality of the physical layer are both determined by the IEEE 802.11 standard. The two major modules whose functionality was adapted to the AutoNet requirements are the interface queue and the MAC layer. In Figure 11.15, the ns2-specific objects are presented that are integral parts of the data link-layer implementation. Data packets are stored within the interface queue before finally getting access to the medium.

11.3.3.2 Modified CARISMA Traffic Simulator

The CARISMA traffic simulator was modified (Figure 11.16) in order to fulfil the requirements of the coupled simulation environments. Its main new task is to periodically receive data concerning changes in-vehicle behaviour from the network side, realise them, and send new node positions back to the network simulator.

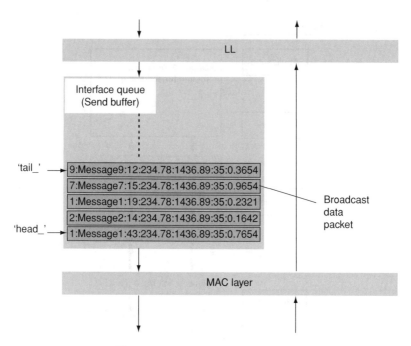

Figure 11.15   Interface queue in ns2.

One exemplary function of the modified CARISMA simulator is to reflect driving behaviour changes subsequent to a received local danger warning. Within this context, the most important information for the traffic simulator is the location of an accident and the IDs of the vehicles that have been warned. CARISMA then assumes that all drivers of these vehicles still try to get to their destination, but intend to choose a new route not directly passing by the location of the accident. The traffic simulator chooses routes according to the shortest path policy. Hence, paths weighted with high factors are not likely to be chosen as parts of a new route.

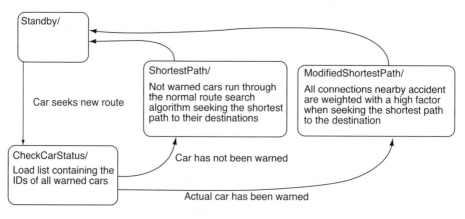

Figure 11.16   Modified traffic simulator.

11.3.3.3   Simulator Coupling Controller

In order to set up the AutoNet simulation environment enabling the analysis of impacts of vehicular networks on traffic flows, a feedback loop must be installed to ensure that both simulators report their status and results to each other. For example, CARISMA has to transmit the geographic scenario size and the number of vehicles to ns2 in the beginning of a simulation run. Apart from that, it must periodically inform ns2 about the actual vehicle positions, whereas the traffic simulator relies on being notified about the geographic location of the danger zone and the vehicles which are currently aware of it. Last, ns2 needs up-to-date information about network connectivity.

There are several hurdles to overcome in order to establish a high performance connection between traffic simulator and network simulator. First of all, the update frequency between the simulators must be determined. A high frequency guarantees an exact alignment, but also reduces the overall performance of the simulation environment. If the traffic simulator sends all vehicle positions to the network simulator every millisecond, for example, it will take a very long time to compute even small scenarios with only a few hundred vehicles. On the other hand, using larger update intervals of several seconds would cause unacceptable failures. In the event that a vehicle receives a warning message and its driver intends to drive around the danger zone, for instance, the traffic simulator should be notified as soon as possible to take into account this change of behaviour. Hence, if the update frequency is too low, a real-time simulation of the interaction between communication networks and traffic flow cannot be put into effect. For example, the time-discrete traffic simulator CARISMA updates vehicle positions and directions every second, so the highest possible update frequency between both simulators is one data transfer per second of virtual simulation time. This interval turns out to be a good trade-off between optimum simulator alignment and computational efficiency, since the interaction between communication networks and traffic flow does not have to be investigated on a millisecond basis like in an intersection collision warning system.

One further important issue is the adaptation of possibly different coordinate systems that need to be considered by simulator coupling. For example, the point of origin in the ns2 coordinate system is located in the left bottom corner of the scenario, whereas the one within CARISMA is located in the left upper corner. Hence, some coordinate conversions have to be implemented in order to ensure flawless interoperability.

The third important challenge of interlinking the two simulators is the difference in-vehicle mobility representation. Usually, every communicating vehicle controlled by traffic simulator is represented as a network node from the network simulator's point of view. It is crucial that vehicle positions and the locations of network nodes are exactly aligned to each other during the whole simulation. For example, CARISMA simulates acceleration and deceleration of vehicles, whereas the network nodes in ns2 are only able to move at constant speed. As a consequence, CARISMA's level of positioning accuracy is slightly higher than the one applied in ns2. In order to couple a traffic simulator with a network simulator, three alternatives can be used:

1. The first approach was to run both simulators exactly in parallel and send absolute vehicle positions from the traffic simulator to the network simulator every second, which is visualised in the very left chart of Figure 11.17. The idea was to set the network nodes in the network simulator to new positions after each time step and thus guarantee exactly

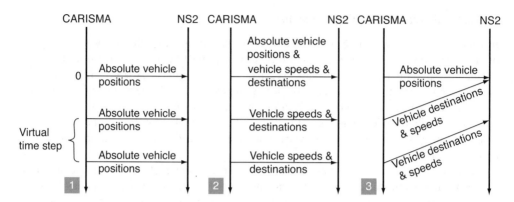

Figure 11.17   Three different approaches for aligning vehicle movement in CARISMA and ns2.

equal vehicle positions in both the traffic simulator and the network simulator. As a result, network nodes do not continuously move, but 'jump' from one point to the other. This involves several severe problems, however. First, the mobility model of network simulators typically is not attuned to accepting absolute node positions. For moving a node from one point to the other within one time step, it requires a geographical destination and data regarding the velocity at which the node is supposed to move. Second, the network simulation severely lacks in accuracy due to the fact that nodes do not move continuously, but remain at a fixed position for a whole time step. Connectivity between vehicles would be estimated unrealistically, because the network topology would not change within a whole second. In a real-world scenario, however, the structure of AutoNet ad-hoc networks permanently changes due to vehicles connecting or disconnecting when driving through an urban environment. Therefore, this approach does not meet the requirements for simulator coupling: the representation of vehicle mobility in the network simulator resulting from this first model can be seen in the very left part of Figure 11.18. The black line represents the path of a vehicle, simulated by the traffic simulator. The black, cross-like symbols mark the locations where a position update is sent to the network simulator. Grey crosses symbolise the node positions at synchronisation times as seen from the network simulator's point of view. Geographical locations of vehicles are identical in both simulators only during synchronisation times, because network node positions do not change before the next update.

2. The second approach works as follows: only for initialising network node positions are absolute positions transmitted. The traffic simulator then estimates, based on current vehicle speeds and directions, where nodes are probably located after one time step. This

Figure 11.18   Resulting representations of vehicle mobility for the example of ns2.

information is sent to the network simulator as vehicle destinations and velocities. The approximation is rather rough, however, since velocity is wrongly assumed to remain constant during the whole interval. Also, bending of roads is not taken into account by this approach. As a consequence, a considerable vehicle positioning error occurs after each time step, which is visualised in the middle part of Figure 11.18. Network nodes move continuously in this case, but their positions in the network simulator and the 'real' vehicle locations in the traffic simulator do not comply with each other. Hence, this mechanism is also unsuitable for simulator coupling.

3. The third approach to synchronise vehicle positions in both simulators turned out to be the best one: simulators do not run exactly in parallel any more, but the traffic simulator is one second in virtual simulation time ahead of the network simulator. Not until finishing a complete simulation step does the traffic simulator send the actual vehicle positions to the network simulator, which interprets them as node destinations. Thereby, positioning errors can almost be completely avoided. Varying vehicle speeds in the traffic simulator are basically taken into account as well: at the end of a simulation step, the traffic simulator computes average velocities of the vehicles and passes them to the network simulator. The resulting representation of mobility in the network simulator is shown in the right part of Figure 11.18: the path a network node moves along fits the original path in the traffic simulator very well. The only drawback is the shift in virtual simulation time, which means that events simulated by the traffic simulator reach the network simulator one second later.

After considering issues like different update frequencies, coordinate systems and vehicle positioning methods, the overall architecture and communication protocol of the simulator coupling component can be designed. One of the main goals within this context is to provide a high performance connection, which is quickly adaptable to new requirements. For the example of coupling CARISMA and ns2, a separate communication module has been attached to both ns2 and CARISMA in order to make the data transport functionality as independent from the specific simulators as possible. Thus, the software developed will easily be applicable for other simulators in future projects as well. Figure 11.19 depicts the communication modules as flow charts. Boxes represent states of the algorithm described, whereas arrows account for modelling possible transitions between these states. If a label is attached to an arrow, the label acts as a prerequisite for the transition to be run through.

The functionality of the models will be briefly described in the following. Basically, the two modules are running in parallel. Both work in completely different ways, however. The first one, which is visualised on the left side of Figure 11.19, can best be described as a client, sending and requesting pieces of information to and from the second one, which acts as a server.

Since the traffic simulator CARISMA contains the server module (right side of Figure 11.19), it must be started before the ns2 client application. After establishing a TCP connection, the client sends a predefined simulation time step to the server, which thereby gets to know the simulator synchronisation time period. After successful reception of data, the server sends back an 'ok' as an acknowledgment to the client, which remains in the waiting state until reception of the 'ok' message. By implementing this algorithm with a 'send and wait for acknowledgment' characteristic, message collisions between the modules can be avoided. As a next step, the client requests the geographic scenario size, the number of network nodes and their initial positions from the server. By sending the first 'step' command to the

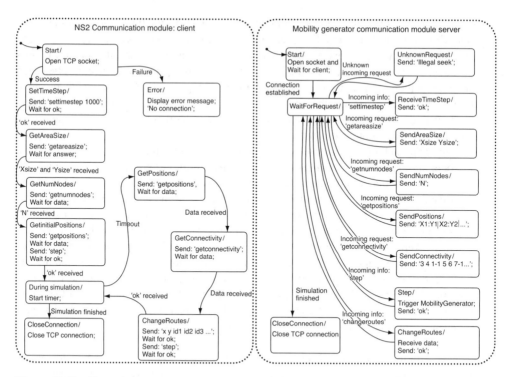

Figure 11.19 Flow chart representing the data exchange mechanism between network simulator and traffic simulator.

server, the client makes the traffic simulator perform a simulation step of the duration defined in the beginning with the help of the 'gettimestep' message. After acquiring the data, the client communication module starts running through a loop periodically requesting new node positions, connectivity information and sending data regarding hazard position and warned vehicles. Also, one 'step' command is sent to the server per run through the loop. The server reacts properly on every of these requests. Should it receive an unknown request string, this is displayed to the user ('illegal seek'). Both modules finish their service as soon as simulation time ends.

11.4 Field Operational Testing Methodology

Although the simulation methodology provides important contributions to the investigation of AutoNets, they have a major drawback: simulations do not reflect the real world. For example, there are several phenomena in traffic flows that cannot be explained in a rational and traceable way. Moreover, simulations do not reflect heterogeneity with respect to different vehicle brands and types. This means they neither consider typical limitations and restrictions in the vehicles, nor do they take into consideration interoperability with respect to different features and functional sets of each automotive key player. Finally, simulations cannot be used to measure user acceptance and user experience. However, such factors are crucial for choosing the respective feature set of cooperative applications as well as for their successful introduction into the market.

In order to overcome these issues and to address the related problem statements, field operational tests (FOTs) are an important research methodology. FOTs aim to test and evaluate real applications in real-world scenarios. Such testing can be seen as the next step towards market introduction that follows intensive simulations studies. This is particularly important since FOTs are usually very expensive and time consuming in their preparation: in fact, the development and setup of FOT applications and the architecture required has to rely on sound results of the simulations in order develop respective applications and mechanisms systematically and selectively. Therefore, FOTs have to fulfil four important criteria, which are crucial for their success:

- *Real system components.* It is important that the system being developed should be as close as possible at the final product. Hence, both real hardware components and technologies as well as real software systems should be used, which should be close to existing standardisation activities. In the event standardisation of respective technologies is not yet finalised, the participating partners of the FOT should actively drive and control respective standardisation activities in order to harmonise the overall system.
- *Real vehicles and traffic.* The testing performed by an AutoNet-based FOT should take place in real-world scenarios, that is they should be performed with real vehicles in real traffic scenarios in cities, rural areas or motorways. This requirement ensures two important factors: (i) AutoNet applications are embedded into real traffic scenarios and, thus, inherently include all facets of traffic and user experiences; (ii) real tests identify situations or system states, which may not be identified in the development phase. For example, an application may require additional system states that may occur in real traffic scenarios, but not in simulated traffic scenarios.
- *Include all stakeholders.* The partners of an FOT have to comprise all stakeholders that will be relevant for the final productive applications of the cooperative system. It is important that their requirements, expectations, experience and also the calculations of all stakeholders are considered beforehand. This way, the interaction among the relevant partners is fixed and the market introduction of respective applications can be organised very efficiently.
- *Large and heterogeneous fleet.* It is also apparent that AutoNets and the cooperative applications must be supported and agreed by all vehicle manufacturers available in the marketplace. This is necessary to ensure interoperability of the different AutoNet applications. Moreover, it allows adaptation of the functionality set of the AutoNet application to the different limitations of the vehicle manufacturers. In order to generate statistical and meaningful results, the vehicle fleet used for testing should comprise a significant number of vehicles to generate representative results.

So far, a number of larger FOTs have been started in Europe. Examples include the European projects DRIVE[6] and EuroFOT,[7] and the German project simTD (see Appendix A).

---

[6] DRIVE C2X homepage: http://www.drive-c2x.eu.
[7] EuroFOT homepage: http://www.eurofot-ip.eu.

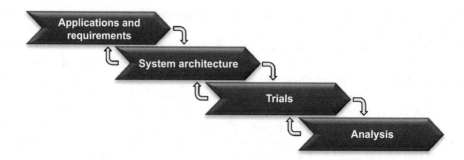

Figure 11.20   Phases of an FOT.

FOT projects have in common that they require a long and intensive planning phase due to their high financial costs and the number of partners involved in the FOT. For example, the simTD FOT comprises 18 partners with an overall budget of around 69 million Euros. The methodology of an FOT in general comprises four partly overlapping phases as illustrated in Figure 11.20: specification of (AutoNet-based) applications and their requirements, system architecture development, testing in the trials and final benchmarking in the analysis phase. It is important to mention that there will be an intensive interaction between the different steps during their overlapping phases. In the following sections, we will emphasise on the four phases.

11.4.1   Applications and Requirements

In the first step of an FOT, the partners have to agree on a set of AutoNet applications to be considered throughout the project. Since many stakeholders are typically involved, it is important that every partner suggests a set of applications that fits their expectations of the project results. From this overall set of suggested applications, a subset has to be chosen in an objective way. This objective decision can be achieved using, for example, the analytic hierarchy process (AHP) [3]. The number of applications being chosen obviously depends on the resources and the budget of the partners participating in the FOT project. For example, in simTD 21 applications were chosen, implemented and evaluated, which are summarised in Figure 11.21.

After this decision process, the chosen applications need to be specified in several iterations:

- *Application description.* First, the application needs to be described in an abstract way, that is the purpose needs to be defined, the basic functionality must be specified, and it should be outlined which prerequisites are needed in order to deploy the application.
- *Application specification.* In a second step, the functionality of the applications has to be specified in detail. This comprises the feature set as well as the respective function blocks, technologies, and protocols being used.
- *Requirements.* Based on the application specification, the requirements of the applications have to be derived. These requirements are needed to develop the system architecture in a way that the applications can be deployed.

Traffic	Driving and Safety	Additional services
Monitoring of traffic and complementary information/basic functions • Data collection in the infrastructure side • Data collection by the vehicle • Identification of road weather • Identification of traffic situation • Identification of traffic events/incidents **Traffic (flow) information and navigation** •Foresighted road/traffic information • Rpad works information system • Advanced route guidance and navigation **Traffic management** • Alternative route management • Optimized urban network usage based on traffic light control • Local traffic-adapted signal control	**Local danger alert** •Obstacle warning • Congestion warning • Road weather warning • Emergency vehicle warning **Driving assistance** • In-vehicle signage/traffic rule violation warning • Traffic light phase assistant/Traffic light violation warning • Extended electronic brake light • Intersection and cross traffic assistance	**Intersection access and local information services** • Internet-based usage of services • Location-dependent services

Figure 11.21 Applications chosen, developed, and tested in simTD. Reproduced by permission of simTD research project.

- *Validation considerations.* For every application, the validation prerequisites need to be specified. This is an important factor to evaluate the functionality of the application itself as well as its economic impact, which can be analysed in the last step of the project. In order to validate an application, validation methods, metrics and measurands for the application have to be defined. The methods – which are sometimes called *models* – are needed to determine the necessary measurands that have to be measured during the testing phase in order to derive the respective metrics.
- *Test Cases.* For each application, a number of test cases have to be specified in order to generate the measurands (and, thus, to calculate the metrics). For cooperative AutoNet

applications, tests are typically divided into two classes: (i) functional tests addressing the behaviour of the application, and (ii) non-functional tests, which aim at application-unspecific metrics (e.g. economic or traffic-related factors).

The results of this specification process are the input for the second phase of an FOT, the definition of the system architecture.

11.4.2   System Architecture

The system architecture provides the technical basis for the FOT. It covers the process from the specification of the architecture to the implementation of its functionality and the final integration into the overall system. This way, the system architecture also has an 'integrating factor': it brings together the different partners of the FOT since they have to collaborate with each other in order to meet their specific requirements within the overall objective of the FOT.

Obviously, the requirements and measurands defined in the 'applications and requirements' phase are one important input factor for the specification of the system architecture, which need to be consolidated beforehand. The requirements already comprise the expectations of the different stakeholders (which are typically reflected in the applications). However, there are several other factors that need to be considered, which are illustrated in Figure 11.22 [2]. Technical factors include the properties of available hardware (and software), but also their technical limitations. Since parts of the functionality already exists with different partners, these systems should be considered in the system architecture and, thus, need to be compliant and interoperable among each other. Typical examples are the limitations of automotive compliant hardware operate at temperatures between −40°C and +80°C (interior electronics) or even +125°C (powertrain electronics), resulting in limited computation and storage capabilities. Examples for software limitations may include the use of particular real-time operating systems, which do not provide the full feature set and libraries necessary for the chosen applications. Another important factor for the system architecture specification are costs. FOTs typically comprise a significant number of communicating units. Hence, the costs for these

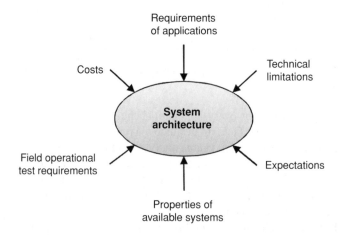

Figure 11.22   Applications chosen, developed, and tested in simTD.

units have to be reduced due to multiplier effects. Finally, there are additional requirements that are necessary to perform the field operational tests. For example, the system architecture has to provide respective mechanisms in order to perform the tests, to manage the vehicles, the units being developed, and many other factors.

It is apparent that several input factors for a system architecture contradict each other. Therefore, the final set of requirements needs to be agreed and thus the feature sets of the applications need to be adapted accordingly in case a requirement cannot be met by the system architecture. Hence, a real cooperation is necessary between the people specifying the applications and their requirements, and the team that specifies and develops the system architecture.

The result of the system architecture specification is a set of required subsystems together with their interactions with each other, which reflects the overall system being developed. An approved tool for the specification is the unified modelling language UML,[8] which describes the subsystems, their interactions and interfaces in a standardised and well arranged way. Moreover, this model-driven description maintains a consistent view on the system and supports consistent and interoperable code generation for the implementation.

Based on this first specification, the subsystems need to be refined and developed in several follow-up activities.

- Each subsystem has to be structured from a hardware point of view.
- Each hardware has to comprise software. Software comprises the operating system being used, the system components together with their interfaces, and the applications.
- The communication protocols between the subsystems as well as the programming interfaces (APIs) need to be defined.
- A development process for each hardware unit (or subsystem) must be specified. This process describes the development of the required hardware, and the software development stages as well as the prerequisites and tools that will be used for the implementation. It is important that the processes for the different subsystem should be harmonised as much as possible, because the functionality of an application may distribute over several system components. This way, all developers have to agree on these development processes.
- Setup of the required hardware and implementation of the system components and applications for each subsystem.
- Gradual integration in a contrawise manner, that is integration of applications and system components, integration of the software components to the target hardware unit, integration of the hardware units within the subsystem, and finally integration of the subsystems in order to finalise the overall system.

Due to the complexity of cooperative systems, testing is an important issue throughout the complete development process. Testing allows the identification of potential faults at an early stage, which helps to reduce cost and efforts since software bugs can be fixed much more easily in the early development cycles. Moreover, testing helps to find inconsistencies between the interfaces early in the overall system. In fact, a test-driven software development process [1] can be very useful to develop such a complex system. Note that 'testing' in development

---

[8] UML documents can be accessed for free at the Object Management Group (OMG) homepage: http://www .omg.org/spec/UML/.

terminology is different from the test cases specified in the applications and requirements phase. Test cases are developed for the trials, whereas testing in the development terminology basically refers to unit testing and testbenches in order to validate the functionality of the respective software or hardware component.

Several comprehensive organisational activities also need to be addressed in the system architecture that are essential for the development as well as for the following trial phase. For example, in complex development environments a change management process together with a process of tracking of requirements, errors and changes needs to be established and tooled. This is important since potential changes (or software errors requiring API modifications) have to be communicated transparently among the implementing parties. Moreover, tools, processes, and strategies need to be developed for the FOT in order to provide software update mechanisms and software configurations for the subsystems, logging and monitoring concepts, and many other aspects.

Finally, the overall system has to be built in the test area. For example, vehicles and traffic signs will be equipped with their respective hardware components, traffic centres and test systems will be installed and commissioned for the trial phase, which follows the system architecture phase.

11.4.3   Trials

The core activity of the trial phase is to prepare and perform the test cases specified during the application and requirement phase in order to generate the measurands for the evaluation and analysis of the applications and the system. This requires several preparation activities for an efficient and successful execution of the test cases. First of all, several tools are required to control and manage the tests. Examples include a test management tool to handle the test cases, for AutoNet-based FOTs a vehicle management system is required in order to schedule the vehicle fleet. Such systems are also part of the system architecture. This means the trials phase has to overlap with the system architecture phase in order to harmonise the different activities – and to include potential requirements necessary for the trials.

Besides those organisational preparations, test cases need to be prepared too. Based on the test case specification, the test cases need to be staged in trials and the trials must be planned and scheduled. The result will be a screenplay for each trial, which describes the activities and correlations of the participating parties, the number of vehicles required, the configuration of the subsystems involved, and the measurands that will be logged for the later analysis. After this preparation phase, the trials can be executed. One trial typically is oriented at the following course of action.

1. Configuration of the entities participating in the trial.
2. Execution of the trial according to the screenplay. (Depending on the test case, this action may also include the instruction of people participating in the trial.)
3. Consistency checks to ensure that the trial was successful.
4. Analysis of the data for this trial.

An important aspect of FOTs is the analysis of user experience. This means it is important to cover both a significant number of vehicles as well as a significant number of different users.

For the trials, there will be a number of professional and instructed drivers combined with a number of experienced drivers without any instruction. This ensures that the trials can be performed efficiently. These drivers should be replaced periodically by other drivers in order to have a sound and driver-independent basis for the evaluation of the trials. Besides the trials, it is important to consider a large number of 'common drivers', which do not have any direct relation with the FOT. These drivers are important to evaluate the user experience and user acceptance of the applications in the everyday life.

For AutoNet-based FOTs, it is important to have a significant number of vehicles including respective drivers performing the different tests. This is one of the highest cost elements since vehicles have to be allocated and fitted with the required software and hardware components, and drivers have to be hired accordingly. From an organisational point of view, three different types of vehicle fleets are appropriate for field operational trials:

- *Partners' fleet.* The vehicles are typically provided by the partners of the FOT consortium. The role of this fleet is to perform the initial tests during the hardware and software development process in the system architecture phase. The partners' fleet also plays an important role for the acceptance tests of the overall system, which approves the system finally developed for being ready for the trial phase. During the trial phase, the partners' fleet will be used to test potential software updates or optimisations, which may be necessary to fix unavoidable software errors occurring in the execution of the test phases. Another important purpose of this fleet is the demonstration of results and milestones during the overall field operational testing. This way, it has to be available from the early phase of the FOT to the end of the project. The number of vehicles depends on the purpose of the project, but will typically be between 10 to 50 vehicles.
- *Internal fleet.* The internal fleet comprises vehicles that are exclusively used to execute the tests according to the respective screenplays. The vehicles will be driven by the hired drivers, that is the professional and experienced drivers as described above, who have to follow the managed and controlled screenplays. Therefore, it is important that the internal fleet must be compliant with statistical requirements that avoid potential side effects by different vehicles on the results. For example, such requirements can be that the vehicles need to be the same size, a similar colour, have similar engines and that the drivers need to be exchanged periodically in order to avoid customisation effects. Depending on the number of trials, the screenplays and the expected results of the FOT, a significant number of vehicles is required for the internal fleet; potential numbers may range from 50 to 200 vehicles for an FOT, which have to be allocated for the trial phase, and which have to be equipped with the respective hardware components.
- *External fleet.* The external fleet is an important basis from which survey results for user experience, and to determine the behaviour of the applications developed in the real world. The external fleet may consist of vehicles from volunteers and employees of companies that are not related to the FOT. These vehicles will be equipped by the FOT partners, the volunteers using the vehicles will get a introduction to the system, and they will use the system afterwards in their everyday life. Therefore the external fleet is not included in the organised trials; vehicles of the external fleet only collect data generated by the applications themselves. Additionally, the drivers have to participate in regular or occasional opinion polls in order to evaluate the user experience as well as acceptance by the drivers. In order

to get a sound basis for the user experience, a significant number of vehicles for the external fleet is required. A review on running FOTs shows that there should be a minimum of 100 vehicles to generate appropriate results.

11.4.4   Analysis

In the final phase of the FOT, the data generated during the trials phase needs to be analysed and benchmarked in order to evaluate the system. Therefore the data needs to be prepared and processed in order to apply the methods and models developed in the applications and requirements phase. The result of this process is a set of metrics that allows each stakeholder to evaluate its expectations on the overall system. Besides the results for functional and non-functional benchmarks, the analysis also has to comprise economic factors with respect to traffic safety, traffic efficiency and – most importantly – cost.

Especially for AutoNets, FOTs have to validate different goals formulated by several government organisations or commissions. Examples include the reduction of accidents, a reduction in people killed in traffic accidents or an increase in traffic efficiency in order to improve the utilisation of the roads. Of particular importance for the partners of the FOT are the expected costs that are required for market introduction. For AutoNet scenarios, major cost factors are the costs for the hardware required in vehicles, the development, operation and maintenance of respective back-end systems, the fitting of the roadside with respective hardware and many other factors. On the other hand, the analysis phase also has to identify potential business models and revenues for such a system, which support the stakeholders for their decision and their strategy for a market introduction.

The analysis phase should be overlapped with the trials phase. This way, the results generated in the trials are validated promptly. In the event that a trial produced unfeasible or corrupt data, the applications can be optimised and the trial itself can be repeated again. Moreover, it is important that the FOT participants have a clear mission before the FOT begins about the metrics for the benchmarks they want to achieve. These metrics need to be specified in the applications and requirements phase, that is in the beginning of the FOT, whereas the validation and benchmarking takes places at the end of the FOT. This means it is important that the partners performing the final benchmarking are also involved in the specification of the different metrics and methodologies during the applications and requirements phase.

An optional and attendent activity in the analysis phase could be an investigation of regulatory limitations for the market introduction of the system being developed and tested during the FOT. Examples for regulatory limitations include privacy aspects, accountability issues and potential bearings in different markets. It is efficient to investigate these limitations within an FOT, because the results are relevant for all partners of the FOT in the same way and, thus, the partners do not have to perform the investigation separately by themselves.

The analysis phase also completes the FOT by an interpretation through consulting the volunteers who are involved in the external fleet as described in the trials phase. These optinion polls generate important results about the user acceptance of different applications as well as their usability and usefulness in everyday life. The results now can be used by every partner of the FOT to integrate the functionality into their vehicles in order to prepare their individual market introduction of a respective function set of cooperative AutoNet applications.

11.5   Summary

The market introduction of AutoNet applications without any preparatory work is an uncontrollable risk. The system must be technically feasible, interoperable, future-proof and cost-effective. Moreover, there are several different stakeholders involved in AutoNets, which makes the market introduction even more complicated. For AutoNets, there are two important methodologies for the preparatory work. In the initial preparatory phase, simulations are an important methodology to evaluate technical aspects, which itself are an important input the design and development of the cooperative system. Therefore, simulations of the driver and vehicle model, the traffic flow model, the AutoNet application model and the AutoNet communication model are able to outline the principles, dependencies and interactions between the different models. In this context, we introduced the four models and their interactions and we presented a simulation environment that covered these aspects. Based on extensive simulation results, the cooperative system needs to be evaluated in real traffic scenarios. In the field operational trials (FOT) methodology, this evaluation is prepared and executed accordingly. In FOTs, the cooperative system is built and tested accordingly in order to generate the results necessary for the market introduction. Besides standardisation activities and measurands for optimising the system itself, FOTs also generate results about economic benchmarks and user experiences, which are a resilient factor to reduce the risk of the market introduction of such a new technology.

References

[1] Beck, K. (2003) *Test-Driven Development by Example. Addison Wesley.*
[2] Stübing, H., Bechler, M., Heussner, D., May, T., Radusc, I., Rechner, H. and Vogel, P. (2010) simTD: A Car-to-X System architecture for field operational tests. *IEEE Communications Magazine*, Volume 48, Issue 5, May 2010.
[3] Saaty, T. L. (2001) *Fundamentals of Decision Making and Priority Theory.* RWS Publications.
[4] Bazza, A. L. C., Klügl, F., Schreckenberg, M. and Wahle, J. (2000) *Decision Dynamics in a Traffic Scenario.* Hermes Science Publications.
[5] Liu, Y., Özgoner, Ü. and Ekici, E. (2005) Performance Evaluation of Intersection Warning Systems Using A Vehicle Traffic and Wireless Simulator. Technical Report, The Ohio State University.
[6] Ohio State University (2005) Simulator Development Overview. http://www.ece.osu.edu/
[7] Takagi, H. and Kleinrock, L. (1984) Optimal transmission ranges for randomly distributed packet radio terminals. *IEEE Transactions on Communications*, vol. 32, no. 3.
[8] Kosch, T. (2005) Situationsadaptive Kommunikation in Automobilen Ad-hoc Netzen. PhD Thesis, Munich University of Technology.
[9] Benjamin, D. (2005) Could 802.11p spell the end for cellular in the automobile? *http://www.abiresearch.com/products/insight/348*
[10] Krauss, S. (1998) Microscopic Modeling of Traffic Flow: Investigation of Collision Free Vehicle Dynamics. PhD Thesis, Universität zu Köln, Germany.
[11] De Waal, C. (2005) BonnMotion: A mobility scenario generation and analysis tool. http://web.informatik.uni-bonn.de/IV/Mitarbeiter/dewaal/BonnMotion.
[12] Krajzewicz, D., Hertkorn, G., Rössel, C. and Wagner, P. (2002) SUMO (Simulation of Urban Mobility) – an open source traffic simulation. *Proceedings of the 4th Middle East Symposium on Simulation and Modelling (MESM 2002).*
[13] Schiller, J. (2003) *Mobile Communications*, 2nd Edition. Addison Wesley.
[14] Zeng, X., Bagrodia, R. and Gerla, M. (1998) GloMoSim: A library for parallel simulation of large-scale wireless networks. *ACM SIGSIM Simulation Digest*, Volume 28 Issue 1.

[15] Chung, J. and Claypool, M. (2005) GloMoSim. NS by example. http://nile.wpi.edu/NS/.
[16] Almeroth, C. K., Belding-Royer, E. M., Jardosh, A. and Suri, S. 2005 Real-world environment models for mobile network evaluation. *IEEE Journal on Special Areas in Communications – Special Issue on Wireless Ad hoc Networks.*
[17] Perkins, C., Belding-Royer, E. and Das, S. (2003) Ad hoc on-demand distance vector (AODV) routing. RFC 3561, Internet Engineering Task Force.
[18] Bayliss, M. (2005) A simplified vehicle and driver model for vehicle systems development. *Proceedings of the Driving Simulator Conference 2005.*
[19] Treiber, M., Hennecke, A. and Helbing, D. (2000) Congested traffic states in empirical observations and microscopic simulations. *Physical Review* E 62 (2).
[20] Bellomo, N., Coscia, V. and Delitala, M. (2002) On the mathematical theory of vehicular traffic flow I: fluid dynamic and kinetic modelling. *Mathematical Model and Methods in Applied Sciences*, Vol. 12, No. 12 (2002).
[21] Kerner, B. (2004) *The Physics of Traffic.* Springer Verlag.
[22] May, A. (1990) *Traffic Flow Fundamentals].* Prentice Hall.
[23] Hartenstein, H. and Laberteaux, K. (2010) *VANET – Vehicular Applications and Inter-Networking Technologies.* Wiley.
[24] Franz, W., Hartenstein, H. and Mauve, M. (2005) *Inter-Vehicle-Communications – Based on Ad Hoc Networking Principles.* Universitätsverlag Karlsruhe.
[25] Schroth, C., Strassberger, M., Eigner, R. and Eichler, S. (2006) A framework for network utility maximization in VANETs. *Proceedings of the third ACM International Workshop on Vehicular Ad Hoc Networks (VANET) 2006.*
[26] Eichler, S., Ostermaier, B., Schroth, C. and Kosch, T. (2005) Simulation of car-to-car messaging: Analyzing the impact on road traffic. *Proceedings of the thirteenth Annual Meeting of the IEEE International Symposium on Modeling, Analysis, and Simulation of Computer and Telecommunication Systems (MASCOTS) 2005.*
[27] Schroth, C., Dötzer, F., Kosch, T., Ostermaier, B. and Strassberger, M. (2005) Simulating the traffic effects of vehicle-to-vehicle messaging systems. *Proc. 5th International Conference on ITS Telecommunications.*

12

Markets

As outlined in the chapters above, the latest V2X communication technologies already allow for the realisation of a wealth of promising AutoNet application scenarios. In fact, first services have already been introduced to the market: numerous vehicle manufacturers, suppliers and telecommunication service providers already offer a number of telematics applications. Examples include BMW Assist[1] (BMW), HD Traffic[2] (TomTom and Vodafone) and the Go-Box tall system[3] (ASFINAG). However, relevant stakeholders such as vehicle manufacturers, third-party telematics services providers, insurance companies or manufacturers of personal navigation devices (PNDs) still predominantly offer proprietary technical equipment that solely supports their respective applications. Most of the existing services have been created independently from each other, thus relying on different technical standards, system architectures and business models.

However, the individual design and operation of telematics services entails a number of challenges:

- First of all, the lack of standardisation with respect to communication protocols as well as the structure and semantics of messages to be exchanged among vehicles and between vehicle and infrastructure components prevents application interoperability. This incompatibility limits the benefit delivered to the users of such services.
- Second, only very few of the AutoNet applications – particularly in the field of traffic safety and traffic efficiency improvements – are believed to be successfully introduced in the market as stand-alone applications for financial reasons: the costs for the respective vehicle systems and the associated hardware and software for the roadside infrastructure domain can barely be compensated by the fees charged to end customers. AutoNet applications that build upon comprehensive roadside infrastructure (e.g. data exchange with roadside stations or with

---

[1] BMW AG Homepage: http://www.bmw.com/com/en/insights/technology/connecteddrive/assist_1.html.
[2] TomTom Homepage: http://www.tomtom.com/hdtraffic.
[3] ASFINAG Homepage: http://www.go-maut.at/go/default.asp.

Automotive Internetworking, First Edition. Timo Kosch, Christoph Schroth, Markus Strassberger and Marc Bechler.
© 2012 John Wiley & Sons, Ltd. Published 2012 by John Wiley & Sons, Ltd.

traffic signs) are expected to require high initial investments into wireless access points and proper integration into actual legacy systems. In all these cases, the possibility of running different applications based on a homogeneous technical roadside infrastructure represents a key prerequisite for economic profitability. Network effects play an additional important role: many of the uses aiming to increase vehicle driver safety merely through low-latency ad-hoc communication among vehicles will suffer from a low penetration rate of wireless onboard units during the initial start phase of a potential market introduction. As comprehensive studies and simulations show, it would take several years to reach a penetration rate of only about 15% – even if 50% of all new vehicles are adequately equipped with respective AutoNet communication technology.

- Finally, the 'closed shop' design of current offerings prevents third party providers from easily contributing new AutoNet-related applications or application extensions to the customers. The scope of most existing applications thus remains mostly limited to a specific static function that can be changed or extended very little over the lifecycle of the underlying hardware infrastructure.

Many technical, organisational and economic challenges still need to be addressed to enable an efficient market for telematics services that operates across the boundaries of proprietary onboard units, specific vehicles and backe-nd systems.

The goal of this chapter is threefold. In the first section, we describe the current state of the market for vehicular telematics services. For this purpose, we revisit the major drivers that create the technical potential as well as the economic demands for V2X communication services. Also, we describe the most relevant stakeholders, their capabilities as well as economic interests. In the second section, we systematically elaborate on major challenges that still prevent the emergence of an efficient market. We thereby incorporate a financial, a technical and a customer benefit perspective. In the third and last section, we argue in favour of an open and modular platform for value co-creation through a flexible ecosystem of various stakeholders as a solution to the identified challenges. We conclude with a brief discussion of different current initiatives that have already started designing holistic platforms rather than developing proprietary island solutions.

12.1   Current Market Developments

The idea of interconnected vehicles [4] was born more than ten years ago when research projects such as inter-vehicle hazard warning (IVHW) [1] were initiated and systems like OnStar[4] were made available to customers [2]. In these early days, a number of technological and economic hurdles prevented a major breakthrough of interoperable telematics services at that point of time. Since then, tremendous advances in both dimensions have opened up new opportunities. The following two paragraphs are devoted to an elaboration of technological enablers that create a market 'push' and economic drivers that trigger substantial market demands ('pull').

---

[4] ONSTAR Homepage: http://www.onstar.com/us_english/jsp/digital_transition.jsp.

12.1.1   Technological Push

Information and Communication Technology (ICT) has advanced significantly during the past few years. Various barriers to the market introduction ([4]) of numerous telematics applications have thus been minimised significantly, or completely removed.

Affordable and powerful hardware for vehicles. Programmable microcontrollers have become smaller, cheaper and more powerful, thus accelerating the deployment of embedded electronic devices within vehicles. Vehicle manufacturers and their suppliers were enabled to exploit their potential and implement various electronic control units and communication systems for their vehicles, such as CAN busses and MOST rings to exchange data among the control units in a reliable fashion. A wealth of information is today processed and gathered by these controllers and exchanged via communication bus architectures, building a tremendous foundation for a multitude of innovative telematic services.

Advances in wireless techologies. A second major driver was the 'revolution' of wireless communication in the past two decades. Wireless communications hardware as well as associated services enjoy great popularity and have become a mass market product. Communication protocols and back-end infrastructure today allow for fast data connections at affordable prices and – at least in urban regions – sufficient network coverage. While only GSM and WAP-based (wireless application protocol) mobile communication with low transmission rates of about 9.6 Kbps was available until the late 1990s, the UMTS standard today alllows for data rates of between 384 Kbps and 7.2 Mbps, thus facilitating the implementation of services that require higher bandwidths. Future communication standards such as LTE (Long Term Evolution) [5] are expected to increase data rates even more – a downlink speed of up tp 250 Mbps has already been achieved in field tests. Combined with a reduction of data transmission latency, LTE allows for device plug and play functionality, and a simplified technical architecture, resulting in ever decreasing setup and operating costs.

Expansion of Internet-based Technologies. The wide adoption of the Internet and its improved mobile accessability represents an additional major technological driver. Simple structured content (e.g. in HTML or XML format), but also complex enterprise-class services (e.g. specified by WSDL interfaces, retrievable via UDDI-based registries, and connected via the SOAP standard) can be accessed globally and at affordable prices. The wide range of services and information available via the Internet, combined with powerful back-end infrastructures, constitute an invaluable resource that calls for utilisation by telematics services.

12.1.2   Economic Pull

In addition to the technological enablers discussed above, a number of economic factors create increased demand for telematics services: first, as illustrated in several simulations [6], dedicated telematics services have the potential to improve traffic safety and, thus, contribute to substantial socio-economic benefit [7]. In the course of the *eImpact* project, for example, scientists quantitatively evaluated the safety effects of different intelligent vehicle safety systems, and thereby assumed a theoretical vehicle equipment rate (the percentage of vehicles running a specific service) of 100%. Also, the calculations were based on the number of accident fatalities and injuries projected for the year 2020 in Europe (20,791 and 8,73695 respectively). The simulations result in a 4.5% decrease of fatalities and a 2.8% decrease of injuries regarding the overall road traffic.

A second kind of added value concerns traffic efficiency: some of the telematics services described in the chapters above provide drivers with real-time information regarding current traffic conditions, propose alternative routes in the case of traffic jams, or generally deliver recommendations regarding the most efficient driving behaviour. With increasing oil prices, ever growing traffic densities particularly in urban areas, as well as a rising awareness of the need for environmental protection, new and innovative traffic management solutions are urgently required.

While technological advancements open up opportunities for the market introduction of a magnitude of innovative interoperable services in the vehicular environment, the economic environment creates a strong demand for them. A multitude of different stakeholders have already managed to successfully enter the emerging market for telematics services with their focused offerings, others are still heavily investing and hope for future revenue opportunities [6].

12.1.3   Stakeholder Analysis

There are a several stakeholders involved in the different AutoNet scenarios. The different stakeholders have different requirements and expectations for such a system. Figure 12.1 shows the four key domains of those key stakeholders, namely vehicle manufacturers and suppliers, service providers, public sector players, and – most important – the users who will benefit from this system.

The following sections discuss the key stakeholders in this market, their capabilities as well as economic interests. It is important to mention that not all stakeholders will be involved in each scenario. Instead, it depends on the use which stakeholder is needed. As a result, the list of potential stakeholders mentioned in the following gives an impression of the variety of

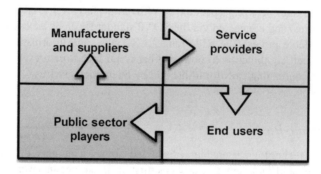

Figure 12.1   Key stakeholder domains for AutoNet markets.

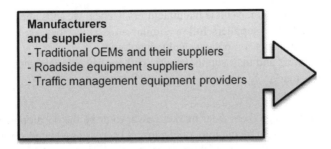

Figure 12.2 Manufacturers and system suppliers domain.

the different parties – and it does not claim to be complete. It is also important to mention that the economic interests of different stakeholders are not necessarily compatible with each other. This means it is very important to find a deployment strategy in which the different expectations of the stakeholders are agreed among each other. As illustrated in Figure 12.1, all stakeholders have to play together in order to introduce and roll out AutoNets.

12.1.3.1 Manufacturers and System Suppliers

Figure 12.2 depicts the key stakeholders in the domain of manufacturers and system suppliers. This domain basically consists of traditional vehicle manufacturers and their system (and component) suppliers, roadside equipment suppliers and traffic management equipment providers.

Vehicle manufacturers and their suppliers aim at a number of objectives regarding the market for telematics services. Vehicle manufacturers expect the opportunity to substantially improve the safety of their vehicle fleets. Applications such as local danger warnings, crossing collision warnings or lane change warnings that are based on real-time data exchange between vehicles as well as between vehicles and back-end infrastructure have the potential to reduce accidents and thus injuries of their customers. Applications that monitor diverse vehicular control units and transfer the information to back-end services owned by the respective vehicle manufacturer may even be able to contribute to an improved understanding of the exact circumstances of accidents. They may also allow the derivation of information about the role different components of a vehicle play in such situations in the long run. Vehicle manufacturers may be enabled to modify their products' construction according to these insights and improve their fleet's safety even more. Services in the field of 'infotainment', on the other hand, hope to allow vehicle manufacturers to improve their brand images, and differentiate themselves from competitors through the provision of innovative features. Finally, vehicle manufacturers are keen on developing novel revenue streams by collecting fees from users of telematics services or other business partners.[5] The suppliers provide the vehicle manufacturers with several types of components need to build a vehicle. The components are typically developed according to

---

[5] We will discuss the users' willingness to pay as well as alternative revenue architectures in the course of this chapter.

the requirements defined by the vehicle manufacturers. Hence, we subsume both the suppliers and vehicle manufacturers: suppliers follow similar aims since they are very interested in equipping vehicles with their components and technology.

Vehicle manufacturers and their suppliers possess three key competencies as players in the emerging telematics market:

1. They are able to leverage their sheer market power to drive the development of necessary standards, across different technology platforms and across associated vendors. The smart application of lobbying may allow them to convince governmental institutions to reserve required frequency bands, to support standardisation activities for specific communication protocols and potentially even to subsidise the market introduction of certain services, such as by granting tax advantages to early customers.
2. The second major capability of vehicle manufacturers and their suppliers concerns their technology leadership. Most vehicle manufacturers and suppliers maintain considerable research and development departments, which are able to develop rich telematics services, adequately integrate these services into the vehicles, and make them accessible with well-designed human–machine interfaces. Also, vehicle manufacturers have the exclusive power to access and process the heaps of data available in vehicular control units, which are exchanged via respective communication technologies with vehicles such as CAN or MOST. No market player but the vehicle manufacturer will have the power to decide which service will be accessible as 'in-dash' (accessible via the central, vehicle-built dashboard huma–machine interface) application and which not. This decision power limits the market power of technology and service suppliers to a certain degree, as will be discussed in the course of this section.
3. Over the decades of market presence, vehicle manufacturers and their suppliers have acquired a unique understanding of end user needs and are thus expected to play a leading role in the marketing and sales of future telematics services. Traditional vehicle supplier companies such as Bosch or Continental may hope for increased sales of system components and terminal equipment.

An important part of this domain are garages, which are specialised to repair vehicles or to equip vehicles with after-market solutions in the area of telematics services. Such garages may be part of vehicle manufacturer companies, they may be associated with specific brands, or they may operate independently of vehicle manufacturers and their brands. Garages are expected to seek additional opportunities for acquiring customers via novel (and back-fitting) telematics services available in the after-market. Electronic applications that recognise technical malfunctions and automatically provide drivers with a recommendation regarding an adequate garage that is as close by as possible, for example, will draw on databases of business partners. Garages increasingly try to establish business partnerships with owners and providers of these services – be it vehicle manufacturers or some third party providers – to open up new revenue opportunities. Core capabilities of such garages comprise a very close customer proximity and a deep understanding of their needs as well as a strong technical expertise. Especially garages with no direct relation with a vehicle manufacturer are specialised for back-fitting different functionalities to vehicles.

Roadside equipment suppliers provide the components necessary for equipping the roadside with AutoNet technology. In general, roadside equipment comprises two core components. First, suppliers of variable (and controllable) traffic signs will provide the components to control the traffic flow according to information provided by traffic management centres. Such components are typically variable message signs or respective traffic lights, which are able to provide information about their functionality and their behaviour. This also comprises respective devices that are able to count vehicles (and determine their density in lanes), and provide this information to respective third parties. The second core component are the so-called 'roadside stations'.[6] However, both have similar aims with respect to the roadside stations are able to communicate with vehicles using the respective communication technology like IEEE 802.11p. Roadside stations are in most scenarios connected to central systems like traffic management centres in order to forward information transmitted by the vehicles (and received by the roadside station) to the central system. An important feature of a roadside station is that it can be connected to variable message signs, which allows the transmission of the information about the connected traffic sign to the vehicles in the vicinity. Other scenarios also assume 'mobile and autonomous roadside stations', which can be used, for example, for a construction area in order to transmit information about the geometry and road characteristics in the construction area.

Roadside equipment suppliers are vitally interested in using existing and accepted standards for information transfers. This is an important precondition to develop interoperable and robust roadside systems at low costs. The suppliers are also vitally interested in offering respective back-fitting solutions for their existing equipment, which makes them more attractive to the operators of such legacy traffic signs.

Traffic management equipment suppliers are the counterpart for back-end systems. They provide the hardware and software components to control the different message signs, but they also provide their knowledge and expertise in processing the data received from the roadside stations. This means traffic management equipment suppliers have similar requirements to roadside equipment suppliers in order to keep the development costs for the respective functionality at a lower level. In addition, traffic management equipment suppliers are also interested in getting as much information as possible from the vehicles in order to improve the accuracy or their models and predicitions.

12.1.3.2   Service Providers

In AutoNet scenarios, there are a number of different service providers, which have to provide different types of services to the end users. As illustrated in Figure 12.3, service providers are classified into information and communication technology suppliers, telematics service provider and content providers.

Suppliers of information and communication technologies (ICT) play a crucial role in the global telematics market. Many of them maintain longstanding realtionships as technology

---

[6] Note that we do not differentiate between the application unit of a roadside station and the communication unit of the roadside station. In fact, both parts may be developed by different suppliers. We do not differentiate them here since both aspects of roadside stations address similar requirements and expectations.

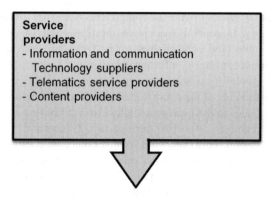

Figure 12.3   Service providers domain.

providers with OEMs already, while others are new to this ecosystem. For AutoNets, there are several key players that provide the ICT services necessary for realising the AutoNet scenarios.[7]

Mobile network suppliers (MNS). MNS provide increasingly powerful communication-based components and other technologies or hosting services to vehicle manufacturers for their further implementation and usage within their products. Examples include onboard units for wireless communication, which are the basis for realising cooperative systems. The MNS' core capabilities comprise the development and production of network components, the ability to propose, drive and offer standard solutions for the benefit and use of all Mobile Network Operators (see below), as well as the technical operation of networks and services [6]. NEC, for example, has developed first onboard units with advanced features in the field of dedicated short-range wireless communication (DSRC) based on IEEE 802.11. In the course of cross-company research projects such as FleetNet, NoW or PRE-DRIVE C2X as described in Appendix A, the modules have been integrated into vehicles and connected to internal controller units and communication architectures. As described in the previous chapters, WLAN-based DSRC supports spontaneous ad-hoc and direct communication of a vehicle with other vehicles or with fixed roadside stations in its vicinity. For message transmission beyond the transmission range of the wireless technology (typically a few hundred metres), the network nodes cooperate and relay the messages on behalf of other nodes [10]. Although field operational tests have shown very promising results in different research projects like PRE-DRIVE C2X and simTD, the technology as well as the services relying on it have not yet been introduced into the mass market due to a number of challenges that will be discussed in the course of the next section.

---

[7]Note that we do not consider internet service providers (ISPs) as ICT suppliers. This is due to the fact that telematics services in many scenarios rely on the Internet. However, they use the Internet 'as is', and ISPs only will play a minor role with respect to expectations and requirements to AutoNets.

The company *Autonet mobile* has been following a less sophisticated approach and offers an onboard unit (OBU) that provides customers with a high quality WiFi connection within the vehicle and in its close vicinity [9]. The module does not even have to be integrated into vehicles but can be bought as an after-market product that can be easily installed in the luggage compartment. The company charges a one-time fee for the router, and offers an associated one-year contract with a monthly fee, depending on the amount of data that is expected to be transmitted via the module. The OBU features powerful antennas to connect to 3G cellular network and relies on some additional functionality for optimal connectivity. The device is expected to work in about 95% of the USA, including all major metropolitan areas. Passengers will be able to connect to the Internet by using any WiFi enabled device such as notebooks or smartphones. Autonet Mobile has also started to team up with manufacturers such as General Motors, Chrysler or Volkswagen, and has eventually managed to establish its module as one of official optional features within the order lists for some of the vehicle types sold by these manufacturers. In early 2009, Cadillac announced the availability of an in-car wireless Internet option for its model CTS sport sedan based on the Autonet mobile technology, thus making Cadillac one of the first car makers to offer wireless Internet in a production vehicle. Interestingly, early market experience indicated that the option attracts families and private users first, as opposed to the mobile business professional one might expect to constitute the initial market. This early integration of a WiFi hotspot into vehicles may be a first indicator for a future in which many services, be they infotainment, navigation or communication-related, are based on systems and infrastructure located 'in the cloud', rather than dependent on computation or data that are captive in the vehicle. Vehicle manufacturers are well advised to pay attention to this development in order to avoid losing their governance over service provision and delivery within their products. We will discuss this trend and potential strategies to address it later in this chapter.

Software platform providers. Software platform providers have a strong interest in expanding their existing products into platforms that play the role of a telematics application operating system on the one side, and attract a significant number of ecosystem partners on the other side to become an increasingly rich automotive de facto standard. Tier One electronics and software providers such as Continental or Bosch already manage the design and implementation of large parts of the vehicular software application environments. For them, it is a natural step to expand the scope of their products towards truly Internet-enabled platforms, which allow them to tap the dynamics of Internet applications for drivers or passengers in the vehicles.

In fact, Continental has recently started marketing its AutoLinQ Internet platform, which builds on Google's open source operating system Android. AutoLinQ represents a system comprising all necessary hardware and software components necessary to enable drivers and passengers in vehicles to use Internet-based applications. The hardware will, among other things, encompass all the parts required to control the respective applications, appropriately display their output and enable them to access Internet services through an integrated 3G wireless communication module. With this extendable platform, Continental plans to follow an always-online approach, offering – among other things – the possibility fo implementing location-based services. As Continental has also gathered substantial pre-existing knowledge in the context of vehicle control units and communication buses, applications requiring technical vehicle status notifications will be made possible, too. The incorporation of the Android platform as an underlying operating system may help the business partners Continental and Google in establishing an agile ecosystem that grows the diversity of services available to its users on its own. Android is an open source environment which many application developers are already familiar with. Numerous independent third party developers are thus expected to provide applications enriching the existing portfolio. The system sets boundaries to its developers' design freedom as it aims to provide a portfolio of applications that are interoperable. This means they interact with each other rather than offering a loose set of constantly new and mutually independent applications. In order to deliver the growing number of telematics services to the end user, the so called Android Market Place can be used. A few software platform providers – among them also Continental – already offer infotainment solutions based on Windows Automotive, a competitive operating and middleware bundle tailored for the automotive environment. While the Windows-based platform is sometimes said to have advantages in terms of performance and the credibility of its ecosystem, the Android environment is expected to meet a more agile market of ecosystem partners [11].

Car stereo suppliers. Companies supplying hifi components for vehicles constitute a further group of relevant suppliers of ICT in the telematics market. Parts of their traditional business model are at risk: vehicle manufacturers increasingly integrate multi-functional application platforms into their vehicles, which often cover services such as road navigation, remote assistance, Internet access and entertainment. All these functions can be operated on one single hardware and software platform. Separate hardware devices for receiving analogue radio signals are – particularly in premium class vehicles – not deployed any more. In order to develop new business opportunities, major car stereo suppliers have started to market and rollout Internet-streaming devices. For example, the Blaupunkt miRoamer system, connects to a cellular phone via Bluetooth to access the Internet, and then uses the Internet radio aggregator service miRoamer to stream tens of thousands of radio stations. First estimates of the amount of data transmitted per month (for an average user) are at about 2GB [12]. The major risk inherent in this product strategy is the uncertainty, whether and when vehicle manufacturers are willing to place these devices in their head units. The critical question here is whether or not the multi-functional hardware and software platforms that are increasingly being deployed in vehicles will leave a niche for a separate car stereo unit.

Mobile network operators (MNOs). MNOs are seeking opportunities to expand their existing business models as the global market for telematic services emerges. In order to move beyond basic voice and text messaging, MNOs therefore increasingly turn to value-added electronic services such as those enabling seamless machine-to-machine (M2M) communication [13]. The use of M2M applications over cellular networks in telematics is steadily rising: industrial use cases include asset tracking (e.g. tracking large containers while in transit), vehicular diagnostics (where, for example, various measurements relating to fuel can be observed in an effort to optimise efficiency and reduce costs) and locating stolen vehicles and cargo in an effort to retrieve them. Frequent messaging over a cellular network provides managers of diverse vehicle fleets with immediate and exact location capabilities [13]. The global network operator Telefonica, for example, offers an M2M application solution that enables data transmission between machines, for both fixed and mobile objects. Services offered include, but are not limited to, cargo tracking, route planning, order management, logistics for road, rail, air and sea, as well as initial auto telematics with driver navigation, safety, vehicle diagnostics, vehicle security, mobility services and traffic information. Still, a multitude of challenges prevents these M2M business models from becoming a breakthrough innovation for the mass market. For example, the insufficiently standardised server-to-vehicle protocols around the world inhibit efficient interregional M2M communication, and has so far limited the revenue potential for mobile operators. In the second part of this chapter, these challenges will be elaborated upon in detail.

Mobile phone manufacturers. Producers of mobile phones have started to play an increasingly important role in the realm of telematics services. Today, manufacturers such as Nokia or Apple still consider vehicular telematics a topic of secondary relevance. First developments, however, denote a significant change: Apple and BMW have teamed up to allow iPhone users to manage and use their devices via the BMW built-in control buttons. With the emergence of an efficient market for telematics services, mobile phone producers and vehicle manufacturers are well advised to interconnect their products even further. The seamless synchronisation and transfer of music, application and calendar data, or emails between vehicles and iPhones bears great potential for both parties. In the near future, dedicated telematics services that are used via mobile phones are expected to enter the mass market. For example, the mbrace system, which was introduced to the public in 2009 [14], already offers subscribers some services that connect their vehicles and smartphones. Mercedes-Benz and its partner Hughes Telematics have set up a platform that lets drivers use their iPhones or BlackBerry devices to lock and unlock car doors, locate their vehicles on a map, and contact roadside assistance. Because the smartphone software sends commands to the car on the data network and through Hughes's servers, the owner can be in Paris and unlock his car in London, UK, for example. Market experts expect a growing number of remote applications that can be operated with the help of smartphones, including the ability to check a vehicle's emissions without having to visit a garage and, for low-mileage drivers, the option of buying cheaper insurance based upon online odometer readings [14].

Personal navigation device (PND) manufacturers. PND manufacturers develop road navigation applications, which belong to the most mature telematics services. Producers like TomTom have brought their devices to perfection as they feature rich navigation services, have fairly smooth–human machine interfaces and come at an affordable price. PND manufacturers now face diverse opportunities: particularly, in the lower-class car segments, these players may seek to replace proprietary in-dash solutions in the long term and provide vehicle manufacturers – in the role of a tier one supplier – with all required hardware and software. From there, they may try to expand the functionality of their devices and develop them towards a platform for a multitude of telematics services that covers a well accepted HMI, connectivity to back-end functionality, a set of key functions such as road navigation, and a scalable computing infrastructure that can also run third party services. For example, TomTom teamed up with Renault, where TomTom supplies PNDs that are seamlessly integrated into Renault vehicles.

In the upper class segments, on the other side, opportunities seem less promising. More and more vehicles are produced with built-in navigation solutions that are part of powerful, multi-functional software and hardware platforms which may make external navigation devices redundant. PND manufacturers may possibly try to replace the vehicle-manufacturer-specific navigation service software and provide vehicle manufacturers with their proven industry standard; they might argue that costs incurring for design and maintenance of the software can be lowered by passing on the in-house deployment of proprietary solutions. A more promising potential strategy would be to team up with other key players in the telematics market and build rich, stand-alone telematics modules in order to exercise increased pressure on vehicle manufacturers. TomTom has already initiated a cooperation [15] with Google to extend its functionality: users may seek and find certain addresses at home on their fixed Computer via Google Maps and seamlessly transfer the target to the TomTom device before leaving the house and starting to drive with the help of the TomTom device. This strategy of building stand-alone telematics modules will be discussed in more detail in the section regarding telematics service providers.

However, pure PND manufacturers such as TomTom and Garmin are currently facing serious challenges to their business models. With the introduction of powerful smartphones with high-resolution touchpad displays, mobile phone manufacturers such as Nokia and Apple are also able to provide telematics services at lower costs – or in case of Nokia even for free in Germany. Hence, we will likely see a fusion of mobile phone manufacturers and PND manufacturers in the mid future.

Other 'niche' ICT players. Other market stakeholders have emerged that offer fully-fledged external computers that can be neatly integrated into vehicles. Such 'in-car PC providers' offer their customers a flexible platform for installing and accesssing all kinds of services. The UK-based company In-Car PC, for example, enables its users to run any sort of Windows application, use a remote desktop or similar technology to remotely access an office PC or network, browse the Internet, send and receive email, use Skye and other VoIP services, view and edit documents, presentations, spreadsheets and other information just as on a normal notebook computer [16]. According to company brochures, In-Car PC mounts touch-sensitive displays in almost arbitrary places, including in-dash integration, removable on-dash placement, sun visor integration, positioning directly in front of the user, completely covert setup as well as headrest integration for rear seat passengers. The company is also the authorised stockist for a number of key MNOs and can supply their car PCs with a 3G Internet connection. Market stakeholders such as In-Car PC have a strong interest in serving the specific needs of individual car drivers or business users such as fleet managers. These frequently have highly proprietary requirements, which can hardly be served by vehicle manufacturers who produce rather standardised vehicles and associated telematic units. A New Jersey based taxi company, for example, uses In-Car PC technology to display in-vehicle advertising within their cars. Each vehicle includes a PC and two headrest integrated screens, which automatically turn on with vehicle ignition and display advertising content to passengers. All PCs include 3G-based Internet connectivity, which allows the advertising content in individual vehicles or the entire fleet to be updated remotely at any time, with no need to manually visit each vehicle. Market participants such as In-Car PC may further build on their strengths in the future: providing telematic units that are exactly tailored to their customers' demands, both with respect to hardware platform, in-car screen mounting, back-end connectivity and installed software services. On the other hand, they face the challenge of missing integration into the vehicular infrastructure. Vehicle manufacturers are unlikely to provide these players with any access to communication bus interfaces (e.g. CAN, MOST) and thus to captive car data. Similarly to manufacturers of PNDs or mobile phones, providers of fully-fledged in-car PC technology ought to consider positioning their products as an open and highly flexible platform for in-vehicle telematic services and thus become a relevant competitor of the OEM-specific in-dash solutions that are still mostly run in a 'closed-sho" manner. The opportunities inherent in open platforms that enable access to a multitude of services also provided by third party providers will be discussed in the next section.

Telematics service providers constitute a key force within the upcoming telematics market. A broad ecosystem of diverse types of transportation service providers, infrastructure service providers or whole enterprises with business services push into the emerging telematics market. There are numerous service providers for different types of transportation and mobility active in the market. Examples include, among many others, public transportation service providers, fleet service providers, freight service providers, mobility providers, emergency service providers or automobile associations. They provide different services for different customers in different situations. Hence, they also have different requirements for AutoNet-based cooperative systems. In the following, we will detail two telematics service providers: emergency service providers and automobile associations.

Emergency service providers, for example, offer continuous support in case of breakdowns or accidents. Some of these services are deployed in close cooperation with vehicle manufacturers, where the service may even be triggered automatically in the event that certain in-vehicle crash sensors are triggered. Also, concierge services are increasingly available and popular for vehicle drivers. These services help their customers, for example, making dinner reservations, ordering flowers, buying tickets to the opera or booking a flight. Insurance companies are also pushing into the market by introducing new business models (e.g. 'pay-as-you-drive'), and by introducing data connections between vehicles and their back-end information systems in order to improve the efficiency of business processes after an accident has occured.

Today, particularly minor car accidents require the submission of a loss report (so-called 'first notice of loss'). In most cases, vehicle drivers lack adequate assistance regarding services such as the arrangement of a tow truck or support with the loss report. Insurance companies are therefore about to offer telematics services that establish a communication link in such safety critical situations and support their customers with respect to their actual needs. Apart from this, insurance companies suffer from bad data quality, media breaks and delayed loss reports, which lead to both high operating costs and loss expenses. As of today, insurance companies mostly use dedicated enterprise information systems to control their claims management business processes. However, available solutions are poorly, if at all, integrated with the physical world of insured objects and persons. As numerous studies indicate, cost savings based on pro-active and electronically integrated claims management can reach up to 15% [17]. Such companies are thus desperately seeking solutions that bridge the existing gap between their back-end information systems (such as the claims management systems) and the insured motor vehicles as well as their customers.

First systems and services are now about to be implemented and tested in order to address this need: an SAP-based prototype [19], for example, comprises an in-vehicle software service that exchanges relevant data with an SAP claims management solution. In the event of a minor accident, the equipped motor vehicle submits a loss report to the insurance company, while drivers benefit from a number of tailor-made, value-added services. Triggered for example by the airbag unit, the in-vehicle insurance application directly places an emergency call that establishes a voice connection to an emergency centre. In addition, the current position of the motor vehicle, personal as well as insurance-specific data about the driver and crash sensor data is transmitted. The loss report can also be enriched with a set of crash sensor data that enables a first assessment of the situation by the insurance carrier. After the claim has been submitted to the claims management enterprise information system, the insurance company uses the transmitted position data of the vehicle to offer additional location-based services to the driver. Services are displayed to the driver using either in-vehicle human–machine interface or via an application for a mobile consumer electronics device (e.g. Mercedes-Benz Bank's iPhone App called 'Sternhelfer'.[8] The insurance application also provides helpful information such as a driver's eligibility for a rental car as well as the arrival time of a requested tow truck. Finally, the transmitted information can be reviewed in the claims management enterprise system from the perspective of claim personnel at an insurance company. As an example, the claim file holds information about the business partners that offer the various third party services.

---

[8] See Sternhelfer homepage: http://www.sternhelfer.de.

Companies such Octo Telematics have already implemented and successfully marketed first solutions that provide basic assistance to end users in their arrangements after an accident has taken place. By equipping vehicles with a separate on-board unit that measures and tracks parameters such as geographic position, driving speed and used routes, the company is enabled to also offer usage-based ('pay-per-use' or 'pay-as-you-drive') insurance premiums: at the moment, most vehicle insurance premiums are calculated simply by type of vehicle, horsepower, sales price, age and sex of the vehicle holder, and depend on place of residence. The Octo Telematics on-board unit that establishes an information link between vehicle and back-end system provides the insurer with a far broader range of tools to optimise its risks and to strengthen its brand. According to its website [18], the company has managed to cooperate with a considerable number of insurance companies and end customers already.

Automobile associations, such as the German ADAC or the European ACE increasingly expand their telematics services. In fact, they will seek opportunities to use diverse telematics services to directly reach out to their members. On the one hand, this will allow them to provide drivers with rich and real-time information regarding traffic or weather conditions. On the other hand, it also enables the associations to immediately interact with their members and offer them a multitude of electronic services and, thus, expand into new business areas. Core capabilities of automobile associations in a future telematics market are mainly twofold:

- Automobile associations may act as lobbying institutions representing the interests of a number of companies that already sell their products in close cooperation with the associations: financial companies (e.g. travel insurances, credit card companies), vehicle rental companies, workshops or travel agencies. Some of the existing associations are expected to play a crucial role when it comes to the negotiation and definiton of open platforms and standards for service interfaces into vehicles, for example.
- The second major capability concerns the diversity of content and services that some of the associations have already built up over the last decades. Even in a global telematics market where services and data become easily available on a global scale and within the vehicles, these close customer relationships can be expected to persist. Automobile associations are well advised to build on their strong user basis and establish themselves as trusted advisors and well-known platforms in a possibly increasingly untransparent market for telematic services.

In addition to the above mentioned telematics service providers already in the market, Internet radio aggregators, tailored weather forecast services, mobile gaming providers, news aggregators, mobile advertisement brokers and many others enrich the growing ecosystem of telematics service providers. Internet giant Google, for example, offers an increasing number of services in this area: as part of a cooperation with BMW, for example, its search engine functionality has been made available via the head unit display of many upper class models.

An increasing competition for the control over the channel between all the above mentioned service providers and the actual end user becomes apparent. Most of the stakeholders still offer proprietary hardware and software solutions to approach their customers: insurance companies, for example, often provide users with separate on-board units that feature communication modules, GPS receivers, displays, processors and more. The installation of separate hardware devices within the vehicle for each of the offered services, however, is impractical and unacceptable from the drivers' perspective. Service providers are thus seeking to enter alliances with vehicle manufacturers in order to integrate their services into vehicles and be granted access to captive vehicle data such as geographic position and diverse vehicle status notifications in order to enrich their service offerings.

The entering into an alliance, however, will probably only be granted to service providers who do not endanger core business areas that are served by vehicle manufacturers. The provision of road navigation services such as Google Maps, for example, may represent a substitute service to the proprietary navigation systems offered by vehicle manufacturers at one point of time. Vehicle manufacturers will supposedly aim to exclude such services from the channel they control (in-vehicle telematic platforms) in order to protect relevant parts of their business. Service providers may then try to form their own alliances and establish a new well-accepted channel into the vehicles. Emerging high performance broadband technologies such as the LTE communication standard, as well as the relentless march of improvements in the field of processing power and energy consumption of mobile devices of ever decreasing size, may facilitate this strategy [20]: very small on-board units may assume the role of a gateway for the provision of services that are actually operated at the back-end. Multiple driver-focused and passenger-focused services and applications might then primarily be accessed in an 'offboard fashion via the cloud' [20], while the provision of energy, captive car data, physical space for mounting the display and other infrastructure becomse less and less important. The challenges inherent in such strategies that rely on the establishment of a single standardised platform, as well as many other important factors, will be discussed in more detail in the subsequent sections.

Finally, *content providers* are the third important key player in the service provider domain. In general, content providers provide the base information necessary to realise the services. In contrast to telematic service providers, the information provided by content providers is of 'general purpose', that is it is not necessarily related to any traffic or cooperative scenarios. As a result, content providers typically have minor requirements on cooperative systems, whereas service providers have stronger requirements on the information of the content providers. The following examples illustrate the different types of content with different relations to AutoNets:

- information about the current weather, including a weather forecast;
- regional points of interests;
- navigation maps;
- traffic information, such as for a specific region.

Attentive readers likely suppose (by the traffic information) that it is sometimes difficult to differentiate between service and content providers. In fact, acquisitions of traffic-related information providers by companies providing telematic services increased recently. For example, both European navigation map providers, Navteq and Teleatlas, were taken over by the PND manufacturer TomTom and the mobile device manufacturer Nokia, respectively. In both acquisitions, the strategic decision was made to reduce the risk of losing partnership with the content provider necessary for realising the respective telematics services.

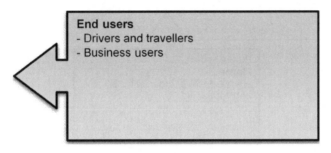

Figure 12.4   End users domain.

12.1.3.3   End Users

Of course, one of the most important stakeholders of cooperative systems are the end users (Figure 12.4), which basically determine the success of the system; in fact, a very positive acceptance of cooperative systems accelerates the market introduction of such systems since vehicle manufacturers and their suppliers are very interested in satisfying the customer's needs as quickly as possible. End users of cooperative systems comprise drivers and travellers as well as business users, which aim at utilising the system commercially [6].

Drivers and travellers are considered private individuals who consume telematic services within the vehicles. Among these stakeholders, safety is considered one of the pivotal criteria when buying a vehicle [6]. A survey conducted by the Cooperative Vehicle-Infrastructure Systems (CVIS) initiative revealed that 'European drivers are willing to fit their cars with new systems if they imply a significant increase in safety; probably even if this means an increase of the car price' [22]. In the context of safety-related telematics services, surveys show that a majority of car drivers assess obstacle warnings as very useful or quite useful, while almost 50% of drivers believe they are very worth or quite worth paying for. Also, a majority of drivers assess road status reports as very useful or quite useful, while about 40% of users believe they are very worth or quite worth paying for [22]. More than 90% of European drivers consider the provision of information about a car travelling on the wrong side of the road (ghost driver) to be most useful (very dangerous situation), followed by a '5 kilometres accident ahead warning notification' and the status information about current traffic flow [6]. Besides services that aim at the prevention of accidents due to unexpected conditions on the road ahead, private users have been found to greatly value applications in the field of infotainment.

Business users use telematics services for professional purposes. This stakeholder group comprises individual business users who are very interested in reducing travel time, as well as whole businesses, which basically provide telematic services to other companies or the public as described above. This means such businesses comprise, among many others, delivery services, transport companies, vehicle rental companies, taxi services, breakdown services or emergency services. Some studies suggest these stakeholders might assume the role of early adopters, as they supposedly had a proportionally high willingness to invest in new technologies, if these technologies entailed advantages for the conduct of their business. As can be seen in the context of the above discussed service of the firm Wireless Car, that equips cars with WiFi hotspots, however, private users were among the first customers.

Figure 12.5 Public sector domain.

12.1.3.4 Public Sector Players

According to Figure 12.5, public sector players comprise public operator organisations as well as several national and international institutional authorities.

Public operator organisations are the public correspondence of business users. Their aim is to provide services to the public in the form of basic services, without the need to maximise profits. Examples include road operators or even public transportation operators. Both may be controlled by public institutions, or may potentially be realised as public private partnerships, which may vary from city to city, and from country to country. In contrast to business users, public operator organisations are in general covered by public authorities. This way, they have more influence on political decisions concerning regulations or telematics services introduction.

Players in the *authorities* sector assume imortant institutional roles as they set the boundary conditions for a telematics services marketplace. They may facilitate the establishment of standards, and subsidise the efforts of specific market participants in order to accelerate market introduction processes. Stakeholders include, but are not limited to, the European Parliament and Commission, national legislative organs, the Police but also entities such as local traffic management centres and other traffic and infrastructure authorities. The interests pursued by governmental stakeholders are mainly threefold: as briefly outlined in the introductory part of this chapter, they have a macro-economic interest in the prevention of traffic injuries and fatalities. Applications such as Local Danger Warning or Crossing Collision Warnings – which are based on real-time data exchange between vehicles – are considered to help governments to reach their ambitious goals, such as cutting the yearly number of fatalities from around 50,000 to around 25,000 within a time frame of ten years in Europe. Besides safety aspects, public authorities also aim to increase traffic efficiency and to thus limit carbon emissions as well as to save macroeconomic opportunity costs incurred from time wasted through traffic congestions. Besides these two kinds of interest, governmental authorities also aim to stimulate their countries' national economies through facilitating technology leadership.

Public sector players are able to support the emergence of an efficient market for telematic services by supporting the often long-winded progress of standards development. As laid out in the next section of this chapter, one of the main challenges still preventing many technologies and associated applications from entering the mass market is the lack of standards that are deployed across module and vehicle boundaries. Governmental bodies can establish incentive structures and accompany standardisation processes in a way that avoids uncooperative behaviour of single players, for example. Secondly, public sector players may subsidise research-related as well as industry initiatives that contribute to an accelerated market introduction of specific applications or technologies. For this purpose, the European Commission has announced a number of action plans for the deployment of telematic infrastructures across Europe and has already funded a significant number of research projects with the strong participation of industry leaders. Among others, the Commission proposed the promotion of deployment of advanced driver assistance systems and safety and security related ITS systems, including their installation in new vehicles (via type approval) and, if relevant, their back-fitting in used vehicles, as well as the definition of specifications for roadside-to-roadside, vehicle-to-roadside and vehicle-to-vehicle communication in cooperative systems [21]. Appendix A gives a profound overview of the different standardisation and regulatory bodies. Recently, several Field Operational Tests (FOTs) have been initiated both on national and international levels to extensively test and evaluate proper functioning of developed services across different firms.[9] The facilitation particularly of inter-company research and development activities is of special relevance: the setup of initiatives where different companies with their individual economic interests closely interact and jointly develop interoperable services has proved to be very difficult. The relevance of such initiatives as well as their inherent challenges will be discussed extensively in the final section of this chapter.

The next chapter is devoted to an analysis of the remaining challenges on the path to the creation of an efficient global marketplace for telematics services.

12.2   Challenges

Technological challenges still hamper the market introduction of many telematic services: intermittent connectivity to broadband wireless access technologies, caused by the vehicles' high mobility and the fragmentary coverage of mobile networks, represents a major challenge – it may affect the reliability of the actual message transmission or even interrupt the consumption of services. To ensure end-to-end quality of service even in situations with limited or temporarily interrupted connectivity, dedicated in-vehicle functionality for priority-based queuing and rescheduling of message delivery are required. In the event that the available network bandwidth is not high enough to transmit all the data packets waiting to be sent within the vehicle, the in-vehicle communication component must label urgent messages appropriately and deliver them prior to all other queued messages.

Message exchange between back-end services and vehicles as well as the invocation of in-vehicle applications also require a reliable communication protocol that allows for the discovery and identification of vehicles as unique mobile nodes. Existing approaches still struggle with issues such as scalability and privacy protection. A further critical long-term technical challenge concerns the overall scalability of the communication architecture. According to official tables of the German Federal Statistical Office from January 2008, there are about 41 million cars on German roads [19]. The technical infrastructure for operating the different vehicle-to-vehicle and vehicle-to-roadside services will have to process, transmit and store an enormous amounts of data. Complex interfaces of back-end (e.g. enterprise-class) services

---

[9] Interested readers are referred to Appendix A, which introduces important activities in the fields of cooperative systems and AutoNets, including current FOT activities.

often require the transmission of numerous parameters and login credentials. Since the number of parameters directly affects the message size, the scalability of a service infrastructure is challenged even further. The communication architecture's security represents another technical challenge since data exchanged between vehicles and back-end services may be highly sensitive. Also, the intrusion of hackers into vehicles must be avoided to ensure road traffic safety. For this reason, scalable algorithms, protocols and infrastructures for message encryption, authentication and the related storage and maintenance of certificates need to be designed and deployed. Besides these and many other remaining technical challenges which have been discussed in the previous chapters, a number of organisational and economic challenges need to be addressed too.

12.2.1 Harmonisation and Standardisation

The lack of harmonisation represents a major barrier to the successful market introduction of many electronic cooperative telematics services. Harmonisation and, thus, standardisation for cooperative AutoNets are necessary in several different areas, as illustrated in Figure 12.6.

On a *hardware and software level*, heterogeneous platforms aggravate the provision of standardised and interoperable systems and services. In fact, most hardware and software providers follow a highly proprietary approach: vendors of mobile insurance solutions (such as the above mentioned company Octo Telematics, which offers a pay-as-you-drive payment scheme) require their customers to install dedicated on-board units in their vehicles. Car stereo suppliers such as Blaupunkt offer dedicated hardware platforms too. Customers have to install yet another device to receive an additional telematic service; in this case, the reception of Internet radio stations. The above mentioned company In-CarPC offers holistic computer-based solutions for in-vehicle deployment. Their product goes one step beyond the above mentioned 'island solutions' as it allows for installing and operating more than one single service. However, the devices have no access to any captive vehicle data (e.g. real-time sensor data within the vehicle) and thus lack true integration into the vehicle infrastructure. Services

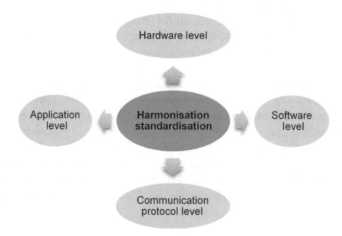

Figure 12.6 Relevant areas for harmonisation and standardisation of AutoNets.

operated on the in-car computer are not necessarily harmonised either, leading to a lack of interoperability. In short, the above mentioned solutions support the operation of a set of specific services, but usually are neither extendible nor interoperable with services operated by other vendors. For this reason, large electronics and software providers have started working on a flexible infrastructure platform that serves as a single basis for operating a multitude of services, potentially provided by many different stakeholders. As outlined above, Continental AG's AutoLinQ platform can be considered a promising step in this direction. A number of research projects have been initiated during the past years as well which aim at defining one single, flexibl, and internationally applicable hardware and software architecture, such as NoW, CALM and AKTIV (see Appendix A). However, as the focus of these and many other projects varied, and since project members followed different design paradigms and had to obey legal constraints that differed from country to country, the resulting architecture proposals were not fully harmonised.

The additional lack of harmonisation on a *communications protocol level* aggravates the market introduction of many services even further. Particularly V2V applications that rely on real-time data transfer among vehicles (e.g. local danger warning applications) require the availability of standardised and comprehensive communication. Without a standardised system, vehicles built by different manufacturers will not be able to communicate with each other, thus heavily deteriorating the value perceived by customers. The identification of an appropriate radio spectrum builds the foundation for a harmonised infrastructue. For this purpose, frequency bands at about 5.9GHz have been protected and reserved in Europe, the US and Japan. Based on this first agreement, diverse network protocols are needed in order to control the actual data exchange between network nodes. During the past few years, both industry and research institues have conceptualised and promoted a wealth of versions of the IEEE 802.11 standard, for example. These versions differ with respect to paramters such as utilised frequency range, data transfer rate, physical transmission range, medium access control functionality, modulation techniques and many other things. The industry-wide agreement on one specific standard (e.g. the IEEE 802.11p standard, see also Appendix A) will be pivotal for the realisation of both robust and efficient data exchange specifically for V2V communication-based cooperative applications.

On an *application level*, cross-manufacturer harmonisation and standardisation has made the least progress. While certain frequency bands and first network protocols could be defined, the specification of application standards that allow for true interoperability are still in their infancy. In the area of mere V2V applications, projects are underway that aim at defining and standardising message semantics, application protocols and smart message dissemination algorithms. However, none of them has been acknowledged as de facto standard so far. In the field of V2Central, V2Internet and V2Private services, proprietary application protocols are in place that do not encompass interfaces to connect to other services. However, an imporant milestone was achieved in the COMeSafety activity and deployed in the project PRE-DRIVE C2X (see Appendix A), which provides a first proposal for the back-end integration and basic service interoperability. This opens up the possibility of integrating the information of back-end services into cooperative AutoNets. Vice versa, this services-oriented approach also allows for including vehicles in business processes by considering vehicles as 'mobile services'.

In short, the lack of standardisation at multiple levels impedes the successful market introduction of numerous promising telematic services. In most cases, service providers have to design, build and operate a separate communication infrastructure, leading to high costs and limited customer benefits. Moreover, such proprietary systems will be available for specific vehicle brands, which aggravates the deployment of cooperative AutoNet applications where a considerable number of vehicles need to contribute. The problem inherent in this situation can be illustrated as follows: if eBay, Amazon.com and all the other Internet-based service providers had to equip their customers with proprietary computers and communications media rather than relying on the availability of standardised computers, operating systems, browsers and the Internet, their businesses would be at stake. In fact, in order to facilitate the emergence of an effcient market for telematics services, a harmonised and standardised system is required.

12.2.2   Life Cycle

The lack of harmonisation and standardisation still impedes the success of the telematic services market which would greatly benefit from application interoperability and the consequent network effect. A further issue that has beleaguered V2X services in the past is that the life cycle of vehicles, which is typically in the range of decades, far outlives the life cycle of consumer electronic devices and associated services. To overcome this issue, design paradigms need to be changed: technology must be installed in the vehicles that allows for (ideally remote) updates at any point of time in the future. This would allow one to add new applications and new value to the customers' vehicles on a regular basis. Road navigation systems that are updated via the Internet in the event of road networks undergoing changes represent the first examples of the implementation of this paradigm. In short, a comprehensive infrastructure for telematics services will have to incorporate the possibility updating existing applications and adding new ones.

In turn, this also results in respective requirements concerning the hardware platforms running the telematic services within the vehicles. For computers, consumer electronic devices, and respective services, it is a general principle that new software and features typically require additional memory and higher processing capabilities, which typically contradict the life cycle of the components used in vehicles. Obviously, it will not be accepted by drivers that they need to replace their complete car in order to upgrade their telematic services functionality.

12.2.3   Costs and Revenues in an Emerging Business Ecosystem

In addition to these organisational issues, economic hurdles still endanger the establishment of cooperative services too. As described in the subsequent paragraphs, the realisation of an efficient market for telematics entails significant expenditures for the setup of an adequate system (capital expenditures, CAPEX) as well as costs for operating activities (operational expenditures, OPEX). Contrary to many existing business models in the automotive arena, these costs can hardly be taken over by one single market player. The huge costs incurred from the development and operation of vehicles' on-board technology, as well as extensive roadside and back-end service infrastructures, rather need to be shared among the different involved stakeholders. In the following, major cost items that incur along a future telematics value chain will be briefly delineated [6].

For marketing and sales purposes, campaigns need to be conducted, advertising needs to be paid for and (personal as well as electronic) sales channels have to be setup and maintained. In the client services stage, ongoing client communication, contract management, life cycle management and invoicing processes lead to considerable costs. Concerning the actual telematics service, one-time costs incur for the design, development, testing and integration of software and hardware modules into the vehicles (e.g. on-board communication modules or software application platforms). Further capital expenditures must be borne for the implementation of roadside stations (e.g. traffic signs with integrated communications modules) as well as the back-end services which support the diverse services. Operational costs are added as services both within vehicles and at the back-end need to be maintained: content needs to be generated and provided on a regular basis, and errors need to be identified and addressed, to name only a few cost items. For the transmission of data via licensed frequency bands, additional cost must be taken into account. As one single player will hardly be able to take over all the cost items, a network of cooperating stakeholders will have to be built up.

Revenues, on the other side, are unlikely to be captured by a single player either. They will have to be distributed among the stakeholders in accordance with the respective value they contribute. To flourish in such an environment, relevant stakeholders will have to form a business ecosystem which is defined as a network of organisations that share a key technological platform [23]. Irrespective of a single organisation's individual strength, all actors in such a business ecosystem are connected and share the success or failure of the network as a whole. As of today, a truly succesful symbiosis of a considerable number of telematics stakeholders has not been developed. As discussed above, most business models still rely on the individual development and operation of individual and thus isolated solutions. This strategy prevents sharing the costs across different players and leads to the above mentioned lack of interoperability. In the course of this chapter, we will discuss examples of early business ecosystems in more detail.

12.2.4 Customer Acceptance

Besides the issue of high initial and ongoing costs for the set-up and operation of a telematics infrastructure, additional challenges exist with respect to customer acceptance. First, many safety-related applications suffer from a limited perceived value for the customer, particularly in an early stage of market introduction. Local danger warning applications that rely on the periodic and real-time exchange of beacon messages between vehicles, for example, only add real value in the case that a considerable number of vehicles are equipped with the necessary on-board units. As illustrated in Figure 12.7, it is expected to take years from the start of an application's market introduction to gain a share of about 20% of all the vehicles on the roads that might be appropriately equipped, allowing the safety application to actually work. As soon as a relevant number of vehicles run the respective application and broadcast the beacons to cars in their vicinity, the application has passed a critical threshold and may attract new customers rather quickly. In the meantime, however, customers would not perceive a clear value from such a service. A possible approach to addressing this network effect issue is to provide customers with services that add value right from the beginning and thus to convince customers to invest in the installation of on-board units. Infotainment applications, for example, may be consumed via already existing cellular networks and are thus immediately available for

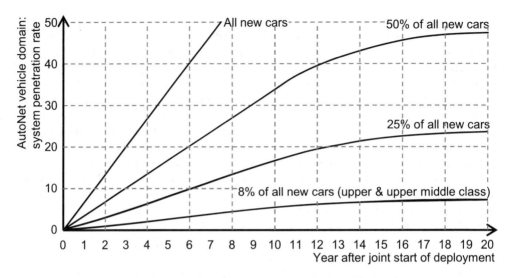

Figure 12.7 Evolution of V2X system penetration with respect to different deployment strategies, based on an analysis in the German Network on Wheels project (NoW).

all customers. On-board units that allow for accessing both cellular networks and WiFi-based ad-hoc networks between vehicles provide customers with the possibility of connecting to back-end services immediately as well as the option to consume ad-hoc AutoNet services in the futue, as soon as a critical number of other vehicles is equipped as well. Safety-related applications would be offered as ancillary services for free in the first place; their perceived value would increase over time corresponding to the overall equipment rate of vehicles.

A further issue concerns the design of applications that exactly meet the prospective customers' needs. In this context, the idea of a *killer application* was born in the past. Vehicle manufacturers have mostly followed a 'closed-shop' approach and tried to set up a clearly defined and limited porfolio of proprietary applications. However, as successful business models have proved in the Internet arena, the provision of an open, but governed platform may be more promising than the installation of a static and limited application portfolio. Facebook, for example, has allowed the mass of Web developers to offer applications via a clearly defined interface to their end users. The company reviews these applications and assures their adequacy for their user base and thus maintains a certain level of quality. Without the creativity of the long tail of developers, their offering would be relatively limited (to the core functionality of connecting to other users and some ancillary functions) and less attractive. Vehicle manufacturers are well advised to follow a similar approach and set up a dynamic content and service architecture that enables individual and personalised service compositions. The governance of such a flexible application platform certainly entails a number of additional challenegs that need to be addressed: the efficient orchestration of business partners, the setup of a content quality assurance system as well the establishment of a content offering and billing infrastructure are required, to name only a few.

A first look in this future direction is the software development kit (SDK) introduced by Volkswagen for their 'App-my-Ride contest'. This contest aims at the development of applications for Volkswagen headunits.[10] With this SDK, Volkswagen opens up the platform of their vehicles for potential future telematic services by providing an application programming interface (API) to external companies and private developers, which provides routines to access a limited set of vehicle parameters. These parameters can be used to develop automotive applications running on Volkswagen headunits.

12.3 Driving the Emergence of a Coherent Business Ecosystem

In order to overcome the challenges discussed above, the diverse stakeholders need to team up and establish a coherent business ecosystem that facilitates harmonisation of services, allows for sharing the required financial investments, and helps unleash the creativity of many diverse players (rather than relying on a stand-alone proprietary solution). In the course of the following section, strategies for the development of a such an ecosystem will be elaborated.

12.3.1   Strategies for the Development of a Modular Business Ecosystem

As described in the previous sections, single players will not be able to carry costs, invent and operate applications that attract a substantial amount of customers, and tackle all the remaining technical challenges. For the successful market introduction of the applications introduced in the course of this book, a modular and open platform is needed that acts as an agile and extendible basis for delivering and using telematics services within vehicles. Modularity comes with a number of advantages [25]:

- *Decentralised design evolution.* The establishment of a set of design rules allows for the parallelisation of design efforts. Rather than requiring a vehicle manufacturer to closely manage the design and operation of all the services offered to the drivers, for example, the creation and enforcement of a set of standards (referred to as design rules in this chapter) allows for an improved division of labour between a multitude of stakeholders. As long as all players comply with the previously agreed rules, each party may focus on its core competencies, while leaving other activities to the business ecosystem partners. An example could be the following:
  - Content providers may specialise in delivering relevant information (which needs to comply with specific standards in order to be consumed by the platform's end users).
  - Vehicle manufacturers might focus on core competencies such as the orchestration of suppliers and the integration of the different components.
  - Third party service providers such as insurance companies might work on innovative business models rather than building proprietary technical infrastructures for operating their services.
- *Complexity reduction.* In a modular environment, systems and infrastructures required for the operation of a specific application can be systematically decomposed into finer granular atomic building blocks, which can then be synthesised into manageable modules. These do not have to be created or redesigned as holistic blocks, but can be dynamically assembled

---

[10] App-my-Ride contest homepage: http://www.app-my-ride.com.

from standardised components. In the field of telematics services, envisioned communications architectures are highly complex and cannot be implemented by a single player or as a single module. The envisioned architecture needs to incorporate a multitude of manageable building blocks that can be handled efficiently.

- *Agility.* Modularity accommodates uncertainty and multiplies available options. Through the encapsulation of parts of a design into mutually independent, but collectively interoperable, modules a portfolio of options is created. Modules are mutually compatible, but they hide internal design information from each other. In this way, the impact of changes in a given design is limited to a defined area as opposed to the currently prevailing, interconnected and monolithic designs. In the field of telematic services, the accomodation of uncertainty plays a key role: it is not yet known which standard for wireless comunications, which message dissemination algorithm or which infotainment services will end up as a de facto standard. The envisioned architecture needs to encapsulate services and technical components to allow for their later exchange if required.

The advantages of a modular platform for telematic services strongly depend on the successful implementation of a set of key design rules that all involved players need to obey. These rules shall be as comprehensive and detailed as necessary to guarantee full interoperability of all different services and as flexible and limited as possible to maximise the remaining design freedom. Many research projects have already tried to design platforms for the flexible provision of interoperable telematic services. Although the projects have lead to major technological advancements, none of them has so far managed to establish a set of design rules that has subsequently been accepted as de facto standard across the whole industry. For this reason, the following paragraphs are devoted to an analysis of different strategies of stakeholders involved in the standardisation process. The goal of this analysis is to get an improved understanding of the interests of the different market participants such as automotive vehicle manufacturers, hardware and software providers, mobile network operators, third party service providers as well as the role intermediary institutions (referred to as standards development organisations or SDOs in the following) may play. In the field of telematic services, research consortia such as NoW, government-funded institutions such as COMeSafety, industry alliances such as Genivi,[11] or standards organisations such as IEEE represent different examples for SDOs. Note that Appendix A gives a detailed description of the example SDOs mentioned here.

As illustrated in Figure 12.8, both user firms and SDOs are the focus of our analysis. In this context, user firms encompass all those companies that comply with a given set of standards in order to be interoperable. The SDO is expected to have an institutional role as it develops and issues design rules and may also enforce their implementation by the user firms. The key question is how user firms may position themselves in an ecosystem of firms that clearly require some standards in order for them to efficiently interact. In general, a user firm has been found to participate 'if the expected benefits from participation are clearly higher than its costs. Moreover, its engagement depends on the interests of all participants, which can be aligned or divergent. While the former dimension is idiosyncratic for each firm, the latter is a global dimension, which influences the strategies of all user firms' [24]. These two dimensions result in four basic strategic options, as shown in Figure 12.9. On a high level, user firms usually

---

[11] Genivi homepage: http://www.genivi.org/.

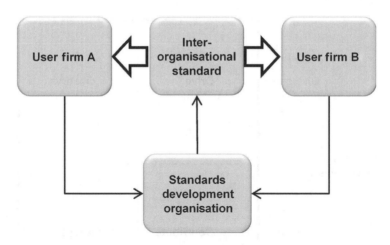

Figure 12.8   Stakeholders in the development of a telematics business ecosystem [24].

follow one of these four strategies according to the following logic: 'If the interests of the actors in an SDO are different from the interests of a particular user and the net benefits from participating are low for this user, then the best strategy for him is only to *observe* the initiative. If the benefits of participating are low, but the interests are aligned, the user should not participate, but simply *adopt* the specifications. If the benefits are high and the interests are aligned, the user should *contribute* to the SDO. Finally, if the interests are in conflict with other actors, but the benefits are high, the user should actively drive some of the SDO work in order to defend his interests' [24].

However, the recommendations regarding the strategic positioning of user firms are over-simplified at such a high level. In the subsequent four paragraphs, the different options are elaborated in more detail.

12.3.1.1   Observe

User firms usually assume observer roles in case the mission, structure or the specifications of an SDO do not meet the interests of their potential contributors. The mission may deviate because the SDO does not cover the telematic application a specific stakeholder is interested in. An SDO might, for example, focus on merely safety-related vehicle-to-vehicle communication applications, while a specific user firm is rather interested in the establishment of a broad telematics platform that allows the offer of infotainment-oriented back-end services. Also, the governance structure might be inappropriate due the lack of a council that grants users the desired voting rights in a specific application arena. If a user firm then expects the costs of proactively changing the situation to be higher than the expected benefits, it will most likely observe the existing SDO's activities rather than contributing to it. Consequently, the firm will also continue to use proprietary solutions for the implementation of its market offerings. The fear of losing the intellectual property inherent in particular solutions represents a further motivation for following an observation strategy. User firms might continue to use proprietary

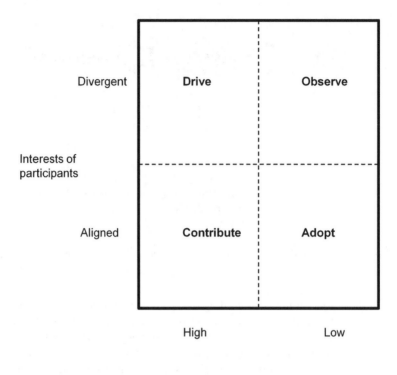

Figure 12.9   Strategies for users during the standards creation process [24].

solutions which they regard as sources of competitive advantage, while simply observing the progress an SDO and its stakeholders make.

User firms have different possibilities to observe an SDO's activities, depending on its respective organisational openness [24]. On the one hand, there are open SDOs such as COMeSafety, which offer public insight into the current development progress. On the other hand, there are more closed institutions that require specific membership status in order to be granted access to specificaiton documents. An example of such a closed institution is ISO, where the contributors have to pay for their membership and where the organisation charges for the standards they produce. In the course of this chapter, we will discuss options for an SDO to adjust governance structure, scope, focus as well as the standards development processes in a way to motivate a critical mass of relevant user firms to assume the roles of active contributors to a given standard.

12.3.1.2   Adopt

According to Loewer, adopt means the following: 'if the mission, structure, and specifications of an SDO already meet the interests of a firm though it expects no benefits from influencing them, it can simply adopt the specifications. In this free-rider stratey, the firm does not get involved in most issues concerning the SDO and just uses the specifications' for the implementation of respective applications [24]. The motivation for a user firm to follow this strategy depends on whether the SDO offers additional services to those members that have actively engaged themeselves in the specification process or not. An SDO may, for example, provide their active members with a preferred access to certain infrastructure or specification components, while charging others a certain fee.

12.3.1.3   Contribute

When the mission, governance structure and specifications of an SDO meet the interests of a user firm, and when the active participation is expected to yield significant benefits in the future, a firm is likely to assume a contributor role. Different factors impact a user firm's perception of the value of proactive involvement. First of all, the establishment of structured and transparent procedures for the development of a set of design rules are crucial. Clear roles and transparent decision making processes keep coordination costs low and allow for the efficient use of a participant's resources. Second, an appropriate policy for treating the individual members' intellectual property plays a key role: According to Loewer, 'two fundamental cases can threaten a firm's contributions. First, the firm might completely lose the property rights for its contributions. This lowers the motivation to contribute, while also fostering free-rider behavior in other firms. [...] Second, the firm might be sued for damages caused by a specification.' [24]. The setup of an IPR (intellectual property rights) policy that allows contributors to retain their existing intellectual property, facilitates the joint development of novel intellectual property, and protects contributors from excessive claims later users could come up with is thus pivotal for a successful SDO. Third and last, incentive systems inherent in an SDO's governance structure may ensure the participants' motivation to take over administrative tasks or promote the industry-wide adoption of specifications: 'The benefit from supporting the administration depends on the incentive system of the SDO. Usually greater adminsitrative responsibilities are coupled with more decision rights.' [24].

12.3.1.4   Drive

User firms usually assume a driving role in the case that the mission, governance structure, or specifications of an SDO do not completely meet their interests, but the expected benefit of participating is high. 'The best strategy then is to drive the SDO actively towards the firm's interests. Generally, this can be done on the three layers of formation, organisation, and development' [24]. If a number of user firms, for example, do not agree with the mission that is defined during the formation phase of an SDO, these players may either decide to drive the reshaping of the mission statement or to found a new SDO with perfectly aligned mission settings. On an organisational layer, user firms may not see their interests properly reflected in the SDO's organisational structure. The user firms might then urge the SDO to adapt the structure, for example, to establish new councils to cover dedicated topics, or to modify the decision making processes. On the development layer, a user firm may find its individual requirements are not being addressed to a sufficient extent: according to Loewer, a firm 'may be lacking proper specifications for its requirements, while no other firms see any benefit in developing such specifications. The the firm has to drive a critical mass of others to participate in the development' [24].

12.3.1.5 Discussion

While user firms may follow one of the above discussed strategies, the SDO's role is to effectively coordinate the development and adoption of design rules and thus to establish them as a standard. Overall mission, governance structure, scope, focus and processes need to be designed in a way to motivate as many of the relevant user firms to actively contribute as possible. As illustrated in Figure 12.10, SDOs may follow one of four major strategies for coordination purposes. In case most of an SDO's user firms have different interests with respect to a certain standard and expect only a little benefit from using it, they are expected to choose the observation strategy as discussed above. If the standard, however, plays a key role as a building block for the development of the overall telematics platform, the 'SDO should impose one specific solution on its members' [24]. In the second case, when user interests are aligned, but the perceived value of contributing is low, user firms are expected to assume an adoption role. 'This is often the case when the members decide myopically, not perceiving the long-term value of the development effort. Here the SDO should support the development in order to lower the costs of contributing and raise the benefit for the [pro-active] participants' [24]. The SDO may, for example, take over the responsibility of developing parts of the over all standard or of promoting the standard within the whole industry in order to unburden the individual members. In the third case, when user interests are aligned, and expected benefits are high, user firms usually voluntarily contribute to an SDO's effort. 'In this case, the SDO has

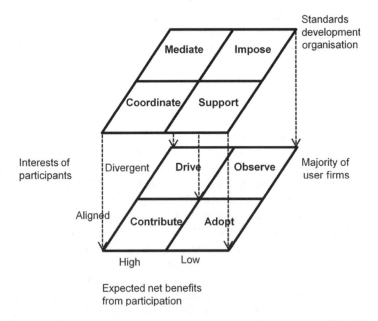

Figure 12.10   Strategies for managing user firms participating in an SDO [24].

to maintain a stable organisational environment for the developing groups. Usually, it should only coordinate the development process without much intervention' [24]. Working groups should then have a lot of freedom regarding the actual development processes and should only be forced to follow certain fundamental rules regarding IPR policy, voting mechanisms and the distribution of knowledge. In the fourth and last case, when high expected benefits are coupled with divergent interests, user firms usually try to drive the SDO into a direction that optimally suits their respective individual needs: 'While some differences are normal and important for motivating the search for innovative solutions, the SDO has to mediate fundamental divergence or conflicts. Otherwise, the integrity of the developed specifications would be threatened' [24].

12.3.2   Early Examples of Telematic Business Ecosystems

So far, numerous initiatives have been formed in order to establish telematic business ecosystems. In the following we will put the focus on two exemplary initiatives in this field: ng Connect and Genivi.

12.3.2.1   ng Connect

The ng Connect program[12] represents an early example of a fully fledged business ecosystem of companies that aim to jointly develop a modular telematics platform. Conceived and founded by Alcatel-Lucent, the program brings together infrastructure, device, application and content companies to create an end-to-end ecosystem with the resources and expertise required to effectively deliver services and applications to both service providers, enterprises and consumers.[13] ng Connect stakeholders have recognised that the provision of innovative telematic services only works by preventing isolated development activities and by bringing together innovative companies that are not generally linked, by integrating concepts from these diverse companies, and by improving time to market: according to the webpage, the next generation of high-bandwidth services arrives a lot faster when you pool the strengths and resources of global leaders and innovators in infrastructure, devices, applications and content.

The program now encompasses companies in the fields of gaming, advertising, automotive transportation, media, computing, consumer electronics, handsets, apps vendors, white goods (vendors of large machines accomplishing routine housekeeping tasks such as washing machines) and even retailers. All these companies contribute to a joint, comprehensive solution, and benefit from the advantages of a cross-discipline business ecosystem in the following way:

- *Primary research and solution concept development.* Program members share costs incurring from primary research activities, for example regarding the customers' willingness to pay for specific services, market forecasts or the implications of the 4G wireless communication evolution. Besides primary research, program members are provided with resources

---

[12] ng Connect homepage: http://www.ngconnect.org/.

[13] According to the program's website, member firms include: 4DK Technologies Inc., Alcatel-Lucent, Atlantic Records, BUZZMEDIA Inc., chumby, Connect2Media, Creative, dimedis, FISHLABS, GameStreamer, Gemalto, HP, Intamac, Kabillion, Kyocera, LearningMate, MediaTile, QNX Software Systems, R360, RebelVox, Samsung, SIGNEXX, Total Immersion, TuneWiki, V-Gate, Words & Numbers.

for the development of solution concepts: mechanisms for developing or sharing product requirements to enable pre-integrated solutions, state-of-the-art solution concept lab facilities, a dedicated solution concept software development team, program management resources, an end-to-end service provider network line-up, including fixed and wireless networks, for assessing end-to-end solution requirements, and finally facilities for demonstrations to target customers, the media and other parties.

- *Integration, validation and customer showcases.* The ng Connect program further motivates its members to contribute to a common standard by providing them with state-of-the-art facilities for jointly integrating and validating devices and services over real-world network and service deployment scenarios. In this way, participating companies may hope for reduced time to market and deployment costs. As demonstrating an integrated service offering is pivotal to seizing market leadership and customer mindshare, the program further incentivises their members by granting them access to software development and engineering support services in showrooms where they can effectively present to customers.
- *Sales and marketing.* Finally, members are supported in their efforts in bringing their products to market. According to the homepage of ng Connect, the ng Connect Program brings a strong collective voice to the market, allowing members to benefit from a level of sales and marketing visibility and reach they could not achieve on their own. ng Connect offers support regarding the joint development of sales and marketing tools, premises at trade shows, executive conferences, and marketing events, marketing and sales training and many others.

In November 2009, the ng Connect program revealed the 'Long-Term Evolution (LTE) Connected Car', a concept car that offers a range of telematics applications in the fields of entertainment, infotainment, security and driving-related services. Video-on-demand services provide customers with a wealth of video content, including movies, recorded TV programs and user-generated content. Multi-player games can be played within the vehicle, in other vehicles, or with players anywhere in the world. A comprehensive audio library becomes accessible to the customer that is not limited to data stored within the vehicle, but can draw on music resources available somewhere on the Internet. The concept car also offers a data connection to home automation and security systems for managing climate control systems or lighting in the home of the driver. A built-in WiFi hotspot allows all passengers within the vehicle to utilise WiFi enabled in-vehicle devices and to access all other services available in 'the cloud' (e.g. social networks or weather forecast websites). An eCommerce platform provides passengers with access to an application store with numerous 'apps' to buy right from the vehicle. Augmented navigation services increase the convenience of immediate, real-time GPS updates with point-of-interest overlay and integrated location-based services. Continuous broadband connectivity via the emerging 4G wireless communications standard LTE enables real-time traffic, weather and road condition alerts to increase driver safety and to allow for permanent technical vehicle health monitoring. Finally, the concept car can be considered a mobile sensor that transmits data regarding status, location and road conditions to central services that shares relevant information with other drivers.

While still at the concept stage, the LTE Connected Car underscores a few critical factors for designing and successfully bringing to market future telematic services: first and foremost, it emphasises the relevance of cross-industry collaboration and the increased focus on establishing ecosystems of different value system stakeholders such as service providers, content providers, network carriers and vehicle manufacturers. Rather than relying on controlling the whole efforts of conceptualising, building and operating complex telematics services, all involved parties should pursue ways to create and manage relationships among companies that make up a connected vehicle ecosystem. Second, the concept car unleashes the potential of storing data and applications 'in the cloud' rather than solely in the vehicle. Media content and application functionality may be stored and operated at back-end services that are accessed via always-on broadband network connectivity (e.g, the upcoming LTE standard) [20].

12.3.2.2 Genivi

The Genivi alliance was officially launched in 2009, heralding a new era of cooperation among automakers, suppliers and technology providers in the interest of streamlining the development and support with a particular focus on in-vehicle infotainment (IVI) or IVI products and services. Comprising members such as the BMW Group, Wind River, Intel, GM, PSA, Delphi, Magneti-Marelli and Visteon, Genivi represents a growing business ecosystem that addresses the increasing costs and complexity of developing IVI services. In-vehicle infotainment is a rapidly changing field within the automotive industry and includes diverse services such as music, news, multimedia, navigation telephony and Internet services. The alliance members joined forces to create a standardised, modular and thus extendible open source software platform that establishes a consistent foundation upon which automobile manufacturers and their partner suppliers can add their differentiated applications and services to create competitive IVI products. This platform aims to significantly reduce the cost of developing devices and services, and to accelerate the pace at which new telematics applications can be developed, and will allow for new business models to emerge in the in-vehicle infotainment market.[14] In order to successfully coordinate this ecosystem and to motivate its members to contribute to the joint platform, a dedicated governance structure has been established, which addresses the following issues:

- *Strategy.* In terms of strategy, the Genivi organisation accounts for developing the long range and annual strategic plan, taking decisions regarding the roadmap for alliance deliverables, identifying and implementing liaison relationships with other alliances and determining the timing and form of alliance expansion.
- *Technical integration.* At a technical level, the organisation develops and monitors the technical working plan of the alliance. This includes the gathering of requirements, specification development, enforcement of IPR policies and processes, testing and release of reference implementations and adoption and compliance activities. The organisation's activities aim to optimise each single stakeholder's motivation to contribute, and to prevent its members from just observing the progress other fellow members make.
- *Marketing.* Similar to the ng Connect program, Genivi tries to bring a strong common voice to the market, allowing members to benefit from a jointly developed alliance marketing plan, and from a council that manages internal and external communications and facilitates programs that bring greater awareness to the alliance and its deliverables.

---

[14] See Genivi homepage: http://www.genivi.org/.

12.4 Summary

This chapter provided an overview of different stakeholders in the emerging market for telematic services and products. We elaborated on their respective interests, strengths and potential weaknesses. We also analysed some of the challenges that often prevent these stakeholders from successfully introducing their products to the market: lacking cross-vendor standardisation, life cycle management, financial issues and customer acceptance. Subsequently, we postulated the establishment of coherent business ecosystems to tackle these challenges. For this purpose, we analysed the interplay of companies and a standards development organisation in situations where a set of common design rules is required to bring an integrated service or product offering to market. Finally, two examples of early, promising telematics business ecosytems were demonstrated: ng Connect and Genivi.

At the moment, most market players still follow a closed-shop approach and offer proprietary platforms for the provision of telematic services for their customers, entailing all the disadvantages mentioned above: huge development costs that cannot be shared among business partners, limited resources (only within the boundaries of the respective firm) for tapping into innovation potential, lacking interoperability with other vendors' services (thus delimiting the power of economic network effects), and a lack of extendibility. The first results achieved by the emerging business ecoysystems such as ng Connect and Genivi are promising, on the other hand, and underscore a number of critical factors for designing future telematics services. First of all, ecosystem management will become pivotal for vehicle manufacturers. Service providers, content provides, network operators and vehicle manufacturers will have to create and manage ecosystems of innovators around a core technology platform. Second, vehicles increasingly assume the role of mobile nodes that pro-actively sense and transmit information to back-end services and also receive data and application functionality via always-on broadband communication networks. In fact, service providers will need less and less space within the vehicle to provide users with their services, spurring the trend towards 'vehicle-to-cloud' functionality.

References

[1] DEUFRAKO Project Consortium (2002) IVHW System Concept and Issues Relevant for Standardization.
[2] Chatterjee, A., Kaas, H-W., Kumaresh, T. .V and Wojcik, P. J. (2002) A road map for telematics. *The McKinsey Quarterly*, No. 2.
[3] Car-to-Car Communication Consortium website (2010). http://car2car.org
[4] Lasowski, R. and Strassberger, M. (2009) Market introduction and deployment strategies. In: Moustafa, H. and Zhang, Y. *Vehicular Networks: Techniques, Standards, and Applications*. CRC Press.
[5] Sesia, S. and Toufik, I. (2009) *LTE – The UMTS Long Term Evolution – From Theory to Practice*. John Wiley & Sons.
[6] Birle, C. et al. (2009) CoCar Feasibility Study: Technology, Business and Dissemination. Public Report by the CoCar Consortium.
[7] Malone, K., Wilmik, I., and Noecker, G., Rossrucker, K., Galbas, R. and Alkim, T. (1984) Socio-Economic Impact Assessment of Stand-Alone and Co-Operative Intelligent Vehicle Safety Systems (IVSS) in Europe. eImpact, Technical Report.
[8] Matheus, K., Morich, R., Paulus, I., Menig, C., Luebke, A., Rech, B. and Specks, W. (2005) Car-to-car communication – market introduction and success factors. *5th European Congress and Exhibition on Intelligent Transport Systems and Services*.
[9] Autonet Mobile (2010) The only wi-fi spot designed for your car. http://www.autonetmobile.com/.

[10] Festag, A., Baldessari, R., Zhang, W., Le, L., Sarma, A. and Fukukawa, M. (2010) CAR-2-X communication for safety and infotainment in Europe 2010 *NEC Technical Journal, Special Issue: ITS Safety and Security*, Vol.3, No.1.

[11] Hammerschmidt, C. (2010) Continental puts android internet platform on wheels. http://eetimes.eu/showArticle.jhtml?articleID=217800032.

[12] Engadget (2010) Blaupunkt shows off miRoamer-powered internet car radios. http://www.engadget.com/2009/01/13/blaupunkt-shows-off-miroamer-powered-internet-car-radios/.

[13] Ronn, I. (2010) How mobile operators can generate revenue from industrial telematics. http://www.totaltele.com/view.aspx?ID=334782.

[14] Spinelli, M. (2010) Mercedes-Benz launches new apps-based telematics service. http://www.popsci.com/cars/article/2009-11/mercedes-benz-launches-new-apps-based-telematics-service.

[15] TomTom (2010) The shortest route between Google Maps and TomTom. http://www.tomtom.com/page/tomtom-on-google-maps.

[16] In-Car PC (2010) In-Car PC website. http://www.in-carpc.co.uk/internet.htm.

[17] Accenture (2002) *Unlocking the Value in Claims*. Accenture – Insurance Solution Group.

[18] Octo Telematics (2010) Octo Telamtics website. *http://www.octotelematics.com/*

[19] Baecker, O. and Bereuter, A. (2010) Technology-based industrialization of claims management in motor insurance. *Tagungsband MKWI 2010, Universitaetsverlag Goettingen, Multikonferenz Wirtschaftsinformatik 2010*.

[20] Koslowski, T. (2009) LTE Concept Car Shows New Vehicle-to-Cloud Opportunities. Gartner Research Paper.

[21] European Commission (2008) Communication from the Commission. Action Plan for the Deployment of Intelligent Transport Systems in Europe. European Commission, Technical Report.

[22] RACC Automobile Club (2008) D.DEPN.4.1a Stakeholder Utility, Data Privacy and Usability Analysis and Recommendations for Operational Guarantees and System Safeguards: Europe. CVIS, Technical Report.

[23] Moore, J. F. (1996) *The Death of Competition: Leadership and Strategy in the Age of Business Ecosystems*. Wiley.

[24] Loewer, U. M. (2005) *Interorganisational Standards. Managing Web Services Specifications for Flexible Supply Chains*. Physica Verlag.

[25] Clark, K. B. and Baldwin, C. Y. (2000) *Design Rules. Vol. 1: The Power of Modularity*. MIT Press.

13

Impact and Future Projections

Cooperative systems for traffic scenarios in general, and AutoNets in particular, are one of the key technologies for future mobility. AutoNets provide both vehicles and infrastructure entities with precise information about their environment and the behaviour of other vehicles in a way that would not be possible with autonomous sensing of the environment by vehicles with their sensors only. This digital context empowers new driving assistance applications and traffic control algorithms, which sustainably improve traffic safety and traffic efficiency. Both traffic safety and traffic efficiency are two of the most important aspects of future mobility. Mobility is expected to further increase in the future, resulting in more accidents on the road. Cooperative systems help to reduce accidents and their severity, because with the exchange of information among the entities dangerous situations can be detected earlier and respective actions can be taken in time. Increased mobility also manifests in more vehicles using current streets. The improved traffic efficiency using cooperative systems is considered one of the key factors that will help the existing road infrastructure handle the additional vehicles, especially in megacities. Moreover, traffic efficiency helps to reduce fuel consumption, because knowledge about the current situation on the road empowers both vehicles to optimise their powertrain strategy to the current traffic conditions, and road operators to optimise traffic flows on the roads.

We also showed that there are many stakeholders involved in AutoNets, covering the three domains of AutoNet mobile domain, AutoNet infrastructure domain, and generic domain. Besides the stakeholders addressed in this book, there are several 'indirect' stakeholders that are interested in AutoNets. One example are politicians, which may be very interested in the successful market introduction of new and innovative technology like AutoNets for several reasons, such as increasing the locational advantage of their region. A second example are students and researchers, for whom AutoNets open up a new research area, resulting in potential spin-offs for universities with novel and profitable solutions for the different domains of AutoNets. However, we also showed that the stakeholders have different expectations of and requirements for the AutoNets, which are sometimes even contradictory. This means that it is very important to balance and harmonise the different expectations in order to define the final system ready for the market introduction. Here, we want to summarise four of the most

Automotive Internetworking, First Edition. Timo Kosch, Christoph Schroth, Markus Strassberger and Marc Bechler.
© 2012 John Wiley & Sons, Ltd. Published 2012 by John Wiley & Sons, Ltd.

important aspects from our point of view, namely business models, standardisation, market introduction and scalability.

Business models. At this very moment, it is not clear what a business model for new services may look like. This is due to the fact that the different stakeholders have different business models in mind, and from a long-term perspective, the stakeholders want to recoup their investments. For example, vehicle manufacturers want to sell value-added assistance functions and services based on AutoNets, suppliers want to sell (standardised) hardware and software solutions to the vehicle manufacturers, traffic management centres want to collect value-added data from AutoNet-enabled vehicles to improve their traffic flow predictions (and sell the more accurate predictions to customers), whereas third party providers may want offer their commercial services for AutoNet use cases. On the other hand, the public wants to have less pollution, drivers are interested in cheap solutions for improving their mobility, whereas road operators have a vital interest in increasing the capacity of their roads with minor investment in the road infrastructure. As a result, the different business models must be combined and adapted appropriately in order to generate an attractive product for customers, where every contributing stakeholder recoups their investment.

Besides vehicle manufacturers, stakeholders operating in the same domain also have to 'open up' their often proprietary systems using respective interoperable service interfaces. This is of particular importance for traffic management centres in order to enable a European-wide harmonised system, which does not end at the borders of other neighbouring traffic management centres. Here, the vision would be to have one single interface to a marke place of European-wide traffic-related data with fixed contracts among the participating providers and customers, where all traffic management centres throughout Europe contribute.

Standardisation. One of the most important prerequisites for the deployment of AutoNets and the development and introduction of AutoNet-based applications are standards. This is of particular importance for cooperative systems, to guarantee interoperability of the system. For AutoNets, there are several levels of interoperability: from a component perspective – mainly provided by vehicle manufacturer suppliers – different hardware and radio platforms as well as the software implementing the communication protocols must be compatible and interoperable. For AutoNet-based use cases, the cooperative applications must be interoperable as well. Therefore, the data provided by different vehicles must be consistent and of a standardised quality, and the cooperative applications have to operate consistently in every vehicle. Moreover, interoperability is required among the service providers of the three AutoNet domains. This helps to reduce both costs and complexity of the system, because one standardised solution is available for all entities of the AutoNet, and no different variants need to be created for different vehicle manufacturers. However, the standards also have to allow the stakeholders to create their unique solutions for AutoNet-enabled applications in order to differentiate them from other competitors. This is of particular importance for the automotive domain, because vehicle manufacturers are very interested in developing systems tailored to the expectations of their typical customers in order to create an additional value for their customers. This way, the same AutoNet application will likely look and feel completely different in vehicles of different brands and types.

The basis for this compatibility and interoperability are valid and sound standards, which are accepted and adopted by every stakeholder of the system. Such standards guarantee well-functioning cooperative AutoNet applications across different vehicle types and vehicle brands, including all stakeholders of an AutoNet system.

Market introduction. One of the major questions of the past few years has been the creation of a roll-out scenario of AutoNet-based cooperative systems. Experts of all domains discussed and simulated thousands of scenarios to decide which equipment rate of the vehicles would be necessary for which type of AutoNet-based application: Whereas some services will work with an equipment rate of 10% for some efficiency services, some robust safety services require more than 90% of AutoNet-enabled vehicles for proper functioning (and for a significant acceptance of the drivers). Moreover, it is still not completely known which drivers are willing to pay for different AutoNet-enabled applications.

This is a difficult question because, especially in the beginning, hardware and software for AutoNet-enabled vehicles will cost additional money – and only a few applications will be available in the beginning. Roadside infrastructure operators have to face a similar challenge: AutoNet-enabled infrastructure such as intelligent and communicating traffic lights will be far more expensive, whereas revenue in terms of efficient traffic flows will greatly depend on the equipment rate of AutoNet-enabled vehicles, and useful and accepted services by drivers also vitally depend on a significant equipment rate of roadside infrastructure in megacities all over Europe and all over the world. This also chimes with the drivers' expectations, because drivers usually do not want to pay for a system when buying a vehicle, when first applications of the system will only be available for use years later.

Scalability. From a technical perspective, scalability will be a major concern for the future that need to be adressed by all stakeholders. Scalability covers almost every aspect of AutoNets, as shown by the following examples:

- Service in back-end systems must be available for potentially millions of vehicles.
- Bandwidth in the wireless networks must be able to serve all the vehicles and some parts of the roadside infrastructure elements.
- Computation power and highly efficient algorithms must be available in traffic management centres.
- Communication protocols must be highly flexible, operating in sparse environments with very few participants (e.g. vehicles) up to very dense environments such as a traffic jam with potentially thousands of vehicles located in one place.

This last point indeed is an open topic, which still needs to be addressed by research. This is of particular importance since AutoNet applications should work in both sparse and dense environments with the same quality. It is also of particular importance for the long-term vision, because there will be only very few AutoNet-enabled vehicles in the beginning, which will continuously increase over the next few decades. This means communication technology and communication protocols must be highly flexible, scalable and future-proof in order to be used in over the next few years.

We also showed in this book that both AutoNet stakeholders as well as research address the issues mentioned above. Harmonisation activities such as COMeSafety currently bring together the different stakeholders and also drive the standardisation activities. Ongoing field operational tests like simTD or DRIVE C2X allow a peek at what such a system may look like, and how potential uses will be accepted by drivers (and stakeholders). Moreover, vehicle manufacturers have also shown the first prototypes of the vision of cooperative vehicles, as illustrated by the following example: in March 2011, BMW presented the concept study *BMW Vision ConnectedDrive* at the 2011 Geneva Motor Show, which gives an insight into BMW's future ideas for a connected vehicle (Figure 13.1). This concept study extrapolates the principle of intelligent networking of driver, vehicle and the outside world into the future, where the vehicle will be transformed into a fully integrated part of the networked world. This networked vision consists of three themes: safety, infotainment and convenience.

- Safety primarily focuses on driver assistance systems for active safety, where the system passes information relevant to safety on to the driver. An interesting aspect here is the presentation and warning of the driver using light effects, here orange/red fibreoptic strands run below and along the cone of vision. Potential uses incorporate a three-dimensional head-up display, displaying information relevant for the journey. In the event of approaching warnings, the information is displayed to the driver right in time with high accuracy and in a way that the driver will not be distracted from the hazard ahead. Using vehicle-to-vehicle

Figure 13.1   BMW vision ConnectedDrive. Reproduced by permission of BMW Group.

Figure 13.2 Safety view of BMW vision ConnectedDrive. Reproduced by permission of BMW Group.

communication, the vehicle is able to 'look into the future' and 'to see around corners'. Figure 13.2 gives us a peek at this scenario.

- An infotainment zone defines a communication level between the driver and the passenger and also spatially promotes active social exchange and the encounter between the two. Here, a passenger information display is mounted in the instrument panel, which serves as a gateway to the passenger's world of entertainment. Features include an emotional browser, which is highly customised to the driver and his or her preferences. Uses include the selection of mood-based playlists, or intelligently filtered location-based information.

- The third level comprises the vehicle itself and focuses on the communication levels of the passengers and the outside world. No matter whether navigation and traffic information or mobile Internet, this level picks up information relevant to the convenience of the passengers and channels it to the interior of the vehicle, or it transmits information to the outside world. The idea is that the vehicle acts as your own personal concierge, in which all your mobile devices are seamlessly integrated while interacting with the vehicle. This makes trip planning pretty easy, while up-to-the-minute traffic information provides the driver with highly precise information for the best (even multi-modal) travelling.

Such concepts – as well as the ongoing activities – show that the automotive industry and its stakeholders have a strong belief in AutoNets, and that they are facing the challenge to create a harmonised system including all relevant stakeholders. This means we are very confident that there will be AutoNet-enabled vehicles on the road in the near future. Although it may take a while to have a sufficient equipment rate required for several AutoNet applications, it will be a success for all the stakeholders, and – what is most important – for the drivers who will be supported in their mobility in a safe, efficient and convenient way.

A

Appendix

A.1 Standardisation Bodies for AutoNets

On the international floor, three organisations with complementary scopes are responsible for standards: the International Organisation for Standardisation (ISO), the International Electrotechnical Commission (IEC) and the International Telecommunication Union (ITU). They have formed the World Standards Cooperation to coordinate their activities and the implementation of international standards. ISO is additionally cooperating with local standardisation development organisations (SDO) like CEN in Europe, ensuring the equivalence of International and European standards whenever possible.

In Europe, three SDOs are responsible for the development of standards relevant for AutoNets:

- The *European Committee for Standardisation* (Comité Européen de Normalisation – CEN) is responsible for all sectors excluding electrotechnical aspects.
- The *European Committee for Electrotechnical Standardisation* (Comité Européen de Normalisation pour l'Electrotechnique – CENELEC) is responsible for standardisation of electrotechnical issues.
- The *European Telecommunications Standards Institute* (ETSI) is responsible for the standardisation in the field of telecommunications and related technologies.

A document ratified by one of these organisations 'carries with the obligation, to be implemented at national level, by being given the status of a national standard and by withdrawal of any conflicting national standards'. This fact implies that manufactures with an offer on the European market need to be compliant with these standards.

A.1.1 ETSI

The European Telecommunications Standards Institute (ETSI)[1] develops globally-applicable standards for Information and Communications Technologies (ICT) including fixed, mobile,

---

[1] ETSI homepage: http://www.etsi.org.

Automotive Internetworking, First Edition. Timo Kosch, Christoph Schroth, Markus Strassberger and Marc Bechler.
© 2012 John Wiley & Sons, Ltd. Published 2012 by John Wiley & Sons, Ltd.

radio, converged, broadcast and Internet technologies. ETSI is an independent non-profit organisation with more than 700 member organisations from over 60 countries all over the world. ETSI is located in Sophia Antipolis, France, and features a centre for testing and interoperability. ETSI standards are available free of charge.

ETSI collaborated with the European Conference of Postal and Telecommunications Administrations (CEPT) and the European Commission on the allocation of radio spectrum specially usable for AutoNets in Europe. In 2007, ETSI established the Technical Committee on Intelligent Transport Systems (TC ITS). Within TC ITS, five working groups (WG) cover different aspects of ITS:

- WG1 – Application requirements and services.
- WG2 – Architecture, cross-layer and web services.
- WG3 – Transport, network.
- WG4 – Media and medium-related issues.
- WG5 – Security.

Each technical committee establishes and maintains a work programme, consisting of work items whereby each work item needs to be supported by at least four ETSI members. An ETSI work item is the description of a standardisation task, and normally results in a single standard, report or similar document. The technical committee approves each work item, which is then formally adopted by the whole membership. A technical committee usually hands responsibility for a work item to a small group of experts, led by a rapporteur. The resulting document is referred to as an ETSI deliverable.

A.1.2   CEN

The European Committee for Standardisation (CEN)[2] views itself as áa business facilitator organisation in Europe, removing trade barriers for European industry and consumersá. CEN comprises 31 national members developing voluntary European standards which are also accepted as national standards in each of the 31 member countries.

CEN Technical Committee 278, established in 1991, standardises in the field of AutoNets and supports the following aspects:

- Communication between vehicles and road infrastructure.
- Communication between vehicles.
- Vehicle, container, swap body and goods wagon identification.
- In-vehicle human–machine interface as far as telematics is concerned.
- Traffic and parking management.
- User fee collection.
- Public transport management.
- User information.

[2] CEN homepage: http://www.cen.eu.

A.1.3   ISO

ISO (International Organisation for Standardisation)[3] is a network of the national standardisation institutes of 161 countries, headquartered in Geneva, Switzerland. While ISO itself is a non-governmental organisation (NGO), many of the members are governmental institutions in their respective countries, or mandated by their government.

The technical committee most relevant for AutoNets is TC204 on ITS. With a particular focus on communications, working group WG16 is defining the CALM (communications access for land mobiles) set of standards. This system covers all types of communication scenarios (direct and multi-hop V2V, V2R, V2Internet) and scopes (unicast, multicast and geocast) on any physical communication medium (cellular, long-range, medium-range and short-range communication networks) – to be used simultaneously – and all types of applications. The networking layer supports both IP (IPv6) and non-IP (FAST) communications. TC204 WG16 is subdivided into the following sub-working groups:

- SWG 16.0: CALM Architecture.
- SWG 16.1: CALM Media.
- SWG 16.2: CALM Network Protocol.
- SWG 16.3: Probe Vehicle System.
- SWG 16.4: Application Management.
- SWG 16.5: eCall.
- SWG 16.6: CALM non-IP mechanism.
- SWG 16.7: Lawful intercept and security.

These sub-working groups are working on the following items:

- NP 11915: defines the usage of IEEE 802.11-based WLAN technologies for ITS applications within the CALM architectural framework (without IEEE 802.11p/WAVE, which is considered separately).
- NP 29284: specifies event-based vehicle probe data as an extension of ISO 22837.
- CD 21215: defines the usage of 5.8GHz to 5.9GHz communication technologies within the CALM architectural framework for V2V and V2R (called CALM M5).
- DIS 21217: defines the CALM architectural framework for the network, MAC and physical layers.
- DIS 24100: specifies basic principles for privacy in probe data systems.
- DIS 24103: defines the media adapted interface layer, an abstraction or virtualisation layer with network protocols and air interface definitions for a variety of wireless communication technologies. It is open for future extensions.
- DIS 21210: specifies how IPv6-based communication protocols are to be used within the CALM architectural framework (DIS 21217).
- TS 25114: specifies the management of probe data collection in a consistent way with SAE J2735.

---

[3] ISO homepage: http://www.iso.org.

A.1.4   IETF

The Internet Engineering Task Force (IETF)[4] is an open international community dealing with the evolution of the Internet architecture and (co-)operation within the Internet.

The actual technical work of the IETF is done in different working groups, which are organised by topic into several areas (e.g. routing, transport layer issues, security, mobility, etc.). In general, the IP-based protocols used in AutoNets are standardised in RFCs (request for comments), which are defined in the respective working groups of the IETF. Examples include Mobile IPv6 and network mobility, which were specified at the MIPv6 working group and the NEMO working group, respectively.

A.1.5   IEEE

IEEE (Institute of Electrical and Electronics Engineers)[5] is organised as a non-profit organisation. Within IEEE, the Standards Association (IEEE-SA) is developing industry standards for a broad range of industries. IEEE-SA cooperates with IEC, ISO and ITU.

Of special relevance for AutoNets is the IEEE 802 LAN/MAN Standards Committee. which has published the well known standards for the Ethernet family, for Token Ring and for Wireless LAN, as well as for Wireless PAN, Wireless MAN, Bridging and Virtual Bridged LANs. The WLAN standards are developed by the 802.11 Working Group. The IEEE standards relevant for AutoNet systems are the following:

- *IEEE 802.11p* for communication in the AutoNet ad-hoc network bomain.
- *IEEE 1609* family of standards, also known as wireless access in vehicular environments (WAVE), define an architecture with system services and interfaces for secure vehicle-to-vehicle and vehicle-to-roadside communications, together with the management functions and device modes for system operation.
  - *IEEE 1609.2* defines security services for applications and message management with message encryption and authentication as well as secure message exchange mechanisms for AutoNets. Details can be found in Chapter 9.
  - *IEEE 1609.4* is built on IEEE 802.11p and specifies the upper MAC layer with multiple channel operation.
  - *IEEE 1609.5* defines communication management services.

A.1.6   Car2Car Communication Consortium

In 2004, the Car2Car Communication Consortium (C2C-CC)[6] started as a non-profit organisation and has grown ever since, now conprising all major European vehicle manufacturers, several suppliers, research organisations and other partners. The overall objective of the C2C-CC is to implement cooperative intelligent transportation systems to further improve

---

[4] IETF homepage: http://www.ietf.org.
[5] IEEE homepage: http://www.ieee.org.
[6] C2C-CC homepage: http://www.car-to-car.org.

road traffic safety and efficiency by means of inter-vehicle communication and vehicle to roadside infrastructure communication and applications. The technological focus for communication is on an ad-hoc network for vehicles based on 5.9GHz, which features low latency communication and geo-routing.

The consortium contributes to the European standardisation bodies, including associated validation and certification processes. It works closely with the European standardisation organisations, in particular ETSI TC ITS, in order to achieve commonly agreed European standards for Intelligent Transportation Systems. An important objective is to support a joint adoption of these standards within the European automotive industry.

A.2   Research Projects on AutoNets

In this section, we provide an overview of research projects that have addressed key aspects of AutoNets. We start with early projects laying the foundation for the current activities in this area. In the past few years, there have been numerous national and international projects focusing on AutoNet technology and applications. Obviously, we cannot introduce all of these projects; instead, we will focus on a few. Besides these projects, several national projects like the Italian *Infonebbia*, the Dutch *SPITS*,[7] the Portugese *DRIVE-IN*[8] and the French *SCOREF*,[9] the US projects *VSC*,[10] and *CICAS-V*,[11] or the Japanese projects *ASV*,[12] *AHS*[13]/*Smartway* and the *Internet ITS Consortium* have provided important results. There have been even more national and other projects on an international level (like the European projects *COM2REACT*[14] or *GEO-NET*[15]) that we cannot name here and may not even be aware of.

A.2.1   Early Activities

In research, connected vehicle technology has been an important topic for a long time. One of the first systems featuring V2R communication was the 'Wolfsburger Welle' project from 1981 to 1983 [2]), connecting vehicles with traffic lights through infrared communication. V2V and V2R communication technologies were an important topic in the research programme 'Prometheus' (1987 to 1995). The main goal of Prometheus was to develop driverless vehicles, which was done mainly by onboard sensor and actuator systems, but also by communication among vehicles and between vehicles and infrastructure components.

To address the issues of cooperative intelligent transportation systems, a lot of preparation work was done in several national and international projects – in the European Union, the

---

[7] SPITS: Strategic Platform for Intelligent Traffic Systems; https://spits-project.com/.

[8] DRIVE-IN: http://drive-in.cmuportugal.org/.

[9] SCOREF: Système COopératif Routier Expérimental Francais; http://blog.inria.fr/scoref/.

[10] VSC: Vehicle Safety Communications.

[11] CICAS-V: Cooperative Intersection Collision Avoidance System.

[12] ASV: Advanced Safety Vehicle.

[13] AHS: Automated Highway System.

[14] http://www.com2react-project.org/.

[15] GEO-NET: http://www.geonet-project.eu/.

United States of America, Japan and other parts of the world. The following examples will give an impression on these activities and their main concerns:

Chauffeur II (1996 to 2002). The Chauffeur II project was funded by the European Union. The overall goal of Chauffeur and its successor Chauffeur II was to augment the utilisation of existing roads. Chauffeur II therefore proposed a telematics-based vehicle control system to increase the density of freight traffic in a safe manner. The project developed a radar system to link trucks electronically. This system allows a leading truck to control following trucks in a dense lane [3].

CarTALK 2000 (2001 to 2004). The EU project CarTALK 2000 aimed at the development of new driver assistance systems using inter-vehicle communication [4]. Inter-vehicle communication in CarTALK 2000 was based on self-organising ad-hoc radio network technology [5].

MoTiV (1996 to 2000). MoTiV was funded by the German Federal Ministry for Education and Research (BMBF). One part of MoTiV was the project 'Mobility in Connurbations' [6], which aimed at increasing vehicular safety and mobility by combining intermodal means of transportation. The successive research programme on mobility and transportation dealt with intelligent transportation, sustainable mobility, environmental protection and increased safety on the road using vehicle-to-vehicle communication.

INVENT (2001 to 2005). The BMBF project INVENT addressed various applications by combining traffic on the road, information and communication technologies. Example applications are driver assistance systems using inter-vehicle communication to warn vehicles about dangerous incidents or road conditions ahead [7]. Moreover, this information will be used to develop traffic management systems to dissolve – or even entirely prevent – congestions on the road [8].

PATH (since 1986). PATH has been supported by the California Department of Transportation, USA.[16] Its focus had two basic aspects: (i) advanced transportation management and information systems, and (ii) advanced vehicle control and safety systems. In order to distribute traffic and sensor information, PATH uses dedicated short range communication (DSRC) technology to communicate with a generic computer device, which can be located either at the roadside or within a vehicle.

DRiVE and OverDRiVE (1998 to 2002). The DRiVE project (dynamic radio for IP services in vehicular environments) [9] and its successor OverDRiVE [10] were funded in the IST (information society technology) programme of the European Union from 1998 to 2002. The objective of DRiVE was to enable spectrum-efficient high-quality wireless IP services for multimedia applications in a multi-radio vehicular network environment. In order to achieve this objective, DRiVE intends the convergence of cellular and broadcast networks. The convergence is realised by optimising the inter-working of different radio systems (GSM, GPRS, UMTS, DAB and DVB-T) in a common dynamically allocated

---

[16] PATH homepage: http://www.path.berkeley.edu.

frequency range. The OverDRiVE project extends DRiVE by providing mechanisms for spectrum sharing between the communication systems using a dynamic allocation of the spectrum according to the actual load of the network. Moreover, OverDRiVE aims at the development of a vehicular router to provide multi-radio access to a moving intra-vehicular network.

COMCAR (1999 to 2002). The COMCAR project (communication and mobility by cellular advanced radio) [11] was part of the UMTSplus project funded by the German ministry BMBF. The objective was the design and implementation of a mobile communication network to provide mobile, asymmetric and interactive IP-based services in vehicles. Therefore, existing and upcoming communication systems like GSM, GPRS, HSCSD, UMTS, DVB-T and DAB were optimised and integrated in a vehicular communication network. COMCAR basically deals with communication aspects of the lower layers such as spectrum efficiency of the supported networks. It also provides flexible mechanisms for the quality of service support in such heterogeneous communication environments. A mobile middleware supplemented this quality of service support, which allows adaptive multimedia applications to react to changes in the communication environment in a user-specific manner.

FleetNet (2000 to 2004) and Network on Wheels (NOW) (2004 to 2008). Compared to communication in wired IP-based networks, the communication architecture of FleetNet differs fundamentally in many aspects. Examples are the addressing of the vehicles, the routing of data in the context of mobility and the medium access for the wireless transmission. Figure A.1 shows the protocol architecture of FleetNet [12]. It comprises the lower three layers for inter-vehicle communications, namely FleetNet physical layer (FPHY), FleetNet data link control layer (FDLC), and FleetNet network layer (FNL). FleetNet does not specify a transport layer; instead, FleetNet applications immediately use the FNL for their communications. The FleetNet communication architecture also specifies control and management planes for both FDLC and FNL, which are not depicted in Figure A.1. The management planes control the respective layers. For example, it adapts the transmission power to the number of communicating vehicles in order to minimise interferences.

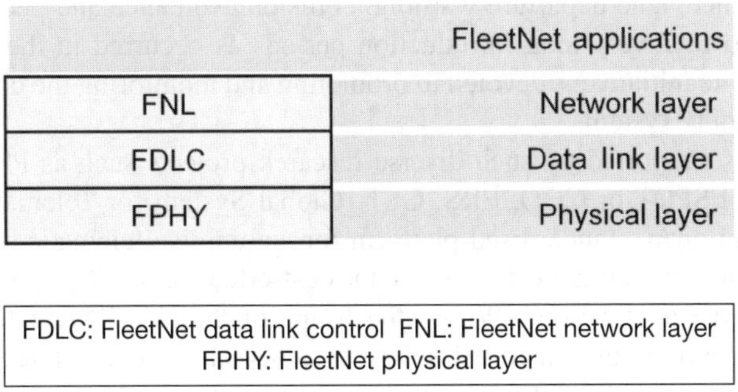

	FleetNet applications
FNL	Network layer
FDLC	Data link layer
FPHY	Physical layer

FDLC: FleetNet data link control FNL: FleetNet network layer
FPHY: FleetNet physical layer

Figure A.1   FleetNet communication architecture.

Although the outcomes of those early projects were promising, they suffered mainly from the the poor capacities of available wireless communications channels. These constraints on the overall transmission rates obliged the engineers to design interconnected systems that were very much focused on one specific application. This in turn made a successful market introduction of the developed systems impossible, considering the inherent deployment problem of cooperative systems (as we discuss in detail in Chapter 12) and the costs for integrating wireless communication technology into both vehicles and infrastructure systems.

With new wireless technologies available and the associated tremendous increase of wireless bandwidth at the beginning of the new century, the idea of interconnected vehicles experienced a global renaissance. Both cellular and wireless ad-hoc peer-to-peer networks promise to be capable to carry a variety of different applications. With lessons learned from the early stages and having in mind the economical constraints with respect to system deployment costs, in particular wireless ad-hoc communication based on the widely deployed IEEE 802.11 wireless LAN standard seem to be the most promising technology and have been strongly pushed by research, industry and politics. Decisions from both US and EU regulation authorities on the usage of an effectively protected frequency spectrum for intelligent transportation systems gave a further boost to this development.

There is now a broad consensus that the key to achieving the benefits of AutoNets lies in a common and standardised means of communication between the components participating in the AutoNets. These components can be located either in a back-end infrastructure, in dedicated roadside entities, in vehicles or in mobile devices. In the following, we will present some of the more recent projects and activities in this field. These projects are far from complete, but they give an impression of the scope vehicular networking research has covered.

A.2.2   The eSafety Initiative

eSafety is a joint initiative of the European Commission, industry and other stakeholders. It aims at accelerating the development, deployment and use of so-called 'intelligent integrated safety systems'. Such systems are expected to use information and communication technologies to increase road safety and reduce accidents on Europe's roads, as well as to reduce environmental pollution to a minimum. An important goal of this initiative is to facilitate and accelerate the introduction of innovative technologies in the automotive environment such as braking assistance, lane departure warning, collision avoidance and pedestrian protection systems. To avoid relatively long introduction periods as occurred in the case of anti-lock braking systems, the initiative is devoted to promoting and monitoring the deployment of new, intelligent road safety systems.

The initiative is supported by such diverse research projects such as PReVENT, EASIS, GST, CVIS, SAFESPOT or COOPERS. GST (Global System for Telematics) targetted the development of an open, standardised platform for automotive telematic services. Agreeing on a common environment is expected to ease the cost-effective development, deployment and spread of innovative applications. Within the frame of the EASIS (electronic architecture and system engineering for integrated safety systems) project, the integration of today's heterogeneously developed safety systems into one comprehensive network of so called integrated safety systems was promoted. The standardisation of a robust in-vehicle electronic

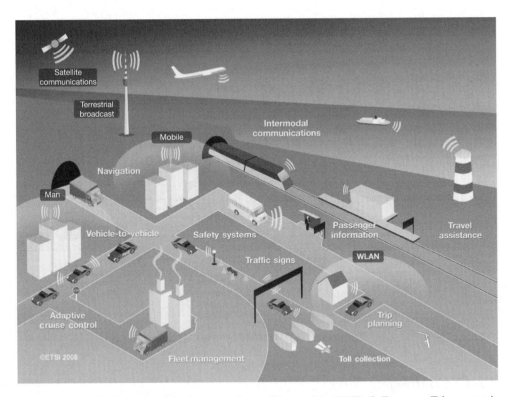

Figure A.2 Vehicular communication scenarios as illustrated by ETSI. © European Telecommunications Standards Institute 2008. Further use, modification, redistribution is strictly prohibited. ETSI standards are available from http://pda.etsi.org/pda/.

architecture was strongly pushed by this project as well, to both improve the quality and reduce development times of future telematic services. PReVENT, a third project integrated into EU's eSafety initiative, was launched to enhance road safety by developing and demonstrating cutting-edge technologies for preventing road traffic casualties. INSAFES, SASPENCE, WILLWARN, SAFELANE and INTERSAFE are among the various subprojects of PReVENT.

A number of graphical illustrations provide an overview of connected vehicle scenarios. The Technical Committee *Intelligent Transportation Systems* (ITS)[17] at ETSI has illustrated different usage scenarios and technologies (see Figure A.2). These usage scenarios are as diverse as traffic safety and wireless payment. Usually, certain technologies would be used in each scenario.

---

[17] The term Intelligent Transportation Systems (ITS) refers to the application of information and communication technologies to the transportation sector, both vehicles and infrastructure for road, maritime and air traffic. AutoNet systems thus make up a subset of ITS. They have seen growing research interest over the last couple of years, as can be seen by the number of sessions organised and the number of papers published at the world's largest ITS event, the yearly ITS World Congress (http://www.itsworldcongress.org).

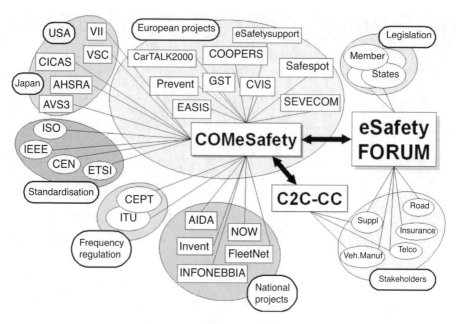

Figure A.3 COMeSafety overview diagram.

A.2.3 COMeSafety

COMeSafety[18] was a specific support action of the European Commission within the sixth Framework Programme. It started in January 2006 with a duration of four years. COMeSafety supported the eSafety Forum with respect to all issues related to AutoNet communication. It provided a platform for other projects and stakeholders both for the exchange of information and the presentation of results. Regular electronic newsletters and publications at major conferences and press events complemented the dissemination efforts. For European and worldwide harmonisation, liaisons were established and workshops were organised. COMeSafety thus provided an open integrating platform, aiming for the interests of all public and private stakeholders to be represented.

Consolidated results and interests were submitted to the European and worldwide standardisation bodies. COMeSafety actively supported the process of European spectrum allocation for AutoNets by participation in ETSI and CEPT technical groups. This was a common concern of all the related projects and regarded as a basic requirement for a successful operation providing the expected impact on road safety. Relevant ISO and IEEE work was also considered. With liaisons with all relevant stakeholders, the provision of information and preparation of strategic guidelines COMeSafety supported directly the eSafety Forum on the items of cooperative systems for road safety and traffic efficiency. Figure A.3 summarises the corresponding parties involved in COMeSafety together with their interactions.

---

[18] COMeSafety homepage: http://www.comesafety.org.

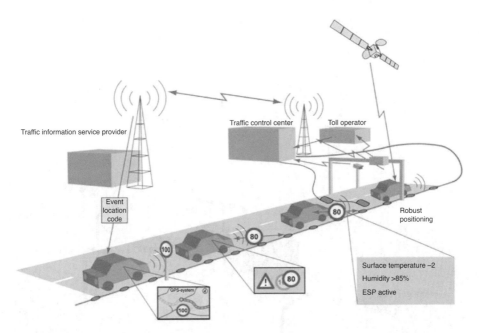

Figure A.4   COOPERS overview diagram. Reproduced from COOPERS research project.

A.2.4   COOPERS

The integrated project COOPERS (cooperative systems for intelligent road safety)[19] was a European research and development activity within the sixth Framework Programme by the European Commission, starting in February 2004 with a duration of four years. It focused on 'cooperative traffic management' between vehicles and traffic infrastructure, trying to support a more harmonised development between the automotive industry and infrastructure operators.

Figure A.4 shows the expected deployment scenario of COOPERS, which also includes the different stakeholders participating in the project. The validation and test drives of the COOPERS system concept were performed on public motorway sections in France, Belgium, The Netherlands, Germany (Berlin and Bavaria), Austria and Italy.

A.2.5   CVIS

CVIS (Cooperative Vehicle-Infrastructure Systems)[20] started as a major European research and development project in February 2006 for the duration of four years. It was dedicated to provide an implementation of the ISO CALM set of standards (see Section A.1.3 where all types of applications would use IPv6 (including network mobility (NeMo)), operating over a variety of different communication media. Only for time-critical safety applications, a direct usage of 5.9GHz communication via the CALM FAST stack was allowed. CVIS was thus not

---

[19] COOPERS homepage: http://www.coopers-ip.eu.
[20] CVIS homepage: http://www.cvisproject.org.

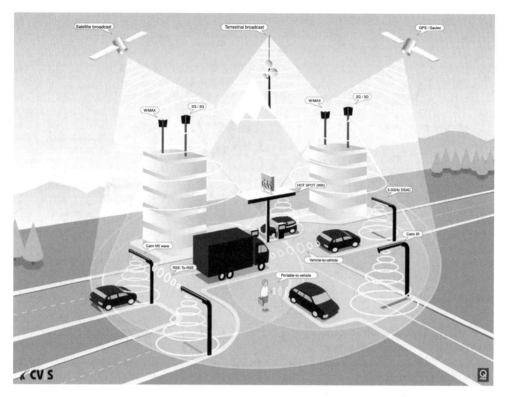

Figure A.5 CVIS overview diagram. Reproduced from CVIS research project.

focused on one particular application or technology, but rather on the underlying architecture and infrastructure to allow vehicles to communicate with each other and with the roadside infrastructure. It was dedicated to a seamless system design, developing a secure and open service framework with the capability to deal with and integrate data from distributed sensors of the CVIS vehicles and roadside equipment. Data from the vehicles would be taken to affect the operation of traffic control systems directly. Individual vehicles would gain access to theiInformation shown on road signs via the newly established wireless communication links. In addition, they would receive accurate information from the traffic infrastructure to calculate the fastest route to their destination. Among the uses demonstrated were such diverse examples as approaching emergency vehicle warning or tracking of hazardous goods shipments (Figure A.5 provides an overview of the fields addressed by CVIS).

Test sites were created in in six European countries: Belgium, France, Germany, Italy, The Netherlands, Sweden and the United Kingdom. Interoperability in the CVIS system between the different entities and different types of roadside systems was achieved by the implementation of a mobile router, supporting IPv6 and a wide range of communication media, including cellular networks, short-range wireless technology and infrared communication. The technical components developed in the project were tests in real-world scenarios, that is integrated in real control units. Making use of this set up, user acceptance, privacy, security, liability, public policy needs, business models and roll-out possibilities were analysed.

The results of the technical validation of the ISO CALM standards were fed back into the European and global standardisation bodies.

A.2.6 SAFESPOT

SAFESPOT[21] was an integrated research project starting in February 2006 with a duration of four years. The objective of SAFESPOT was on AutoNet safety applications based on cooperation of intelligent vehicles with intelligent roads. A so-called safety margin assistant was developed, detecting potentially dangerous situations in advance extending drivers' awareness. This assistant, providing warnings to the driver, was based on AutoNet technology, more specifically on V2V and V2R communication. It therefore used information provided by the AutoNet, fused with on-board sensor information.

The technical development included applications as well as several enabling technologies:

- V2V and V2R communication using ad-hoc networking.
- An accurate relative positioning.
- Local dynamic maps.
- Wireless sensor networks.

A dedicated subproject called BLADE (business models, legal aspects and deployment) dealt with legal, business, and deployment organisational aspects. SAFESPOT applications and technologies were validated in test sites located in France, Germany, Italy, The Netherlands, Spain and Sweden.

A.2.7 SeVeCom

SeVeCom (Secure Vehicular Communication)[22] was another EU-funded project in the sixth Framework Programme, running in parallel to the three aforementioned integrated projects. It focused on security and privacy for AutoNets of different communication types. It defined a security architecture and addressed three major aspects:

- Threats, such as bogus information, denial-of-service attacks and identity cheating.
- Requirements like authentication, availability and privacy.
- Operational properties, including network scale, privacy, cost and trust.

A.2.8 GeoNet

The EU-funded project GeoNet[23] develops AutoNet technologies to allow vehicles to exchange information beyond their immediate surroundings and line-of-sight. It started in February 2008 with a duration of two years. GeoNet enables the dissemination of information to destination vehicles in a defined geographic area using IP-based technology. The protocols are designed

[21] SAFESPOT homepage: http://www.safespot-eu.org.
[22] SeVeCom homepage: http://www.sevecom.org.
[23] GeoNet homepage: http://www.geonet-project.eu.

to be reliable and scalable. TCP/IP is used as a unifying layer between different wireless technologies and various applications. For scalability reasons, GeoNet deploys IPv6. This way, the approach used in GeoNet is similar to the Internet integration solution developed in FleetNet, where the FleetNet adaptation layer within the FleetNet network layer is responsible for translating IP addresses into geographical positions (cf. figure A.1).

It is worth mentioning that this approach will allow for both IPv6 and non-IPv6 communications. This will, effectively, open the door for the development of new applications, which require data to be transmitted to explicit geographical areas. Thereby, GeoNet is specifying, implementing, validating and standardising the necessary functional blocks in coordination with European projects, ETSI and the Car-to-Car Communication Consortium (see Section A.1.6).

A.2.9   FRAME, E-FRAME

E-FRAME[24] (starting in May 2008 with a duration of three years) and its predecessor FRAME (starting in July 2001) are part of a series of projects funded by the European Commission. A project called KAREN produced the initial European ITS Framework Architecture, taken up by two projects called FRAME-S and FRAME-NET and then further developed by FRAME and E-FRAME. The Framework Architecture builds on a set of user needs and their respective functionality.

The two more recent FRAME projects also added tools and other support aids to assist in using the framework architecture. Today, this framework architecture is usually simply referred to as FRAME'. The FRAME Forum was established in October 2004 to provide support for the architecture, driven by many of the initial users.

E-FRAME specifically addresses the expansion of the framework architecture to include support for the deployment of cooperative systems. Additionally, E-FRAME provides advice for the development and for operational issues of a particular implementation. Using this support, interested stakeholders can use the extended architecture as a 'tool' to implement their respective view of cooperative ITS.

A.2.10   VII and IntelliDrive

VII (vehicle infrastructure integration), which is now called IntelliDrive, is a programme of the US Department of Transportation. According to Wikipedia, VII 'is an initiative fostering research and applications development for a series of technologies directly linking road vehicles to their physical surroundings, first and foremost in order to improve road safety. The technology draws on several disciplines, including transport engineering, electrical engineering, automotive engineering, and computer science. VII specifically covers road transport although similar technologies are in place or under development for other modes of transport. [...] VII targets improvements in both safety and efficiency.'[25]

---

[24] FRAME, E-FRAME homepage: http://www.frame-online.net.
[25] http://en.wikipedia.org/wiki/Vehicle_infrastructure_integration.

VII was renamed IntelliDrive,[26] which is considered a multi-modal initiative aiming on interoperable wireless communications among vehicles, road infrastructure and mobile devices. As in the European initiatives, Intellidrive wants to leverage the capabilities of wireless technology to increase safety and reap ecological benefits of smarter transport systems. Besides development of technologies and applications, their potential benefits and costs are analysed. IntelliDriveSM applications provide connectivity in the following scenarios:

- Between vehicles to enable crash prevention.
- Between vehicles and the infrastructure to enable safety, mobility and environmental benefits.
- Between vehicles, infrastructure and wireless devices to provide continuous real-time connectivity to all system users.

With wireless capabilities, Intellidrive wants to enable vehicles to have full awareness of hazards and situations around a vehicle. Based on this, drivers will receive advice on school zones, sharp ramp curves or icy road sections. IntelliDrive also addresses and wants to equip bicycles and pedestrians. To improve mobility, the transportation system would gain (anonymous) access to real-time data provided by thousands of vehicles, supporting transportation managers in monitoring and managing transportation system performance. Traffic signals or transit operations would be adapted according to the current requirements. Travellers, in turn, would receive real-time information about the traffic conditions and would thus be enabled to make informed travel and route decisions.

A.2.11   Travolution

Travolution is a project of the vehicle manufacturer Audi together with other companies and research institutes, which is based on communication between vehicles and traffic lights. Therefore, 25 traffic lights in the city of Ingolstadt were equipped with communication technology in order to communicate the status of traffic lights. This information can be used to optimise the vehicles' behaviour at traffic lights, for example to reduce acceleration or to implement green light speed optimal advisory assistance functionality. Travolution also optimises the control of traffic lights, which reduces the waiting time of vehicles at traffic lights. According to Audi, an area-wide coverage of such a system shows the potential to reduce the emission of carbon dioxide in cities of about 15%.

Besides reducing carbon dioxine emissions, Travolution also implements interesting features for automatic payment at gas stations or parking decks. Therefore, vehicles are able to communicate with infrastructural components mounted at the respective locations.

A.2.12   Aktiv

The German research initiative Aktiv (adaptive and cooperative technologies for intelligent traffic, funded by German Federal Ministry of Economics and Technology) brings together 29 partners from the automobile domain, telecommunication domain, software companies and

---

[26] IntelliDrive homepage: http://www.intellidriveusa.org/.

research institutions. The goal of Aktiv is to improve both traffic safety and traffic flow in the future. The parters are working together to design, develop and evaluate the following issues:

- New driver assistance systems.
- Knowledge and information technologies.
- Solutions for efficient traffic management.
- C2X communication for future cooperative vehicle applications.

Therefore, the Aktiv initiative addresses three areas: traffic management, active safety and cooperative cars. The goal *traffic management* (Aktiv-VM) is to utilise interaction of intelligent vehicle systems and intelligent infrastructure units in order to improve the management in road scenarios. *Active safety* (Aktiv-AS) is focused on the improvement of traffic safety by supporting the drivers with various assistance systems, for example for emergency braking, continuous lateral control, intersection assistance and vulnerable road users. Finally, *cooperative cars* (CoCar) investigated the use of UMTS for future cooperative transport systems. From a technological perspective, the basic idea was to distribute vehicular messages via cellular networks using their inherent broadcast mechanisms. Based on this idea, potential application scenarios were identified, together with a simulation of the expected network as well as the development of a business case for a market introduction of such a system.

The successor of CoCar, calles CoCarX, continues and extends this idea by applying the respective mechansims to the upcoming LTE communication technology. Whereas the work in CoCar did not have any relationto related projects nor to the ongoing standardisation issues in this field, CoCarX had the mission to create a system architecture which was compatible to the system architecture developed in simTD. This way, CoCarX can be seen as an extension to simTD (cf. Section A.2.14 by addressing the cellular communication area for being used in C2X communication scenarios. CoCarX started in 2010.

A.2.13   PRE-DRIVE C2X

PRE-DRIVE C2X (preparation for driving implementation and evaluation of C2X communication technology, 2008 to 2010) is a preparation project funded by the European Commission[27] with an overall budget of 8.4 million Euros. Its goal was to prepare a large-scale European field trial for vehicular communication technology. Based on relevant scenarios and the resulting applications of the European COMeSafety architecture (cf. Section A.2.3) for V2X communication systems, the project developed a detailed specification for such a system and a functionally verified prototype. The prototype is robust enough to be used in future field operational tests. This guarantees easy implementation and quick market introduction. This prototype will be verified and evaluated so that a working system is available at the end of PRE-DRIVE C2X. It will be the basis for all further European activities in the field of cooperative systems. In particular, PRE-DRIVE C2X addresses the following challenges:

Field trials. Specification and development of test equipment and a management centre for testing and evaluation. Selected use cases generate additional requirements and adaptation of the PRE-DRIVE C2X system.

---

[27] PRE-DRIVE C2X homepage: http://www.pre-drive-c2x.eu.

Security. In PRE-DRIVE C2X, the COMeSafety architecture specification will be extended by additional security components.

Back-end Services. Automotive-related services running in the Internet will have a significant impact on the system architecture. They are considered in the PRE-DRIVE C2X system architecture as well.

Since the life cycle of road vehicles is considerably longer than that of electric and electronic components, particular care will be taken to ensure technological persistence in PRE-DRIVE C2X. The operability over a long period of time that extends at least until 2025 has to be considered. This makes it necessary to monitor all relevant upcoming trends in the field of vehicular communication and to anticipate potential emerging technologies. It also requires active participation to relevant standardisation activities. An important precondition for such activities is a common and agreed system architecture. Therefore, the architecture specification in PRE-DRIVE C2X was performed in close cooperation with COMeSafety in order develop a widely accepted system architecture for cooperative systems.

Besides the preparation issues, PRE-DRIVE C2X also developed an integrated simulation model for cooperative systems that enables a holistic approach for estimating the expected benefits in terms of safety, efficiency and environment. This includes all tools and methods necessary for functional verification and testing of cooperative systems in laboratory environment and on real roads in the framework of a field operational test. Moreover, tools and methods developed in PRE-DRIVE C2X are applied to a prototype system to verify its proper functioning and to make an impact assessment including a user acceptance test. PRE-DRIVE C2X strives to further pave the way for Europe-wide implementation of V2X communication technology and to contribute to the next leap forward in active safety.

A.2.14 simTD

The German research project simTD (safe and intelligent mobility, test field Germany)[28] started in 2008 with a period of four years. Its goal is to shape tomorrow's safe and intelligent mobility through researching and testing V2X communication and its applications. simTD will put the results of previous research projects on V2X communication into practice. For this purpose, realistic traffic scenarios will be addressed in a large-scale test field infrastructure around the Hessian city of Frankfurt. The project will also pave the road for the political, economic and technological framework to successfully set up V2X networking. In this context, simTD pursues the following principle objectives:

- Increased road safety and improved efficiency of the existing traffic system using V2X communication.
- Definition and validation of a roll-out scenario for identified applications for scientific questions through practice-oriented experiments and field operational tests.
- Consolidation of V2X applications from the categories of traffic efficiency, driving and safety, and value-added services.

---

[28] simTD homepage: http://www.simtd.org.

- Definition, analysis, specification and documentation of the applications that will be developed and tested. This also includes the resulting requirements for the overall system for selected applications and tests.
- Development of test and validation metrics and methods in each phase of the overall system development in order to enable measurement and evaluation of the results.
- Consolidation and harmonisation of requirements from perspective of feasibility and performance as well as their compatibility of requirements.
- Verification of applications and requirements within the context of individual milestones.

To achieve these objectives, the simTD consortium comprises several German automotive manufacturer, telecommunication companies, the Hessian state government and renowned universities and research institutions have partnered up. The Federal Ministry for Economics and Technology, the Ministry for Education and Research as well as the Ministry for Transport, Building and Urban Affairs are funding and supporting the project.

References

[1] Schiller, J. (2003) *Mobile Communications*, 2nd Edition. Addison Wesley.
[2] Hoffman, G. and Zimdahl, W. (1988) Recommended speed indication within the motor vehicle – A contribution to fuel saving through the "Wolfsburg Wave" information system. *Proceedings of the Symposium on Energy Efficiency in Land Transport.*
[3] Martini, S. and Murdocco, V. (2001) A study on lateral control for automated driving on heavy truck vehicles. *Proceedings of the 2001 IEEE Intelligent Vehicle Symposium.*
[4] Reichardt, D., Miglietta, M., Moretti, L., Morsink, P. and Schulz, W. (2002) CarTALK 2000 – Safe and comfortable driving based upon inter-vehicle-communication. *Proceedings of the 2002 IEEE Intelligent Vehicle Symposium.*
[5] Borgonovo, F., Capone, A., Cesana, M., Fratta, L., Coletti, L., Moretti, L. and Riato, N. (2003) Inter-vehicules communications: A new frontier of ad-hoc networking. *Proceedings of the 2nd Mediterranean Workshop on Ad-Hoc Networks.*
[6] Bundesministerium für Bildung und Forschung (BMB+F) (2002) Mobility in Conurbations. Public Brochure.
[7] Breitenberger, S., Bogenberger, K., Hauschild, M. and Laffkas, K. (2003) Extended floating car data – an overview. *Proceedings of the 10th World Congress on Intelligent Transport Systems and Services.*
[8] Michaelsen, M. and Ress, C. (2004) Innovative hybrid routing services – INVENT traffic network equaliser approach. *Proceedings of the 1st International Workshop on Intelligent Transportation.*
[9] Christodoulides, L., Sammut, T. and Tönjes, R. (2001) DRiVE towards systems beyond 3G. *Proceedings of the 5th World Multi-Conference on Systemics, Cybernetics and Informatics.*
[10] Tönjes, R., Moessner, K., Lohmar, T. and Wolf, M. (2002) OverDRiVE – spectrum efficient multicast services to vehicles. *Proceedings of IST Mobile Summit 2002.*
[11] Keller, R., Lohmar, T., Tönjes, R. and Thielecke, J. (2001) Convergence of cellular and broadcast networks from a multi-radio perspective. *IEEE Personal Communications*, April 2001.
[12] Franz, W., Wagner, C., Maihöfer, C. and Hartenstein, H. (2004) FleetNet – platform for inter-vehicle communications. *Proceedings of the 1st International Workshop on Intelligent Transportation 2004.*

Index

Automotive Internetworking, First Edition. Timo Kosch, Christoph Schroth, Markus Strassberger and Marc Bechler.
© 2012 John Wiley & Sons, Ltd. Published 2012 by John Wiley & Sons, Ltd.